T0188642

Paleontological Data Analysis

# Paleontological Data Analysis

Second Edition

*Øyvind Hammer*
*University of Oslo*
*Oslo, Norway*

*David A.T. Harper*
*Durham University*
*Durham, UK*

This second edition first published 2024
© 2024 John Wiley & Sons Ltd

*Edition History*
Blackwell Publishing Ltd (1e, 2006)

The right of Øyvind Hammer and David A.T. Harper to be identified as the author of this work has been asserted in accordance with law.

*Registered Office(s)*
John Wiley & Sons, Inc., 111 River Street, Hoboken, NJ 07030, USA
John Wiley & Sons Ltd, The Atrium, Southern Gate, Chichester, West Sussex, PO19 8SQ, UK

For details of our global editorial offices, customer services, and more information about Wiley products visit us at www.wiley.com.

Wiley also publishes its books in a variety of electronic formats and by print-on-demand. Some content that appears in standard print versions of this book may not be available in other formats.

*Library of Congress Cataloging-in-Publication Data Applied for:*
Hardback ISBN: 9781119933939

Cover Design: Wiley
Cover Images: Courtesy of Øyvind Hammer; Hans Arne Nakrem

Set in 9.5/12.5pt STIXTwoText by Straive, Pondicherry, India

# Contents

# Preface

We wrote the first edition of "Paleontological Data Analysis" in 2005; it was published a year later. Paleontology has changed a lot since then.

In 2005, quantitative data analysis was still a bit of a fringe pursuit in paleontology. Now, most paleontology students are obliged to take courses in statistics and systematics, even programming, and most paleontology papers include at least some basic statistical analysis. We firmly believe this is a *good thing*, and we hope this book will convey the enthusiasm we feel for quantitative analysis. Yes, in our field, the data are often poor, and assumptions of statistical tests are often partly violated. It is important to remember this. Still, if the alternative is just opinion without justification, we prefer the transparency of a more stringent analysis protocol. Another important aspect of paleontological data analysis is the visualization of large and complex data sets. Data analysis should not be overwhelmingly difficult and opaque but should make things simpler and bring a greater clarity to our investigations. Finally, through the process of data analysis, our research questions and the quality of the data are refined, but our hypotheses may change, and new, often unforeseen, directions materialize. Thus, data collection and data analysis combine into an iterative, creative process.

Apart from paleontology becoming more analytical, what else has happened in our field since 2005?

First, the number of data analysis methods relevant to paleontology has absolutely *exploded*. It is almost impossible to keep up. While the first edition of our book could claim to cover most of the methods commonly used in paleontology at the time, we can now only hope to discuss the fundamental techniques, and to sample some of the more advanced and specialized ones. Two areas in particular have expanded in importance: Bayesian analysis and the (rather vaguely defined) class of "machine learning" methods.

Second, new technologies provide new types of data. Hyperspectral imaging, chemical element mapping, XRF core logging, photogrammetry, and aerial (drone) imaging are some examples of recently emerging methods. But, above all, there is one technology that has truly transformed paleontology since 2005: high-resolution tomography with microfocus and synchrotron CT scanning. The analysis of tomographic data will be introduced in chapter 8.

Third, things have evolved on the software side. In the 2006 edition, we used the software PAST, developed by us. This program has been extended and improved ever since, and it now covers an order of magnitude more methods than in 2006. Because PAST is very easy to use and includes most functions required by paleontologists, it is still a cornerstone in this edition. However, we also need to address the spectacular rise of R. This program, or rather programming language, is a general framework that includes community-provided "packages" for almost any statistical method imaginable. The main advantage of R is its total flexibility; the main disadvantage is that it requires programming skills. While this book will not give a full introduction to R, we will provide some examples to give you the general idea. However, R being a rather esoteric programming language, there is recently a trend toward data analysis with a more elegant language called *Python*. We will also give a few examples of Python code for paleontological data analysis.

Finally, we, the authors, have become older and wiser since writing the first edition. Accordingly, in this edition, we have tried to increase the level of precision in many explanations and definitions.

This is an introductory, practical book. It does not provide the full theoretical basis or all the details. We feel (rightly or not) that paleontologists should have a practical, working knowledge of data analysis, without demanding a full understanding of the mathematical foundations. Still, we do recommend that a proper statistician is consulted whenever doubts arise, for the pitfalls are many.

# Acknowledgements

Harper's contribution was supported by a grant from the Leverhulme Trust (GB).

# 1

# Introduction

## 1.1   The nature of paleontological data

Paleontology is a diverse field, with many different types of data and a corresponding variety of analytical methods. For example, the types of data used in ecological, morphological, and phylogenetic analyses are often quite different in both form and quality. Data relevant to the investigation of morphological variation in Ordovician brachiopods are quite different from those gathered from Neogene mammal-dominated assemblages for paleoecological analyses. Nevertheless, there is a surprising commonality of techniques that can be implemented in the investigation of such apparently contrasting data sets.

### 1.1.1   Univariate measurements

Perhaps the simplest type of data is straightforward, continuous measurements such as length or width. Such measurements are typically made on a number of specimens in one or more samples, commonly from collections from different localities or species. Typically, the investigator is interested in characterizing, and subsequently analyzing, the sets of measurements, and comparing two or more samples to see how they differ. Measurements in units of arbitrary origin (such as temperature on the Celsius or Fahrenheit scales) are known as *interval* data, while measurements in units of a fixed origin (such as distance and mass) are called *ratio* data. Methods for interval and ratio data are presented in chapter 4.

### 1.1.2   Bivariate measurements

Also relatively easy to handle are bivariate interval or ratio measurements on a number of specimens. Each specimen is then characterized by a *pair* of values, such as both length and width of a given fossil. In addition to comparing samples, we will then typically wish to know if and how the two variables are interrelated. Do they, for example, fit on a straight line, indicating a static (isometric) mode of growth or do they exhibit more complex (anisometric) growth patterns? Bivariate data are discussed in chapters 4 and 6.

*Paleontological Data Analysis*, Second Edition. Øyvind Hammer and David A.T. Harper.

### 1.1.3 Multivariate morphometric measurements

The next step in increasing complexity is multivariate morphometric data, involving multiple variables such as length, width, thickness, and, say, distance between ribs (Fig. 1.1). We may wish to investigate the structure of such data and whether two or more samples are different. Multivariate data sets can be difficult to visualize, and special methods have been devised to emphasize any inherent structure. Special types of multivariate morphometric data include digitized outlines and the coordinates of landmarks. Multivariate methods useful for morphometrics are discussed in chapters 5 and 6.

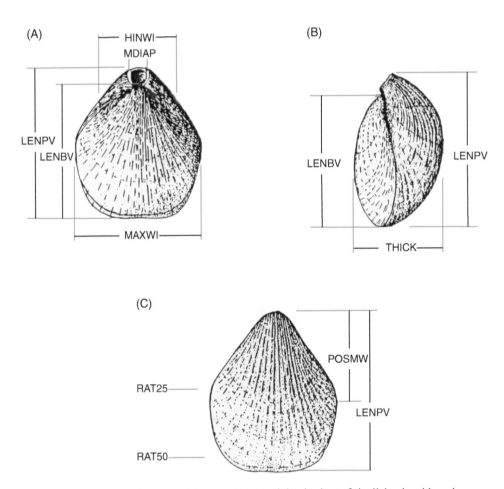

Figure 1.1 Typical set of measurements made on conjoined valves of the living brachiopod *Terebratulina* from the west coast of Scotland. A: Measurements of the length of ventral (LENPV) and dorsal (LENBV) valves, together with hinge (HINWI) and maximum (MAXWI) widths. B: Thickness (THICK). C: Position of maximum width (POSMW) and numbers of ribs per mm at, say, 2.5 mm (RAT25) and 5.0 mm (RAT50) from the posterior margin can form the basis for univariate, bivariate, and multivariate analysis of the morphological features of this brachiopod genus.

### 1.1.4   Character matrices for phylogenetic analysis

A special type of morphological (or molecular) data is the character matrix for phylogenetic analysis (cladistics). In such a matrix, taxa are conventionally entered in rows and
characters in columns. The state of each character for each taxon is typically coded with an
integer. Character matrices and their analyses are treated in chapter 14.

### 1.1.5   Paleoecology and paleobiogeography – taxa in samples

In paleoecology and paleobiogeography, the most common data type consists of taxonomic
counts at different localities or stratigraphic levels. Such data are typically given in an
*abundance matrix*, either with taxa in rows and samples in columns or vice versa. Each cell
contains a specimen count (or abundance) for a particular taxon in a particular sample
(Table 1.1).

In some cases, we do not have specimen counts available; instead, we only know whether
a taxon is present or absent in the sample. Such information is specified in a *presence-
absence table*, where absences are typically coded with zeros and presences with ones. This
type of binary data may be more relevant to biogeographic or stratigraphic analyses, where
each taxon is weighted equally.

Typical questions include whether the samples are significantly different from each
other, whether the biodiversity is different, whether there are well-separated groups of
localities, or any indication of an environmental gradient. Methods for analyzing taxa-in-
samples data are discussed in chapters 5, 9, and 10 (Fig. 1.2).

### 1.1.6   Time series

A time series is a sequence of values through time. Paleontological time series include
diversity curves, geochemical data from fossils through time, and thicknesses of bands
from a series of growth increments. For such a time series, we may wish to investigate
whether there is any trend or periodicity. Some appropriate methods are given in chapter 12.

### 1.1.7   Biostratigraphic data

Biostratigraphy is the correlation and zonation of sedimentary strata based on fossils.
Biostratigraphic data are mainly of two different types. The first is a single table with
localities in rows and events (such as the first or last appearance of a species) in columns.

Table 1.1   Example of an abundance matrix.

|              | Spooky Creek | Scary Ridge | Creepy Canyon |
| ------------ | ------------ | ----------- | ------------- |
| *M. horridus*  | 0  | 43 | 15 |
| *A. cornuta*   | 7  | 12 | 94 |
| *P. giganteus* | 23 | 0  | 32 |

10 cm

Figure 1.2   A muddy-sand community from the Early Jurassic. Community reconstruction is based on the presence of and partly the relative abundance of taxa. After Benton and Harper (2020), modified from McKerrow (1978).

Table 1.2   Example of an event table.

|  | *M. horridus* FAD | *A. cornuta* FAD | *M. horridus* FAD |
| --- | --- | --- | --- |
| Spooky Creek | 3.2 | 4.0 | 7.6 |
| Scary Ridge | 1.4 | 5.3 | 4.2 |
| Creepy Canyon | 13.8 | 13.8 | 15.9 |

FAD, first appearance datum.

Each cell of the table contains the stratigraphic level, in meters, feet, or ranked order, of a particular event at a particular locality (Table 1.2).

Such event tables form the input to biostratigraphic methods such as ranking-scaling and constrained optimization.

The second main type of biostratigraphic data consists of a single presence-absence matrix for each locality. Within each table, the samples are sorted according to stratigraphic level. Such data are used in the method of unitary associations.

Quantitative biostratigraphy is the subject of chapter 13 (Fig. 1.3).

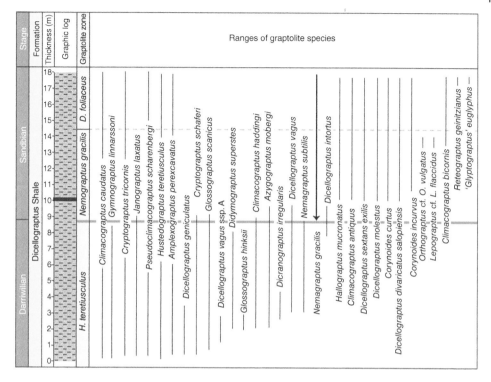

**Figure 1.3** Range chart of fossils, mainly graptolites, through Darriwilian and Sandbian (Middle and Upper Ordovician) strata in southern Sweden. The boundary between the two stages (green horizontal line) is fixed to coincide with the first appearance of the graptolite *Nemagraptus gracilis*. Harper et al. (2022)/Geological Society of London.

## 1.2   Advantages and pitfalls of paleontological data analysis

During the last few decades, paleontology has comfortably joined the other sciences in its emphasis on quantitative methodologies. Statistical and explorative analytical methods, most of them depending on computers for their practical implementation, are now used in all branches of paleontology, from systematics and morphology to paleoecology and biostratigraphy. Over the last 100 years, techniques have evolved with available hardware, from the longhand calculations of the early twentieth century, through the time-consuming mainframe implementation of algorithms in the mid-twentieth century to the microcomputer revolution of the late twentieth century.

In general terms, there are three main components to any paleontological investigation. A detailed description of a taxon or community is followed by an analysis of the data (this may involve a look at ontogenetic or size-independent shape variation in a taxon or the population dynamics and structure of a community) with finally a comparison with other relevant and usually similar paleontological units. Numerical techniques have greatly enhanced the description, analysis, and comparison of fossil taxa and assemblages. Scientific hypotheses can be more clearly framed and statistically tested with numerical data.

The use of rapid computer-based algorithms has rendered even the most complex multivariate techniques accessible to virtually all researchers. Nevertheless, this development has attracted considerable discussion, and critical comments are occasionally made. Such criticisms, if not simply the result of ignorance or misplaced conservatism (which is now rarely the case), are mainly directed at the occasional abuse of quantitative data analysis methods. We will ask the reader to have the following checklist in mind when contemplating the application of a specific method to a specific problem.

### 1.2.1 Data analysis for the sake of it

First of all, one should always have a concrete problem to solve, or at least some hope that interesting and useful, presently unknown, information may emerge from the analysis (the "data mining" approach is sometimes denounced, but how can we otherwise find new territory to explore?). From time to time, we see technically impressive articles with beautiful illustrations that are almost devoid of new, scientifically important content. Technical case studies showing the application of new methods are of course of great interest, but then they should be clearly labeled as such. Do not use a complicated method of data analysis to show something that would be clearer and more obvious with a simpler approach. Shooting sparrows with cannons, or the smashing of open doors.

### 1.2.2 The Texas sharpshooter

The infamous Texas sharpshooter fired his gun at random, hit some arbitrary objects, and afterward claimed that he had been aiming at them in particular. In explorative data analysis, we are often "mining" for some yet unknown pattern that may be of interest and can lead to the development of new, testable hypotheses. The problem is that there are so many possible "patterns" out there that one of them is quite likely to turn up just by chance. It may be unlikely that one particular pattern would be found in a random data set (this we can often test statistically), but the probability that *any* pattern will be found is another matter. This subtle problem is always lurking in the background of explorative analysis, but there is really not much we can do about it, apart from keeping it in mind and trying to identify some process that may have been responsible for the pattern we observe.

### 1.2.3 Explorative method or hypothesis testing?

The distinction between data exploration and hypothesis testing is sometimes forgotten. Many of the techniques in this book are explorative, meaning that they are devices for the visualization of structure in complicated data sets. Such techniques include principal components analysis, cluster analysis, and even parsimony analysis. They are not statistical tests, and their results should in no way be presented as statistical "proof" (even if such a thing should exist, which it does not) of anything.

### 1.2.4 Incomplete data

Paleontological data are invariably incomplete – that is the nature of the fossil record, already pointed out by Charles Darwin and a number of his nineteenth century contemporaries. This problem is, however, no worse for quantitative than for qualitative analysis.

On the contrary, one of the main points of statistics is to understand the effects of the incompleteness of the data. Clearly, when it comes to this issue, much criticism of quantitative paleontological data analysis is really misplaced. Quantitative analysis can often reveal the incompleteness of the data and its importance – qualitative analysis often simply ignores it.

### 1.2.5 Statistical assumptions

All statistical methods make assumptions about the nature of the data. These include both "obvious" assumptions that always apply, such as the samples giving an unbiased representation of the parent population, and more specific assumptions such as the shape of the statistical distribution. Sometimes we are a little sloppy about the assumptions, in the hope that the statistical method will be robust to small violations, but then we should at least know what we are doing.

### 1.2.6 Statistical and biological significance

That a result is statistically significant does not mean that it is biologically or geologically significant. If sample sizes are very large, even minute differences between them may reach statistical significance, without being of any importance in, for example, adaptational, genetic, or environmental terms.

### 1.2.7 Circularity

Circular reasoning can sometimes be difficult to spot but let us give an obvious example. We have collected a thousand ammonites and want to test for sexual dimorphism. So, we divide the sample into small shells (microconchs) and large shells (macroconchs), and test for difference in size between the two groups. The means are found to differ with a high statistical significance, so the microconchs are significantly smaller than the macroconchs, and dimorphism is proved. Where is the obvious error in this procedure?

## 1.3 Software

A range of software is available for carrying out both general statistical procedures and more specialized analytical methods. We would like to mention a few, keeping in mind that new, excellent products are released every year.

A number of professional packages are available for general univariate and multivariate statistical analyses. These may be expensive, but have been thoroughly tested, produce high-quality graphic output, and come with extensive documentation. Examples are SAS and Stata. More specialized, free programs are available for special purposes, such as TNT and MrBayes for phylogenetic (cladistic) analysis, or RASC and CONOP for biostratigraphy.

However, such commercial or special-purpose, stand-alone applications have become much less important in the last 20 years, due to the enormous popularity of the statistical programming environment "R." Being basically a programming language, R gives near-total flexibility, but it also has a relatively high learning threshold. Most new statistical methods that are useful in paleontology are now released as packages for R. It also has very good functions for plotting high-quality figures.

The syntax of the R language is somewhat idiosyncratic and does not adhere to many of the standards of more general programming languages. Partly for this reason, the programming language Python is emerging as a strong competitor to R for data analysis. Python is perceived by many, especially trained programmers, as a more intuitive and elegant language than R, and the number of Python packages for data analysis is rapidly increasing.

In an attempt to lower the threshold into quantitative data analysis for paleontologists, we developed the PAST (PAlaeontological STatistics) software project, including a wide range of both general and special methods used within the field (Hammer et al. 2001). We regard this as a solid, extensive, and easy-to-use package, particularly useful for education but also popular in publication-quality research.

Although most of the examples and figures in this book were made with PAST, we will also provide examples using R and Python.

## References

Benton, M.J., Harper, D.A.T. 2020. *Introduction to Paleobiology and the Fossil Record*. 2nd edition. Wiley-Blackwell, Chichester, UK.

Hammer, Ø., Harper, D.A.T., Ryan, P.D. 2001. PAST: Paleontological statistics software package for education and data analysis. *Palaeontologia Electronica* 4(1), 9.

Harper, D.A.T., Bown, P.R., Coe, A.L. 2022. Chronostratigraphy: understanding rocks and time. In Coe, A.L. (ed.), *Deciphering Earth's History: The Practice of Stratigraphy. Geoscience in Practice*, 227–243. Geological Society, London.

McKerrow, W.S. 1978. *Ecology of Fossils*. Duckworth Company Ltd., London.

# 2

# Statistical concepts

Although a formal introduction to statistics is outside the scope of this book, we need to briefly review some fundamental terms and concepts, especially those concerning the statistical population, the statistical sample, frequency distributions, parameters and models, and hypothesis testing.

## 2.1   The population and the sample

In statistics, the word *population* has another meaning than that used in ecology. The statistical population refers to the complete set of objects (or rather measurements on them) that we are sampling from. In an ideal world, we might hypothetically be able to study the whole population, but this is never achieved in practice. Perhaps the most important purpose of statistics is *to allow us to infer something about the population based on our limited sampling of it*. Contrary to popular belief, statistics can therefore be even more useful for the analysis of small data sets than for vast quantities of big data.

One should try to decide and define what one considers the population. When studying the lengths of fossil horse femurs, the population might be defined as the set of lengths of every preserved (but not yet recovered) femur of a certain species in a given formation at one locality, or of all species in the Pliocene worldwide, depending on the purpose of the investigation. The samples taken should ideally be random samples from the population as defined, and the result of any statistical test will apply to that population only.

## 2.2   The frequency distribution of the population

The statistical population is a rather abstract, almost a Platonic idea – something we can never observe or measure directly. Still, we can define some mathematical properties of the population, with values that can be estimated from our observed samples.

First of all, we must be familiar with the concept of the *frequency distribution*. Graphically, this is a histogram showing the number of measurements within fixed intervals of the measured variable (e.g., Fig. 2.5). We often imagine that the population has an infinite or at least a very large size, so that its (theoretical) frequency distribution can be plotted as a continuous and smooth curve (Fig. 2.1) called the *probability density function* (pdf). The pdf is often normalized to have a unit area under the curve.

The frequency distribution of the population may be characterized by any *descriptive statistic* or *parameter* that we find useful. The distribution is usually reasonably localized, meaning that measurements are most common around some central value, and become rarer for smaller and larger values. The location of the distribution along the scale can then be measured by a *location statistic*. The most common location statistics are the population arithmetic mean, the median, and the mode.

The *population arithmetic mean* or *average* is the sum of all measurements $X$ divided by the number $N$ of objects in the population:

$$\mu = \frac{1}{N}\sum_i X_i$$

The *median* is the value such that half the measurements are smaller than it, or, equivalently, half the measurements are larger than it. For $N$ even, we use the mean of the two central measurements. The *mode* is the most common measurement, that is, the position of the highest peak in the distribution. In rare cases, we have several peaks of equal height, and the mode is then undefined. For symmetrical distributions, the mean, median, and mode will have equal values. In the case of an asymmetric distribution, these three parameters will differ (Fig. 2.1).

Similarly, we can define measures for the *spread* or *dispersion* of the distribution, that is, essentially how wide it is. We could of course just use the total range of values (largest minus smallest), but this would be very sensitive to wild, extreme values (*outliers*) that

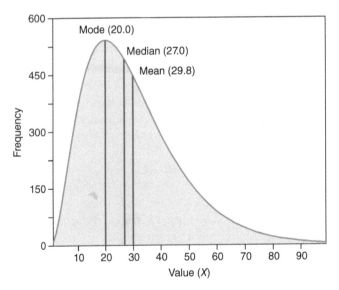

Figure 2.1   For an asymmetric (skewed) distribution, the mean, median, and mode will generally not coincide. The mean will be the parameter farthest out on the tail.

sometimes occur due to measurement errors or "quirks of nature." A more useful measure, at least for symmetric distributions, is the *population variance*, denoted by the square of the Greek letter sigma:

$$\sigma^2 = \frac{1}{N}\Sigma\left(X_i - \mu\right)^2$$

In other words, the population variance is the average of the squared deviations from the population mean. Using the squared deviations ensures that all the components in the sum are positive so they do not cancel each other out. We also need to define the *population standard deviation*, which is simply the square root of the population variance:

$$\sigma = \sqrt{\sigma^2}$$

## 2.3   The normal distribution

Some idealized distribution shapes are of special importance in statistics. Perhaps the most important one is the *normal*, or *Gaussian*, distribution (Fig. 2.2). This theoretical distribution is symmetric and is characterized by two parameters: the mean and the standard deviation. As explained earlier, the mean indicates the position of the distribution along the value axis, while the standard deviation indicates the spread around the mean.

Given a mean $\mu$ and a standard deviation $\sigma$, the bell-shaped continuous normal distribution is defined by an explicit equation:

$$f = \frac{1}{\sqrt{2\pi\sigma^2}} e^{-\left(x-\mu\right)^2/2\sigma^2}$$

where $\pi$ and $e$ are the mathematical constants (this is normalized to ensure the unit area under the curve).

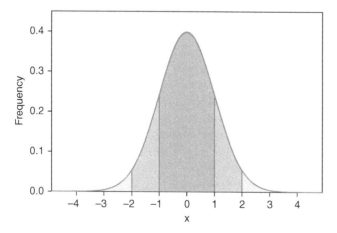

Figure 2.2   Normal (Gaussian) distribution with mean $\mu = 0$ and standard deviation $\sigma = 1$. Vertical lines are drawn at one and two standard deviations away from the mean.

For the normal distribution, about 68% of the values fall within one standard deviation from the mean, 95% fall within two standard deviations, and 99.7% within three standard deviations (Fig. 2.2).

The normal distribution is a theoretical construction, but it is important for several reasons. First, it can be shown that if you repeatedly add together many random numbers (not necessarily from a normally distributed population), the sums will be approximately normally distributed. This is known as the *central limit theorem*. The central limit theorem means that repeated sample means from a population with any distribution will have a near-normal distribution. Second, and probably related to the central limit theorem, it is a fact that many measurements from nature have a near-normal distribution. Third, this distribution is relatively easy to handle from a mathematical point of view. Still, it is important to remember that the normal distribution is only a convenient model. No real populations will be exactly normally distributed.

Many statistical tests assume that the measurements are taken from a normally distributed population. This assumption needs to be thought through in each case. There are many natural processes that can prevent a population from being (nearly) normally distributed, as well as many situations where a normal distribution is not expected, even in theory. For example, waiting times between random, independent events along a timeline are expected to have an *exponential* distribution.

## 2.4 Cumulative probability

Let us assume that we have a normally distributed population (e.g., lengths of fossil horse molars) with mean $\mu = 54.3$ mm and standard deviation $\sigma = 8.7$ mm. From the discussion earlier, we would expect about 68% of the teeth to be within one standard deviation from the mean, i.e., between $54.3 - 8.7 = 45.6$ and $54.3 + 8.7 = 63.0$ mm; 95% would be within two standard deviations from the mean, or between 36.9 and 71.7 mm.

Now what is the probability of finding a tooth smaller than 35 mm? To find the probability of a single specimen having a size between $a$ and $b$, we *integrate* the normal distribution from $a$ to $b$, that is, we use the area under the probability density function from $a$ to $b$. In our case, we want to find the probability of a single specimen having a size between $-\infty$ to +35. Of course, the concept of a negative length of a horse molar is meaningless – the real distribution must be zero for negative values. As always, the normal distribution is only a useful approximation to the real distribution. Be that as it may, we say that the probability of finding a tooth smaller than 35 mm is equal to the area of the normal distribution curve from $-\infty$ to +35 (Fig. 2.3). For any size $x$, we can define the cumulative distribution function as the integral of the probability distribution from $-\infty$ to $x$.

For practical reasons, we often do this by transforming our normally distributed numbers into so-called $z$ values having zero mean and a standard deviation of one and then comparing with the standardized cumulative distribution function that can be looked up in a table (Fig. 2.4).

For our example, we transform our tooth length of 35 mm to a standardized $z$ value by subtracting the mean and dividing by the standard deviation:

$$z = (x - \mu)/\sigma = (35\,\text{mm} - 54.3\,\text{mm})/8.7\,\text{mm} = -2.22$$

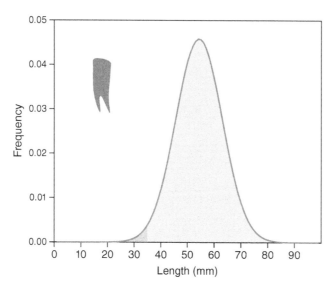

Figure 2.3   A normal distribution (probability density function) of lengths of horse molars, with mean 54.3 mm and standard deviation 8.7 mm. The dark area is equal to the probability of finding a tooth smaller than 35 mm.

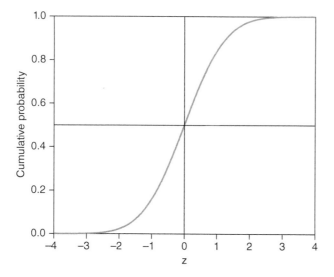

Figure 2.4   Cumulative probability for the standardized normal distribution (integral from −∞). It can be seen directly from this graph, for example, that half the values (i.e., cumulative probability 0.5) fall below the mean (zero), as would be expected from a symmetric distribution.

Note that the $z$ value is *dimensionless*, that is, it has no unit. Using a $z$ table, we find that the area under the standardized normal distribution to the left of −2.22 is 0.013. In other words, only 1.3% of the teeth will be smaller than 35 mm in length.

How many teeth will be larger than 80 mm? Again, we transform to the standardized value:

$$z = (x - \mu)/\sigma = (80\,\text{mm} - 54.3\,\text{mm})/8.7\,\text{mm} = 2.95$$

From the $z$ table, we get that the area under the standardized normal distribution to the left of 2.95 is 0.9984, so 99.84% of the teeth will be smaller than 80 mm. The remaining teeth, or $100 - 99.84 = 0.16\%$, will be larger than 80 mm.

## 2.5 The statistical sample, estimation of distribution parameters

The statistical sample is the actual set of measurements collected with the aim of giving a representative picture of the population. An example of a statistical sample might be the set of widths measured on trilobite pygidia from a given species in a chosen stratigraphic horizon. Often, we have two or more statistical samples to be compared, representing one or more populations.

Population parameters can be estimated from the sample. Thus, an estimate of the population mean can be acquired by computing the *sample mean*, usually denoted by the name of the variate with a bar above it:

$$\bar{x} = \frac{1}{n}\sum x_i$$

Here, $x_i$ are the individual measured values for the sample. By convention, all letters related to samples are written in lowercase in this book.

The sample mean is only an estimate of the population mean, and repeated sampling from the population will give slightly different estimates. By the central limit theorem, these estimates of the mean will be close to normally distributed, with a standard deviation (or *standard error*) of

$$s_e = \frac{\sigma}{\sqrt{n}}$$

Clearly, the standard error of the estimate of the mean decreases as $n$ increases. In other words, we get a more accurate estimate of the population mean if we increase the sample size, as might be expected.

The population variance can be estimated by the *sample variance*, which is denoted by

$$s^2 = \frac{1}{n-1}\sum\left(x_i - \bar{x}\right)^2$$

Perhaps non-intuitively, $s^2$ is defined slightly differently from the population variance $\sigma^2$, dividing by $n - 1$ instead of by $n$. This formulation gives us an *unbiased* estimate of the population variance.

The sample standard deviation is defined as the square root of $s^2$:

$$s = \sqrt{s^2}$$

Even more curiously, even though $s^2$ is an unbiased estimator of $\sigma^2$, $s$ is not an entirely unbiased estimator of $\sigma$, but the bias is small for reasonably large $n$. See Sokal and Rohlf (1995) for details.

---

**Example 2.1**

Using a relatively accurate approximation (Press et al. 1992), we can ask a computer to generate "random" numbers taken from a normal distribution with known parameters. In this way, we can test our estimates of the population parameters against the true values.

We select a mean value of $\mu = 50$ and a standard deviation of $\sigma = 10$ as the population parameters. Figure 2.5 shows histograms of two samples taken from this population, one small ($n = 10$) and one larger ($n = 100$). Note that the sample mean $\bar{x}$ and standard deviation $s$ get closer to the population parameters as the sample size increases, although the numbers will obviously be different each time we run the experiment.

We then take 10 samples, each of size $n = 100$. The mean values are shown in Table 2.1.

The standard error of the estimate of the mean should be $s_e = \sigma/\sqrt{n} = 10.00/\sqrt{100} = 1.00$. This is close to the standard deviation of the estimated means shown in Table 2.1, which is 1.17.

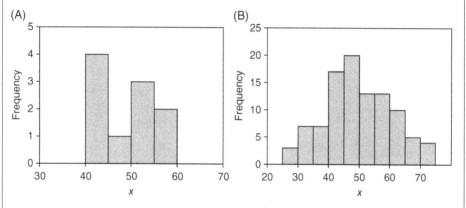

**Figure 2.5** Two samples drawn from a normally distributed population with a mean of 50.00 and a standard deviation of 10.00. A: $n = 10$, $\bar{x} = 48.91$, $s = 6.302$. B: $n = 100$, $\bar{x} = 49.70$, $s = 10.94$.

**Table 2.1** Mean values of 10 samples of size $n = 100$, taken from a normal distribution with mean 50.00.

| Sample | $\bar{x}$ |
| --- | --- |
| 1 | 51.156 |
| 2 | 50.637 |
| 3 | 48.558 |
| 4 | 51.220 |
| 5 | 49.626 |
| 6 | 49.779 |
| 7 | 48.821 |
| 8 | 50.813 |
| 9 | 51.320 |
| 10 | 48.857 |

## 2.6 Null hypothesis significance testing

Consider we have two or more samples (sets of collected data), each of which is hopefully representative of the complete fossil material in the field (the statistical population; section 2.1) from which it was taken. We want to compare the samples with each other and investigate whether the populations they were taken from are "different" in one sense or another. For example, we may want to test whether the mean length of the skull is different in *Tarbosaurus bataar* and *Tyrannosaurus rex* (Fig. 2.6). We have two groups with seven skulls of *T. bataar* and nine skulls of *T. rex*. The lengths of these skulls represent the two samples to be compared (Fig. 2.7). The naïve approach, which is sometimes followed by researchers, is simply to take the mean of each sample and observe that the mean length of the *measured T. rex* skulls, say 82 cm, is different from (larger than) the mean length of the measured *T. bataar* skulls (say 76 cm). But to conclude from this that the two species have different sizes would be entirely misguided! Even if the two samples came from the same population, they would always have different mean values just because of random sampling effects, so no solid conclusion can be drawn from this simple comparison of means.

What we must do instead is to be very precise about the question we are asking, that is, the hypothesis we are testing. First, we need to clarify that we want to compare the means of *T. rex* and *T. bataar* as such, not the means of the limited samples, which are only estimates of the population means. So we spell out the problem as follows: based on the two small samples we have, is it reasonable to assume that the populations they were taken from have different means? This formulation leads to a precise *null hypothesis* to be tested:

$H_0$: the two samples are taken from populations with identical mean values.

For completeness, we should also specify the *alternative hypothesis*:

$H_1$: the two samples are taken from populations with *different* mean values.

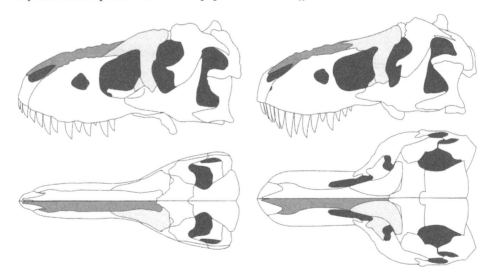

Figure 2.6 Skull of the theropod dinosaur *T. bataar* (left) from the Upper Cretaceous of Mongolia; it is a close relative of *T. rex* from the western USA (right). From Hurum and Sabath (2003).

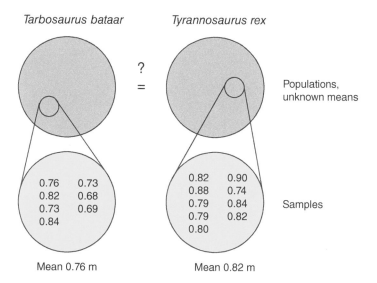

*Tarbosaurus bataar*          *Tyrannosaurus rex*

?
=

Populations,
unknown means

| 0.76 | 0.73 |
| 0.82 | 0.68 |
| 0.73 | 0.69 |
| 0.84 | |

| 0.82 | 0.90 |
| 0.88 | 0.74 |
| 0.79 | 0.84 |
| 0.79 | 0.82 |
| 0.80 | |

Samples

Mean 0.76 m          Mean 0.82 m

**Figure 2.7** The null hypothesis of equality of the population mean lengths of the skulls in *T. bataar* and *T. rex* (top circles) can only be investigated through limited sampling of the populations. The mean lengths in these limited samples (bottom circles) will be different for each collection.

Another alternative hypothesis might be

$H_1$: the *T. rex* sample is taken from a population with a *larger* mean value.

The choice of an alternative hypothesis may however influence the result of the test.

The way forward from this point is a matter of great controversy and endless confusion. In this book, we will mainly discuss the so-called "Null Hypothesis Significance Testing" (NHST). Despite much criticism, NHST is still the most common approach for comparing samples. It is fair to say that although NHST has obvious shortcomings, it is fully consistent, mathematically sound, and even useful. We just have to interpret the results correctly and understand what NHST can do and what it cannot.

By applying an appropriate statistical test (the Student's *t*-test of section 4.3 might be a good choice), we are told that if the null hypothesis were true, the probability of seeing this large a difference between the mean values of two random samples is 2.2%. A standard notation is $p = 0.022$. This is a rather low probability, so we may choose to assume that the null hypothesis is likely to be false and should be rejected. In other words, the sizes of the skulls are different. But what is "low probability"? What if we had calculated $p = 0.13$? The threshold for rejecting the null hypothesis is a matter of taste, but 0.05 has traditionally been a standard choice. In traditional NHST, this *significance level* should be decided before the test is performed, in order to avoid the temptation to adjust the significance level to get the result we wish (in an ideal world of totally detached science, this would not be an issue, but the investigator is normally eager to get a positive result).

If the reported *p* value is lower than the chosen significance level (such as 0.05), we can reject the null hypothesis and state that we have a statistically significant difference between the two species. However, the converse is not necessarily true: if we get a high *p* value, this does not directly allow us to state that the species are probably identical.

Remember from the above that the $p$ value does not state the probability of the null hypothesis being true (although we often use that shorthand); it only states the probability of obtaining the observed (or larger) difference in sample means, if the null hypothesis were true. Therefore, a high $p$ value may be due to an insufficient sample size to detect some small departure from the null hypothesis. It may seem that a possible way out is to reverse the null hypothesis and the alternative hypothesis as follows:

$H_0$: the two samples are taken from populations with different mean values.
$H_1$: the two samples are taken from populations with identical mean values.

Now we could perhaps reject the null hypothesis and thus get statistical support for the alternative hypothesis. The reason that this is not allowed is that the null hypothesis must represent a precise, unique condition. That samples are from identical populations (which they can be in only one way) is an acceptable null hypothesis, but that they are from different populations (which they can be in many ways) is not.

This means that we must be a little careful about the wording if we want to report a high $p$ value. We cannot say that "the species have equal mean sizes $(p = 0.97)$." Instead, we may say, for example, "the null hypothesis of equal mean sizes could not be rejected at $p < 0.05$." This, however, cannot be considered a statistically significant result.

We sum up the traditional steps in carrying out a statistical test as follows:

1) Collect data that are considered representative of the population(s).
2) State a precise null hypothesis $H_0$ and an alternative hypothesis $H_1$.
3) Select a significance level, for example, 0.05.
4) Select an appropriate statistical test. Check that the assumptions made by the test are likely to hold, at least approximately.
5) If the test reports a $p$ value smaller than the significance level, you can reject the null hypothesis.

### 2.6.1 Type I and type II errors

When performing a statistical test, we are in danger of arriving at the wrong conclusion due to two different types of errors, confusingly known as *type I* and *type II*.

A type I error involves rejecting the null hypothesis even though it is true. In our example, assume that the *T. rex* and *T. bataar* did have indistinguishable mean sizes back in the Cretaceous – maybe they even actually represented the same species. If our test rejects the null hypothesis based on our small samples and our selected significance level, making us believe that the samples were from populations with different means, we have committed a type I error. This will happen now and then (in the case of a significance level of 0.05, it will happen in at least 5% of cases), usually when a random sample is not sufficiently representative of the population. Obviously, the risk of committing a type I error can be decreased by choosing a low significance level, such as 0.01. Type I errors are clearly very unfortunate, because they make us state an incorrect result with some confidence.

A type II error involves *not* rejecting the null hypothesis even though it is *false*. In our example, this would mean that *T. rex* and *T. bataar* actually were of different mean sizes, but the test reported a high $p$ value for the null hypothesis of equality. This will also happen occasionally. The risk of committing a type II error can be decreased by choosing a high significance level, such as 0.1. Type II errors are not quite as serious as type I errors, for if the null hypothesis is not rejected, you cannot state anything with confidence anyway. It is just unfortunate that you do not get a positive result.

The probabilities of type I and type II errors can be partly controlled using the significance level, but as the significance level is decreased with the aim of reducing the risk of making a type I error, the risk of making a type II error will increase. However, increasing the sample size will allow us to decrease the probability of making either type I or type II errors.

### 2.6.2 Power

The frequency of committing a type I error when the null hypothesis is true is controlled by the chosen significance level and is called $\alpha$. The frequency of committing a type II error when the null hypothesis is false (and the alternative hypothesis is true) is denoted by $\beta$. The value $1 - \beta$, signifying the frequency of correctly detecting a significant result, is called the *power* of the test. Clearly, we want the power to be high. The power is different from test to test and also depends on the data, the sample size, and the significance level.

### 2.6.3 Robustness

All statistical tests make a number of assumptions about the nature of the population and the sampling of it, but the tests may differ with respect to their sensitivity to deviations from these assumptions. We then say that the tests have different degrees of *robustness*. This term is also used in other fields of data analysis, such as the estimation of model parameters (section 4.6).

### 2.6.4 Effect size

For large samples, the power of the statistical test will be high, and it will pick up even minute departures from the null hypothesis. For example, using large samples of *T. rex* and *T. bataar*, we may find a statistically significant difference in mean skull lengths ($p < 0.05$) of 0.802 and 0.798 m, respectively. Although significant, this minute difference is of no scientific interest. We must therefore always report the actual magnitude of the differences observed, in addition to the $p$ value. This is often called the *effect size*.

### 2.6.5 NHST misunderstandings

NHST is useful for one specific purpose: to check if an observed difference could be due to sampling rather than a true population difference. We might say that, for large sample sizes, NHST is not informative, because the power is so high that any tiny difference between populations will be detected, and the null hypothesis will always be rejected. For small sample sizes, however, when we think we see a considerable difference between the samples, NHST is a good tool to check for nonrepresentative sampling. NHST has been much misunderstood, misused, and applied indiscriminately. In recent years, this has caused many people to argue that it should be abandoned altogether (e.g., Holland 2019).

We believe this is not yet called for, as long as a few things are kept in mind (this is not an exhaustive list):

- The *p* value is not the probability of the null hypothesis being true; rather, it is the probability of obtaining as large a difference as observed in the sample data if the null were true.
- A large *p* value cannot be used to accept the null.
- For large sample sizes, the null hypothesis will nearly always be rejected, so NHST is not very useful.
- A fixed significance level such as $p < 0.05$ can be deceptive and is not required. Many statisticians now suggest abandoning a fixed cutoff point and instead just reporting the actual *p* value.
- The effect size (such as a difference in means) must always be reported, possibly with a confidence interval. This is in fact much more important than the *p* value.

## 2.7 Bayesian inference

There is nothing mathematically wrong with the old-school of statistics and the NHST framework that we have presented so far, often known as the "frequentist" approach, because probabilities are defined through frequencies of occurrence. The theory is solid, and most scientific papers still rely on frequentist methods. However, in the last decades, another way of thinking has been emerging, founded in the work of the Reverend Thomas Bayes (1763) but more fully developed by Harold Jeffreys (1961) and later made practical by developments in numerical methods, mainly an integration method called Markov Chain Monte Carlo (MCMC). This is the school of *Bayesian statistics*. The Bayesians do not work with NHST and its *p* values, but rather talk about distributions and beliefs.

The frequentists are mainly occupied with the probability of the data given a certain parameter value (hypothesis), which is called the *likelihood*. The Bayesians argue that this is usually not what we are really interested in. Rather, we want to know the probability distribution of the parameter (including the probability of a particular hypothesis) given the data. In other words, given the sample data, what confidence can we have in a certain hypothesis for the parameter value, such as the difference in means between two populations being zero?

### 2.7.1 Bayes' theorem

There is no way around it – we cannot introduce Bayesian inference without the Fundamental Creed of the Bayesians, called Bayes' Theorem. The most important application of Bayes' Theorem is the estimation of a model parameter (such as a population mean) given the observed data, in which case the theorem can be written as

$$p(\vartheta \mid \text{data}) = \frac{p(\text{data} \mid \vartheta) p(\vartheta)}{p(\text{data})}$$

In this equation, $\vartheta$ is the parameter of interest. In Bayesian statistics, we usually talk about probability distributions *p*, not single parameter values. Thus, $p(\vartheta|\text{data})$ is the probability distribution of a parameter given the data. This is called the *posterior*, and it is what we want to know (say the Bayesians). The term $p(\text{data}|\vartheta)$ is the probability distribution of the data given a parameter value. This is the *likelihood*, familiar to frequentists. The term $p(\vartheta)$ is the probability of the parameter before the data are collected. This is called the *prior*. The denominator $p(\text{data})$ is called the *evidence* or marginal probability and represents the probability distribution of the data independent of the parameter.

With these definitions in place, we can use Bayes' theorem as a kind of algorithm. Without any data, we start with some belief in the probability distribution $p(\vartheta)$ (the prior). The Bayesians use the term "belief" without shame; in fact, it is endorsed and encouraged on philosophical grounds. The prior can be totally or partially uninformed, or we can guess on, e.g., a normal distribution with some reasonable mean and variance, or we can start with some old idea by Professor Authority that is generally believed. Next, we collect our sample data and compute the likelihood in the classical, frequentist way. Finally, we need the evidence. This is often the most difficult part to compute.

With the prior, likelihood, and evidence, we can easily calculate the posterior distribution of the parameter using Bayes' Theorem. This constitutes our new belief in the distribution of the parameter, updated from the prior after the collection of data. At a later date, we can use this as our new prior, collect more data, and update our posterior accordingly.

### 2.7.2 Markov Chain Monte Carlo

The calculation of the evidence $p(\text{data})$ requires a numerical integration over the parameters, adding up the probabilities of the data under all parameter values. For a single parameter such as the mean, this is fairly straightforward, but in most cases the parameter is multidimensional, and the integration can become very difficult. Moreover, we want to produce a distribution for the posterior probabilities over a range of parameter values, again usually multidimensional, and this is often computationally intractable for both the likelihood and the evidence. These computational difficulties were part of the reason for the delayed adoption of Bayesian methods in data analysis. The development of Markov Chain Monte Carlo (MCMC) methods revolutionized the field. MCMC is basically a numerical integration method, especially well suited for the estimation of posterior probabilities. MCMC traverses through parameter space along a random path designed to sample the distribution evenly. In many Bayesian data analysis programs, the user is required to specify some parameters of the MCMC algorithm, such as the total number of steps (typically millions).

### 2.7.3 What is the point?

When it comes to parameter estimation, Bayesian inference differs from the classical frequentist (maximum likelihood) approach in three main ways. First, the Bayesians try to estimate the probability distribution of the parameter given the data, rather than the probability of the data given the hypothesis. Second, we are allowed to specify a prior, which can be objective or rather subjective, but can be used to moderate the sampling variance of the posterior. Bayes' Theorem also provides a framework for updating our beliefs as new data are collected. Third, the Bayesian approach emphasizes the probability distribution of parameter values, and single values of the parameter (point estimates) are avoided.

This is all good, but it must also be said that Bayesian inference does not always give very different results from a frequentist analysis. For example, it is quite common to specify a uniform (constant or "flat") prior. Since the evidence is also a constant with respect to the parameter value, the posterior will be equal to the likelihood apart from a constant factor. The most probable parameter value in the posterior will therefore be equal to the maximum likelihood value of the frequentists.

### 2.7.4 Bayes factors

The Bayesians also have alternatives to NHST. The Bayes Factor $K$ is the ratio of the "marginal likelihoods" of two competing hypotheses $H_0$ and $H_1$:

$$K = \frac{p(\text{data} \mid H_1)}{p(\text{data} \mid H_0)}$$

The marginal likelihoods must be computed by integrating over all prior probabilities of the parameter values, which can be computationally daunting. However, equations for the Bayes factors have been worked out for a number of special cases. For our *Tarbosaurus* versus *Tyrannosaurus* example described earlier, a Bayesian factor alternative to the $t$-test (Rouder et al. 2009) gives $K = 2.9$, computed as the likelihood ratio of the alternative hypothesis $H_1$ (sizes are different) to the null hypothesis $H_0$ (sizes are the same). This value is larger than 1; hence, $H_1$ is favored over $H_0$, but it is not very large – often, $K = 3$ is cited as the limit below which the evidence for $H_1$ is not substantial.

A useful property of the Bayes factor is that it is symmetric with respect to the two hypotheses. Hence, a very small value, say $K < 1/3$, may be taken as substantial evidence for the null hypothesis $H_0$. This is in contrast with NHST, where the null hypothesis can never be accepted, only rejected.

## 2.8 Exploratory data analysis

So far in this chapter, we have discussed formal procedures for statistical inference: the estimation of population parameters from samples and the testing of hypotheses. Another large area of statistics concerns methods for summarizing and visualizing data sets. In such exploratory data analysis (Tukey 1977), the goal is to discover patterns in order to form new hypotheses. Such open-minded inspection of the data is particularly important when studying highly complex processes particularly suited to paleobiology. Most of this book will describe tools for exploratory data analysis.

## References

Bayes, T., Price, Mr. 1763. An essay towards solving a problem in the doctrine of chances. *Philosophical Transactions of the Royal Society of London* 53, 370–418.

Holland, S.M. 2019. Estimation, not significance. *Paleobiology* 45, 1–6.

Hurum, J.H., Sabath, K. 2003. Giant theropod dinosaurs from Asia and North America: Skulls of *Tarbosaurus bataar* and *Tyrannosaurus rex* compared. *Acta Palaeontologica Polonica* 48, 161–190.

Jeffreys, H. 1961. *Theory of Probability*. 3rd edition. Clarendon Press, Oxford, UK.

Press, W.H., Teukolsky, S.A., Vetterling, W.T., Flannery, B.P. 1992. *Numerical Recipes in C*. Cambridge University Press, Cambridge, UK.

Rouder, J.N., Speckman, P.L., Sun, D., Morey, R.D., Iverson, G. 2009. Bayesian *t* tests for accepting and rejecting the null hypothesis. *Psychonomic Bulletin and Revue* 16, 225–237.

Sokal, R.R., Rohlf, F.J. 1995. *Biometry: The Principles and Practice of Statistics in Biological Research*. 3rd edition. W.H. Freeman, New York.

Tukey, J.W. 1977. *Exploratory Data Analysis*. Addison-Wesley, Reading, Mass.

# 3

# Introduction to data visualization

When you have data, the first thing you do is to plot them. Visualization is perhaps the most important part of data analysis, and many creative methods have been developed for showing data to their best advantage. Since the first edition of this book, there has been a strong trend in scientific publication toward more visually appealing graphics, driven by new software, new high-quality graphic file formats such as SVG, online publication allowing more use of color, and not least a new focus on graphic design principles. This is a very welcome development, but the bar has been raised significantly, and the effort required has increased. We (the authors) are not particularly good at it, but the quality of figures in this book has hopefully improved since the first edition.

We will not go much into graphic design as such, important as it is, but a few basic design principles bear repetition.

## 3.1 Graphic design principles

### 3.1.1 Vector graphics

There are two fundamentally different ways of representing figures on the computer: bitmaps and vector graphics. A bitmap is a collection of pixels, well suited for photographs. Typical bitmap formats are JPG and TIF. The quality of a bitmap image is mainly a function of the resolution (number of pixels per cm) and any data compression applied.

A vector graphic file, in contrast, contains a description of graphic elements such as lines, points, and letters, with their positions, sizes, colors, etc. For scientific data visualization, always use vector graphics, through to the end of the process. This allows object-based editing of the figure in software such as Illustrator, CorelDraw, or Inkscape, and it also provides a vastly superior quality, as the image can be zoomed indefinitely without pixelation. Currently, the two most popular vector graphic file formats are SVG and PDF.

### 3.1.2 Fonts

In figures, always use a "sans serif" font such as Arial or Helvetica. Font sizes smaller than 6 pt are uncomfortable, although this is less of a concern with online publications where the figure can be expanded. Do not mix too many font sizes, but standardize on,

*Paleontological Data Analysis*, Second Edition. Øyvind Hammer and David A.T. Harper.
© 2024 John Wiley & Sons Ltd. Published 2024 by John Wiley & Sons Ltd.

e.g., a header size (e.g., 12 pt), an intermediate size for axis labels (e.g., 10 pt), and a small size for individual elements (e.g., 8 pt).

### 3.1.3 Colors

Only a few years ago, authors had to pay large amounts of money to get their figures published in color. Now, color is usually free, and we should use it, but with some restraint. Highly saturated colors can give a noisy impression. Many plotting packages include predefined sets (palettes) of colors that match well, such as the very attractive "viridis" palette set, which also works well for the color-blind.

Special considerations apply to color gradients used to visualize a continuous range of values, such as temperature or water depth on a map. A grayscale gradient from black to white is a useful option, but color can sometimes show patterns more clearly as well as be more aesthetically appealing. However, a color scale such as the widely used "jet" scale grading from blue through green and yellow to red, can give false impressions of data discontinuities at the color boundaries. Modern color scales such as viridis attempt to avoid such perceptual breaks in the gradient (Crameri et al. 2020).

### 3.1.4 Fills

Many plots include regions such as boxes, circles, confidence intervals, and polygons around points. Such regions can be emphasized and become more attractive when filled in with a low saturation color. Many programs allow semi-transparency, so that such regions can overlap without obscuring each other.

## 3.2 Line charts

One of the simplest types of plots is a depiction of a sequence of data points, usually connected by lines. In paleontology, such line charts typically show the variation of a measured variable such as species richness, body size, fossil density, or some sedimentological or geochemical parameter, through time or stratigraphic level. Typically, time is shown on the horizontal axis, usually running from left to right, while stratigraphic level is shown on the vertical axis, running from bottom to top.

There is some disagreement about the use of line segments between the given data points. Some people argue that there is no information between the data points, and thus connecting them by lines gives a false impression of continuity. However, plotting just the points can make it difficult to see trends and discontinuities. As a general recommendation, we suggest using connecting lines to guide the eye, but also to include dots (or other symbols) at the actual data points (Fig. 3.1). What we do discourage is connecting the dots with a smooth curve, which is a popular option in Excel.

In addition to lines and dots, there are other options for plotting data sequences. We can use bars or steps to emphasize the noncontinuous (discrete) nature of the data. Or we can fill in the area below the curve for a clearer impression (silhouette plot; Fig. 3.1C).

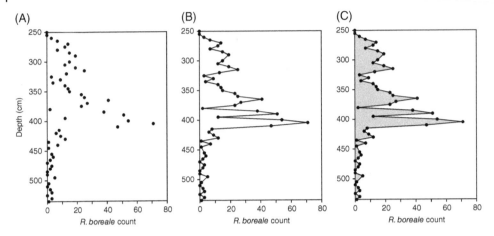

**Figure 3.1** Three alternative ways of plotting counts of a radiolarian through a core: points (A); points and lines (B); and silhouette plot (C).

Showing several curves in the same diagram can easily get cluttered. The *stacked chart* is a type of multiple line plot often used in paleontology and palynology. In such a chart, relative abundances (percentages) are plotted as widths between lines (Fig. 3.2). Although compact, such a plot can be difficult to read because the position of one line depends on the sum of the widths to the left of it, making it jump back and forth (e.g., the gray region in Fig. 3.2).

## 3.3 Scatter plots

A scatter plot shows bivariate data (i.e., *x–y* pairs) as points or other symbols in a Cartesian coordinate system. Scatter plots are used for, e.g., bivariate morphometric data such as length versus width.

In many cases, the data points are divided into given groups according to, e.g., species or locality. Such groups will typically be indicated by different colors or symbols. In addition, two different graphical elements are commonly used to make the groups more visible (Fig. 3.3). The *convex hull* is the smallest convex polygon containing all the points in the group. The 95% *concentration ellipse* is constructed by assuming that the points are sampled from a population with bivariate normal distribution, and the ellipse marks a region within which 95% of the points from this population are expected to fall.

It is possible to plot a third variable using a color scale or a size for each point. A scatter plot where the points are proportional in size to a third variable is called a *bubble plot* (Fig. 3.4).

## 3.4 Histograms

The histogram is the familiar way of presenting a sample distribution as a collection of bars, showing the number of items within consecutive intervals (bins). The main consideration when drawing a histogram is the bin size. Smaller bin sizes may appear to give

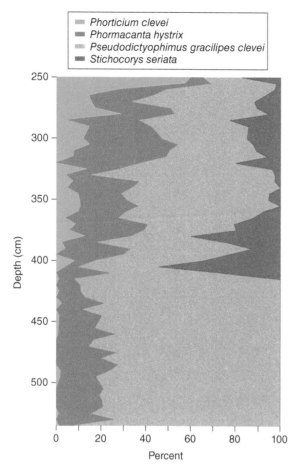

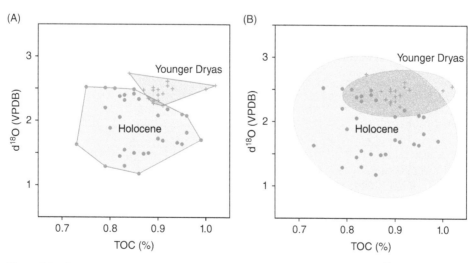

Figure 3.2    Stacked line chart showing percentages of four species of radiolarians through a core. Data from Bjørklund et al. (2019).

Figure 3.3    Scatter plots of total organic carbon (TOC) versus oxygen isotopes ($d^{18}O$) in samples from a core from Andfjorden, Norway. Data from Bjørklund et al. (2019). (A) Holocene (warm) and Younger Dryas (cold) groups marked with convex polygons. (B) Groups marked with 95% concentration ellipses.

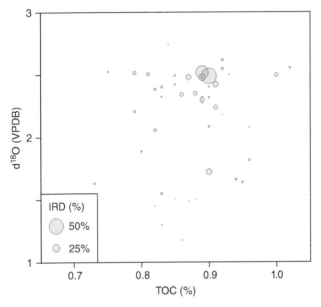

d¹⁸O (VPDB)

IRD (%)

50%

25%

TOC (%)

Figure 3.4   Bubble plot of data from the Andfjorden core (cf. Fig. 3.3). IRD = ice-rafted debris.

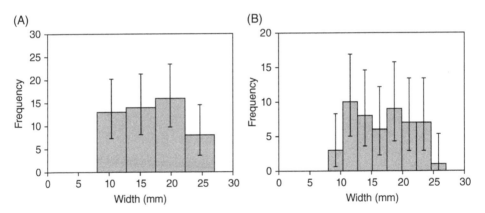

(A)

(B)

Frequency

Frequency

Width (mm)

Width (mm)

Figure 3.5   Histograms of sizes of brachiopods (*Protatrypa*) from the Silurian of Oslo, Norway (*n* = 51). (A) Four bins, as suggested by the zero-stage rule. (B) Eight bins. The 95% confidence intervals become relatively larger as the number of bins increases.

higher resolution, but at the cost of smaller counts within each bin, which increases sampling variance (noise). This is illustrated by the 95% confidence intervals on the counts within each bin (Fig. 3.5). These confidence intervals can be estimated with the Clopper–Pearson method (Clopper and Pearson 1934).

It is common to specify the number of bins manually, but several "objective" methods have also been proposed. For example, the "zero-stage rule" (Wand 1997) gives bin width as

$$h = 3.49 \min\left(s, \mathrm{IQ}/1.349\right) n^{-1/3}$$

where $s$ is the sample standard deviation and IQ is the interquartile range. With this rule, the size of the bins increases with sample dispersion and decreases with sample size.

## 3.5 Bar chart, box, and violin plots

A histogram gives a fairly detailed view of the sample distribution. When several samples are to be presented and compared, more simplified plots are useful.

The means of each sample can be shown in a *bar chart*, optionally including the standard errors of the estimates of the means or their 95% confidence intervals (Fig. 3.6A). In a *mean-and-whiskers plot*, the bars are omitted (Fig. 3.6B).

The *box plot* is a useful way of visualizing a univariate distribution as a sort of abstracted histogram (Fig. 3.6C). A vertically oriented rectangle of arbitrary width is drawn from the 25th percentile (a value such that 25% of the sample values are below it) up to the 75th percentile (a value such that 25% of the sample values are above it). Inside this box, the median is indicated with a horizontal line. Finally, the minimal and maximal values are plotted as short horizontal lines, connected with the box by vertical lines ("whiskers"). In some box plots, presumed outliers in the data are plotted as points outside the whiskers.

The box plot may seem a little arbitrary, but it does summarize a lot of useful information in a simple graphical object. The total range and the range of the main body of values are immediately apparent, as is the position of the median. It is also easy to spot whether the distribution is asymmetric (skewed) or strongly peaked (high kurtosis). Several samples can be plotted in the same diagram without clutter, for easy comparison of their medians and dispersions. The box plot is therefore very popular.

A violin plot (Fig. 3.6D) shows a more elaborate but still stylized representation of the distribution, using smooth "kernel density" functions. The violin plot is often combined with a box plot.

## 3.6 Normal probability plot

It is often of interest to investigate how well a data set can be described by a normal distribution. Together with the histogram, the normal probability plot (Chambers et al. 1983) can be a helpful visual tool for identifying the type of departure from normality, if any (e.g., skew, "fat tails," outliers, or bimodality). The general idea is to plot the cumulative distribution function of the sample against the cumulative distribution function for a normal distribution (Fig. 3.7). If the sample came from a normal distribution, the points would be expected to fall on a straight line. Left-skewed (tail to the left) data will be hollow downwards and right-skewed data will be hollow upwards (U-shaped). Short tails will cause a flattening of the normal probability plot for the smallest and largest values, giving an S-shaped curve. Fat tails will give a steeper plot for the smallest and largest values, giving a "mirrored S" curve.

In one version of the normal probability plot, the sorted observations are plotted as a function of the corresponding normal order statistic medians, which are defined as

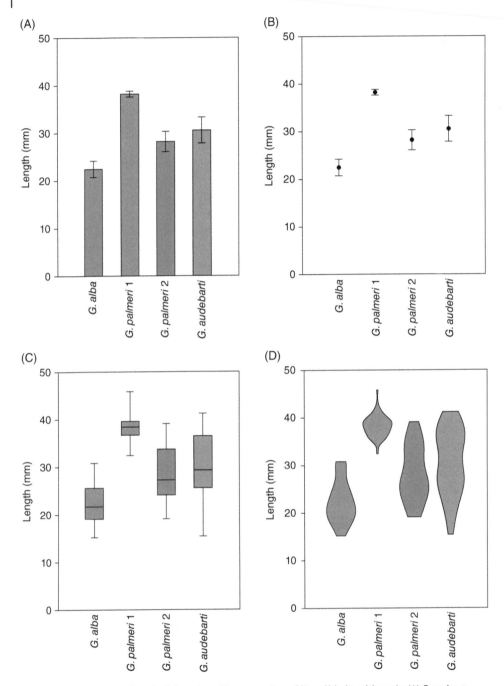

**Figure 3.6** Plots showing shell lengths of four samples of lingulide brachiopods. (A) Bar chart showing means and their 95% confidence intervals. (B) Mean-and-whiskers plot (95% confidence intervals). (C) Box plot. (D) Violin plot. Data from Kowalewski et al. (1997).

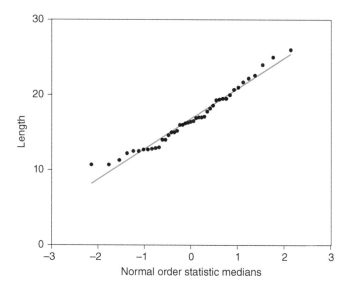

**Figure 3.7** Normal probability plot for the cephala of 43 specimens of the trilobite *Stenopareia glaber* (cf. section 4.3). The line is fitted using reduced major axis (RMA) regression (section 4.6). The data seem to be well fitted by a straight line, as would be expected from a sample taken from a normal distribution. The probability plot coefficient of correlation is $r = 0.986$. However, there is some flattening of the curve for the smallest values, indicating a shortening of the tail at this end.

$N(i) = G(m(i))$, where $G$ is the inverse of the cumulative normal distribution function and $m(i)$ are the uniform order statistic medians, approximated as

$$m(i) = \begin{cases} 1 - 0.5^{1/n} & i = 1 \\ (i - 0.3175)/(n + 0.365) & i = 2,3,\ldots,n-1 \\ 0.5^{1/n} & i = n \end{cases}$$

The normal probability plot is a special case of the quartile-quartile (QQ) plot, for comparing two distributions of any type.

## 3.7 Pie charts

A pie chart is a circular diagram, showing proportions as sectors (Fig. 3.8A). If the central part of the pie chart is blanked out, we get a doughnut chart (Fig. 3.8B), which is possibly less cluttered and also gives room for a text box or icon in the center.

The pie chart is often criticized because the human brain is less capable of judging angles than lengths. A bar chart is therefore easier to read precisely. Still, the pie chart is visually appealing and very compact, and is especially useful when plotted as small icons on geographic maps and stratigraphic logs. The pie chart will not go away anytime soon.

(A)

(B)

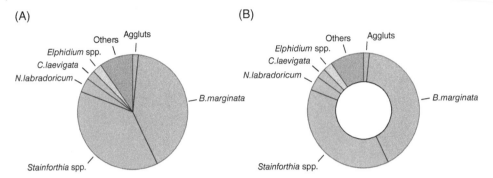

**Figure 3.8** Pie charts of the foraminiferan assemblage at 146 cm core depth (middle Holocene) in the Oslo fjord. (A) Traditional pie chart. (B) Doughnut chart. Data from Hammer and Webb (2010)

## 3.8 Ternary plots

The ternary plot is an odd one. A tri-variate data set can be plotted in three dimensions, but this is difficult to read. However, if we are primarily interested in only the relative proportions of the three values A, B, and C for each item, we can reduce the degrees of freedom from three to two, allowing plotting in the plane.

A good example from ecology is when the three variables are the abundances of three species at different localities (Fig. 3.9). The ternary plot is an elegant way of visualizing such compositional data (Aitchison 1986). For each item (sample), the three values are

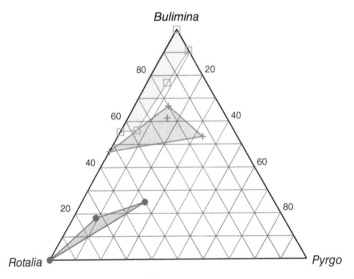

**Figure 3.9** Ternary plot of three foraminiferan genera along a depth gradient in the Gulf of Mexico. Dots (lower left corner): 0–100 m depth. Plus: 100–500 m. Squares: 500–3000 m. The positions of the points reflect bathymetry from the inner shelf (lower left in the triangle) to the abyssal plain (top). Data from Culver (1988).

converted into percentages of their sum. Each variable is then placed at the corner of a triangle, and the sample is plotted as a point inside this triangle. A sample consisting entirely of A will be placed at the corner A, while a sample consisting of equal proportions of A, B, and C will plot in the center.

The ternary plot is familiar to geologists, who routinely use it to plot the relative proportions of three minerals in different rocks, or the proportions of grain sizes such as mud, silt, and sand. It should be just as useful in community analysis. For example, ternary plots have been used extensively to plot the abundances of three suborders of foraminiferans: Rotaliina, Textulariina, and Miliolina. Different environments such as estuary, lagoon, shelf, and abyssal plain occupy specific regions of the ternary plot (Murray 1991).

## 3.9 Heat maps, 3D plots, and Geographic Information System

A heat map is a grid or matrix plot where the values are coded with colors. We will see several examples of heat maps in this book (e.g., sections 6.13, 7.3, 8.2, and 12.1). An important consideration when making a heat map is the color scheme (Crameri et al. 2020). Heat plots can be augmented with contour lines.

Business and news media graphics often add 3D effects to plots such as bar charts and pie charts. This is discouraged in scientific publication as it does not contribute to the communication of the data and is not good visual design practice. More interesting are cases where the data are truly three-dimensional. Modern computer graphics makes it easy to render 3D scenes on flat paper or flat computer screens. Such 3D plots, nevertheless, can be messy and should be used sparingly but are sometimes useful.

Paleontological data are often plotted on geographical maps. For the best results and highest flexibility, a Geographic Information System (GIS) program is often used. Such programs can produce maps at any scale and in any style and combine maps with other graphical elements such as heat maps or pie charts. Popular GIS programs include ArcGIS and the free QGIS.

## 3.10 Plotting with R and Python

While PAST includes a large number of plot types and is easy to use, even higher flexibility can be achieved with general programming languages. Excellent plotting packages are available for both R and Python.

The R core includes some simple plotting functions, but for high-quality, highly configurable graphics the *ggplot2* package is recommended. This package is included in the *tidyverse* bundle:

```
> library(tidyverse)
```

As an example, consider the trilobite cephalon lengths presented in section 4.3:

```
> mydata <- read.table("C:\\data\\stenopareia.txt", sep=" ",
header=TRUE)
```

For ggplot, we need to convert from the original data format with two columns of lengths of the two species, to a data frame with all the lengths in one column, and the species names in a second column. Such data manipulation is often required in R:

```
> md <- gather(mydata, value = 'length', key='species',
na.rm = TRUE)
```

Finally, the ggplot command produces the actual plot (Fig. 3.10A):

```
> ggplot(data = md, aes(x = species, y = length, fill =
species)) +
   geom_boxplot() +
   theme(legend.position = "none")
```

The traditional plotting package for Python is *matplotlib*. For more complex plots with high quality, higher-level libraries such as *seaborn* are available. With the *pandas* package (section 4.3), we read in the data file and plot it with seaborn as follows:

```
mydata = pd.read_csv('C:\\data\\stenopareia.txt', sep=' ')
p = sns.catplot(data=mydata.dropna(), kind="box")
p.set_ylabels('Length (mm)')
```

Here, the dropna() function removes missing values. The resulting box plot is shown in Fig. 3.10B.

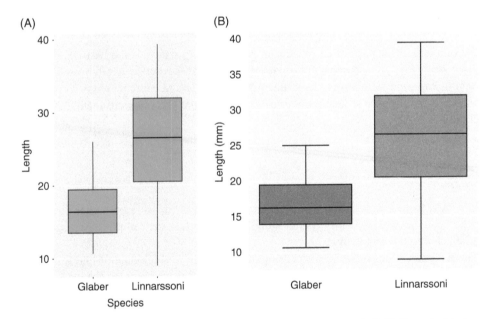

Figure 3.10 (A) Box plot of the trilobite data with the ggplot2 package in R. (B) Plotted with the seaborn package in Python.

**Example 3.1**

Sirius Passet, in the high Arctic, is the most remote of the Cambrian Lagerstätten (Harper et al. 2019). Bed-by-bed collecting from about 11 m of the section has yielded a large collection of fossils. However, the fossils are not evenly distributed through the section or across animal phyla. The pie charts in Fig. 3.11 show the dominance of the euarthropods and the measured section (Fig. 3.12) picks out the horizons with exceptional preservation and the relation of those to other factors.

(A)

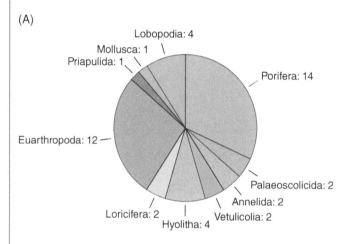

(B)

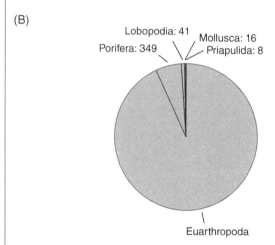

Figure 3.11   Pie charts showing species richness (A) and abundance (B) within each of the main animal groups at Sirius Passet, Greenland.

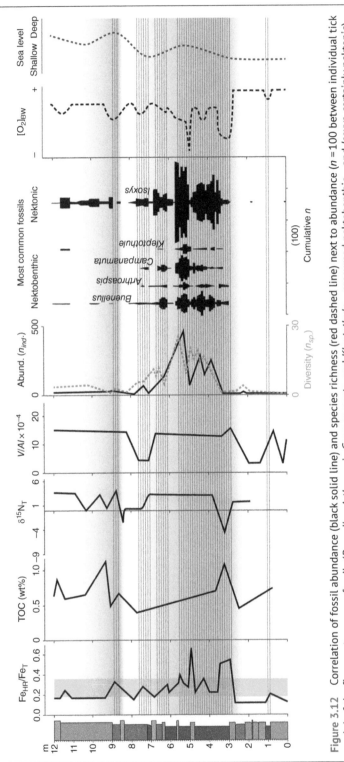

**Figure 3.12** Correlation of fossil abundance (black solid line) and species richness (red dashed line) next to abundance ($n = 100$ between individual tick marks) of the five most common fossils (*Buenellus, Arthroaspis, Campanamuta,* and *Kleptothule* – presumed nektobenthic – and *Isoxys,* certainly nektonic). This plot identifies the horizons characterized by exceptional preservation. Also included in the plot are geochemical trends ($Fe_{HR}/Fe_T$, TOC, $\delta^{15}N_T$, and V/Al) and inferred changes in bottom water oxygen and relative water depths (dashed black lines). Proposed intervals of periodic anoxia in blue with horizontal ruling. Adapted from Hammarlund et al. (2019) and Harper et al. (2019).

# References

Aitchison, J. 1986. *The Statistical Analysis of Compositional Data*. Chapman & Hall, New York.

Bjørklund, K.R., Kruglikova, S.B., Hammer, Ø. 2019. The radiolarian fauna during the Younger Dryas–Holocene transition in Andfjorden, northern Norway. *Polar Research* 38. https://doi.org/10.33265/polar.v38.3444.

Chambers, J., Cleveland, W., Kleiner, B., Tukey, P. 1983. *Graphical Methods for Data Analysis*. Wadsworth.

Clopper, C., Pearson, E.S. 1934. The use of confidence or fiducial limits illustrated in the case of the binomial. *Biometrika* 26, 404–413.

Crameri, F., Shephard, G.E., Heron, P.J. 2020. The misuse of colour in science communication. *Nature Communications* 11, 5444.

Culver, S.J. 1988. New foraminiferal depth zonation of the Northwestern Gulf of Mexico. *Palaios* 3, 69–85.

Hammarlund, E.U., Smith, M.P., Rasmussen, J.A., Nielsen A.T., Canfield, D.E., Harper, D.A.T. 2019. The Sirius Passet Lagerstätte of North Greenland – a geochemical window on early Cambrian low-oxygen environments and ecosystems. *Gebiology* 17, 12–26.

Hammer, Ø., Webb, K.E. 2010. Piston coring of Inner Oslofjord pockmarks, Norway: constraints on age and mechanism. *Norwegian Journal of Geology* 90, 79–91.

Harper, D.A.T., Hammarlund, E.U., Topper, T.P., Nielsen, A.T., Rasmussen, J.A., Park, T-Y., Smith, M.P. 2019. The Sirius Passet Lagerstätte of North Greenland: a remote window on the Cambrian Explosion. *Journal of the Geological Society* 176, 1023–1037.

Kowalewski, M., Dyreson, E., Marcot, J.D., Vargas, J.A., Flessa, K.W., Hallmann, D.P. 1997. Phenetic discrimination of biometric simpletons: paleobiological implications of morphospecies in the lingulide brachiopod *Glottidia*. *Paleobiology* 23, 444–469.

Murray, J.W. 1991. *Ecology and Palaeoecology of Benthic Foraminifera*. Routledge, Oxford, UK.

Wand, M.P. 1997. Data-based choice of histogram bin width. *American Statistician* 51, 59–64.

# 4

# Univariate and bivariate statistical methods

Now that we have introduced concepts such as distributions, populations, samples, and statistical testing, we can start to build a toolbox of useful statistical methods. In this chapter, we will introduce classical statistical procedures for data sets with one or two variables.

## 4.1 Parameter estimation and confidence intervals

In chapter 2, we discussed statistical distributions and their parameters such as mean and variance. We also discussed how the parameters of the population can be estimated from a sample. These estimates have associated sampling errors. Regarding the sample mean, for example, we have seen (section 2.5) that it has an associated standard error, which is the standard deviation of the estimate of the mean under repeated sampling from the population.

Another way of specifying the uncertainty of a parameter estimate is the *confidence interval*. This is an interval computed from the sample, with the property that it will cover the true population parameter in a certain proportion of cases, usually chosen as 95%. Returning to the simulation setup from section 2.5, where we generated random samples from a normal distribution with population mean $\mu = 50$, we now generate 50 samples and compute the sample means and their 95% confidence intervals (Fig. 4.1). In two of the 50 cases (i.e., 4%), we are unlucky and get confidence intervals that do not include the true mean. This is about what we would expect with a 95% confidence level.

For a normally distributed population, we saw in section 2.5 that the estimate of the mean value has associated with it a standard error, given as

$$s_e = \sigma / \sqrt{n}$$

From the standard error, we can give a confidence interval for the mean. Given that about 95% of values are within about 1.96 standard deviations from the mean, we can say that if we take repeated samples from the population, a 95% confidence interval for the true mean is about $[\bar{x} - 1.96s_e, \bar{x} + 1.96s_e]$. Of course, we do not know the true $s_e$ (because we do not know the true $\sigma$), so we must normally estimate the standard error from the sample. A more precise confidence interval for the mean of a normally distributed population with

*Paleontological Data Analysis*, Second Edition. Øyvind Hammer and David A.T. Harper.
© 2024 John Wiley & Sons Ltd. Published 2024 by John Wiley & Sons Ltd.

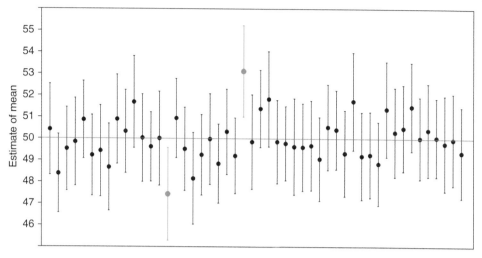

Figure 4.1 Estimates of the means (dots) and their 95% confidence intervals (lines) from 50 samples from a normally distributed population with mean 50 and standard deviation 10. Two of the confidence intervals (red) do not cover the true population mean.

unknown (only estimated) standard deviation is computed using the so-called $t$ distribution:

$$\left[ \bar{x} - t_{\alpha/2, n-1} \frac{s}{\sqrt{n}}, \bar{x} + t_{\alpha/2, n-1} \frac{s}{\sqrt{n}} \right]$$

Here, $s$ is the estimate of the standard deviation, $t$ has $n-1$ degrees of freedom, and $1-\alpha = 0.95$ for a 95% confidence interval.

## 4.1.1 Bootstrapping

Such analytical confidence intervals, usually based on rather strict assumptions about the shape of distributions, have been worked out for many special cases. However, there is also another, very useful and general way of estimating confidence intervals, known as *bootstrapping*. It is simple but may seem rather strange.

Imagine that we have access to the entire population. In this case, we could of course calculate the population mean precisely. But we could also estimate the expected distribution of sample means from a sample of a limited size $n$, as follows (we will initially assume that the population distribution is symmetric).

Randomly draw $n$ individuals from the population. Calculate the sample mean from this random sample. Repeat this many times, say 2000, and draw a histogram of the sample means. Find the point such that 2.5% of the sample means (50 out of 2000) are below it – this is called the 2.5 *percentile*. Also, find the point such that 2.5% of the sample points are above it – the 97.5 percentile. In 95% of the random samples, the sample mean will be between the 2.5 and the 97.5 percentiles, and we take this range as a 95% confidence interval for the mean.

But in reality, we do not have access to the complete population – we have perhaps only one small sample from it, of size $n$. How can we then estimate the population, so that we can simulate repeated drawing of random samples from it? The best we can do is to use the sample distribution as an estimate of the population distribution. So, we draw $n$ individuals at random from the *sample* instead of the population! This is done with replacement, meaning that a given individual can potentially be picked several times. It may sound too good to be true, but this simple approach of (non-parametric) *bootstrapping* does work quite well and is now used in many areas of data analysis (Davison and Hinkley 1997). We will see several examples of the technique throughout the book.

It should be mentioned that this simple method for estimating confidence intervals from the random bootstrap samples (called the quantile method) may be biased. Several techniques have been developed to try to correct for this bias, but they normally give similar results to the quantile method.

### 4.1.2 Credible intervals

Confidence intervals are frequently misunderstood. A 95% confidence interval is constructed to contain the true population parameter (such as the mean) in 95% of the cases under repeated sampling. It is not an interval in which the population parameter will be found with 95% probability (in the frequentist view, the population parameter is not a random variable with a certain distribution, but a single, fixed number). However, in the Bayesian framework, it is possible to construct precisely such a probability interval from the posterior parameter distribution. This is called a *credible interval*.

## 4.2 Testing for distribution

Several statistical tests have been devised to assess whether a sample comes from a population with a certain distribution. Tests for normal distribution are the most common and are often used to check the data prior to applying methods that assume normality. We do not recommend this procedure. First, testing for normality is rarely informative. For small sample sizes, the test will not have sufficient power to detect departure from normality except in extreme cases. For large sample sizes, the power of the test will mean that even the smallest departures from normality will be reported as significant. In fact, no natural populations will have exactly a normal distribution but tests assuming normality will have a certain robustness to this, especially for large samples. Second, testing normality using the same sample as will be used for the following test is considered poor practice. Instead, we suggest considering whether assuming normal distribution is reasonable and perhaps to informally inspect the histogram of the sample. Still, we will briefly describe a couple of commonly used normality tests.

Tests for whether a sample has been taken from a population with normal distribution include the Lilliefors and Anderson–Darling tests (adaptations of the Kolmogorov–Smirnov test of section 4.3). A useful graphic method is the normal probability plot (section 3.6).

### 4.2.1 Shapiro–Wilk test for normal distribution

Among several normality tests, it has been argued (D'Agostino and Stephens 1986, p. 406; Razali and Yap 2011) that the best overall performer for both small and large samples may be the *Shapiro–Wilk* test (Shapiro and Wilk 1965; Royston 1982). The standard algorithm of Royston (1982, 1995) handles sample sizes as small as 3 (although this is not to be recommended) and as large as 5000.

The null hypothesis of the Shapiro–Wilk test is

$H_0$: The sample has been taken from a population with normal distribution.

The test will report the Shapiro–Wilk test statistic $W$, and a probability $p$. As usual, the null hypothesis can be rejected if $p$ is smaller than our significance level, but normality is not formally confirmed even if $p$ is large.

The algorithm for the Shapiro–Wilk test is complicated. Program code in FORTRAN was published by Royston (1982, 1995). For reference, the test statistic $W$ is computed as

$$W = \frac{\left(\sum_{i=1}^{n} a_i x_i\right)^2}{\sum_{i=1}^{n}(x - \bar{x})^2}$$

where $x_i$ are the sample values sorted in ascending order. The $a_i$ are weights that are computed according to sample size. $W$ varies from 0 to 1, with large values for near-normal distributions.

---

**Example 4.1**

Kowalewski et al. (1997) made measurements on living lingulide brachiopods from the Pacific coast of Central America (Fig. 4.2), in order to investigate whether shell morphometrics could produce sufficient criteria for differentiating between species. Figure 4.3 shows a histogram of the lengths of the ventral valves of 51 specimens of *Glottidia palmeri* from Campo Don Abel, Gulf of California, Mexico. The sample seems to have a close to normal distribution. The Shapiro–Wilk test reports a test statistic of $W = 0.971$ and a probability $p = 0.24$, so we cannot reject the null hypothesis of normal distribution at any reasonable significance level. In other words, we have no reason to believe that the sample has not been taken from a normally distributed population.

As an experiment, we also try two other normality tests on this sample. The Anderson–Darling test reports a slightly higher $p$ value ($A = 0.334$; $p = 0.50$) than Shapiro–Wilk, while the Lilliefors test gives a slightly lower $p$ value ($L = 0.092$; $p = 0.35$). This does not change our conclusion (or lack thereof) – we cannot reject the null hypothesis of a normal population distribution.

Figure 4.2 The living linguliform brachiopod *Glottidia palmeri*; the genus differs from the better-known *Lingula* in having internal septa and papillate interiors. It is common in the Gulf of California, where living specimens can be recovered from shallow water. Large samples of this brachiopod are available for statistical analyses. Courtesy of Michal Kowalewski.

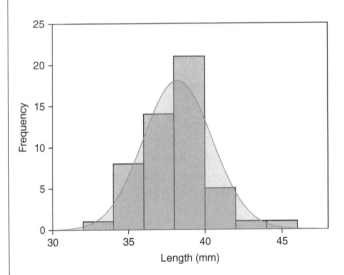

Figure 4.3 Lengths of the ventral valves of 51 specimens of the Recent lingulide *Glottidia palmeri*, Campo Don Abel, Gulf of California, Mexico, with fitted normal distribution curve. The Shapiro–Wilk test cannot reject a normal distribution in this case ($W = 0.971$; $p = 0.24$). Data from Kowalewski et al. (1997).

## 4.3  Two-sample tests

Perhaps the most common type of statistical test is the comparison of two samples of measured data. Many tests have been designed for comparing different parameters, and with different assumptions about the distributions of the populations.

### 4.3.1  Student's *t* test for the equality of means

Student's *t* test is a simple and fundamental statistical procedure for testing whether two univariate samples have been taken from populations with equal means. Hence, the null hypothesis to be tested is

$H_0$: the data are drawn from populations with the same mean values.

The samples should come from close to normal distributions with equal variances, but these assumptions can be somewhat relaxed, especially if sample sizes are large. A variant of the *t* test known as the Welch test does not require equal variances.

Named after its inventor, an employee at the Guinness brewing company who liked to call himself "Student" (his real name was Gosset), this test can be used to test whether, for example, *Tyrannosaurus rex* has the same average size as *Tarbosaurus bataar*, whether a set of stable isotope values at one locality has the same mean as at another locality, or whether a correlation value (section 4.5) is significantly different from zero. The test will report a value for the *t* statistic and a probability value. A low probability indicates a statistically significant difference, as usual.

The *t* test exists in a number of different forms for different purposes. First, there is a one-sample test for comparison with a hypothesized mean value, and a two-sample test for comparing the means of two samples. For the latter, there are really no strong reasons for using the *t* test in its original form, which assumes equality of variance in the two samples. Instead, a slight variant of the test known as the Welch test is recommended because it does not make this assumption. On the other hand, it could be argued that if variances are very different, a statistically significant difference of means should not be over-interpreted.

Another dichotomy is between the one-tailed and the two-tailed test, which have different alternative hypotheses. The two-tailed version tests equality against any departure from equality, while the one-tailed version tests equality against sample A being smaller than (alternatively larger than) sample B. The two-tailed test is by far the most used, but the one-tailed version may be more appropriate in some situations. The one-tailed *p* value is simply half the two-tailed *p* value.

For the classical two-sample *t* test, the *t* statistic is calculated as follows:

$$t = \frac{\bar{x} - \bar{y}}{\sqrt{s^2 \left(1/n_x + 1/n_y\right)}}$$

where $\bar{x}$ and $\bar{y}$ are the means of the two samples and $n_x$ and $n_y$ are the sample sizes. $s^2$ is an estimate of pooled variance based on the sample variances $s_x^2$ and $s_y^2$:

$$s^2 = \frac{\left(n_x - 1\right)s_x^2 + \left(n_y - 1\right)s_y^2}{n_x + n_y - 2}$$

$$s_x^2 = \frac{\Sigma\left(x_i - \bar{x}\right)}{n_x - 1}$$

Note that the *t* value can be positive or negative, depending on whether the mean of *x* is larger or smaller than the mean of *y*. Since the order of the samples is arbitrary, the value of *t* is usually reported without its sign in the literature.

The *p* value is then calculated by the computer (previously a table was used), based on the *t* value and the number of degrees of freedom: $\nu = n_x + n_y - 2$. For details on the numerical procedures, using the *incomplete beta function*, see Press et al. (1992).

The *t* statistic for the Welch test is given by

$$t = \frac{\bar{x} - \bar{y}}{\sqrt{s_x^2/n_x + s_y^2/n_y}}$$

and the number of degrees of freedom is

$$\nu = \frac{\left(s_x^2/n_x + s_y^2/n_y\right)^2}{\left(s_x^2/n_x\right)^2/\left(n_x - 1\right) + \left(s_y^2/n_y\right)^2/\left(n_y - 1\right)}$$

A simple but useful measure of the effect size is Cohen's *d* (Cohen 1988):

$$d = \frac{\bar{x} - \bar{y}}{s}$$

where *s* is the pooled standard deviation. An effect size of $d = 0.2$ may be considered as "small," while $d = 0.8$ may be considered as "large."

Finally, we give the equations for the one-sample *t* test, for the equality of the population mean to a theoretical value $\mu_0$:

$$t = \frac{\bar{x} - \mu_0}{\sqrt{s^2/n}}$$

There are $n - 1$ degrees of freedom.

---

**Example 4.2**

Bruton and Owen (1988) measured a number of samples of illaenid trilobites from the Ordovician of Norway and Sweden. We will look at the length of the cephalon, measured on 43 specimens of *Stenopareia glaber* from Norway (Fig. 4.4) and 17 specimens of *S. linnarssoni* from Norway and Sweden. The box plots are shown in Fig. 4.5. The mean values are 16.8 mm (variance 15.3) and 25.8 mm (variance 64.3) respectively, indicating that *S. glaber* is smaller than *S. linnarssoni*. Is the difference statistically significant, or could it be just a random sampling effect?

Length measurements on animals are typically near-normally distributed, and the box plots do not indicate any dramatic asymmetry in the sample distributions or any

Figure 4.4 The illaenid trilobite *Stenopareia glaber* from Ringerike, Oslo Region. The taxonomy of this group of apparently featureless trilobites is difficult and has benefited considerably from morphometric investigations. Courtesy of David L. Bruton.

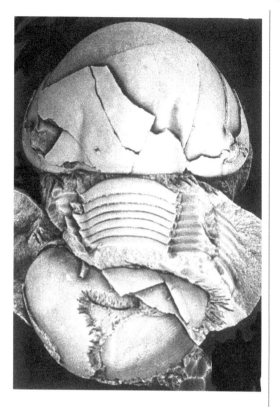

serious outliers. We therefore feel justified in choosing the two-sample *t* test for comparing the two means. Since we do not wish to assume that the variances are equal, we choose Welch's version of the test. PAST reports $t = 4.42$ and $p = 0.00029$, so we can reject the null hypothesis of equality of the means at the $p < 0.05$ level. In other words, we have shown a statistically significant difference in the mean values.

The Bayes factor is 45,100. This very high value can be taken as decisive evidence for the hypothesis that the means are different.

In R, we can use the *t* test function. We first read in the data from a text file with the two samples in two columns, separated by space, and with a header line with variable names:

```
> mydata <- read.table("C:\\data\\stenopareia.txt", sep=" ",
header=TRUE)
```

We now have a data frame called *mydata*, and we call the t.test function for unequal variances:

```
> t.test(mydata$linnarssoni, mydata$glaber, var.equal=FALSE)
        Welch Two Sample t-test
data:  mydata$linnarssoni and mydata$glaber
t = 4.4186, df = 19.082, p-value = 0.0002923
```

```
alternative hypothesis: true
difference in means is not
equal to 0
95 percent confidence interval:
   4.731322 13.242960
sample estimates:
mean of x mean of y
   25.79412  16.80698
```

Here, R reports the basic results of the test, with a *t* value and *p* value. In addition, the sample means are reported, and a 95% confidence interval for the difference in means (R uses an estimate for the confidence interval that does not assume equal variances).

In Python, we first read in the text file with the *pandas* package, and then use the *ttest_ind* function in the *scipy.stats* package:

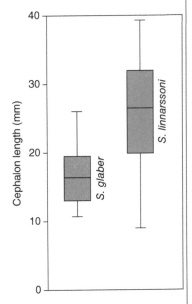

**Figure 4.5** Box plots of the sizes of two Ordovician trilobite species.

```
import pandas as pd
import scipy.stats as stats
mydata = pd.read_csv('C:\\data\\stenopareia.txt', sep=' ')
print(stats.ttest_ind(mydata['linnarssoni'],
mydata['glaber'], equal_var=False, nan_policy='omit'))
```

The function returns the *t* and *p* values:

```
Ttest_indResult(statistic=4.418616089103409, pvalue=
0.0002922556529346143)
```

### 4.3.2  *F* test for the equality of variances

The *F* test is the classic test for equality of variance, assuming the samples are taken from normal distributions. The null hypothesis is

$H_0$: the samples are drawn from populations with the same variances.

Equality of variance may not seem a very exciting thing to test for, but it has some important uses. For example, morphological variance (disparity) tends to reduce during strong selection pressure and can therefore be a useful environmental indicator. The *F* test is also a component in a number of other statistical procedures, most notably the analysis of variance (ANOVA), which will be discussed in section 4.4.

The *F* statistic is simply the larger sample variance divided by the smaller. The probability of equal variances is then computed by consulting a table of the *F* distribution with $n_x - 1$ and $n_y - 1$ degrees of freedom, or using the so-called incomplete beta function (Press et al. 1992).

**Example 4.3**

The sizes of 30 female and 29 male gorilla skulls were measured (data from O'Higgins 1989, see also section 6.11). The variance of the females is 39.7, while the variance of the males is 105.9 (Fig. 4.6). The $F$ statistic is $F = 2.66$, with $p(H_0) = 0.011$. We can therefore reject the null hypothesis of equal variances at a significance level of $p < 0.05$.

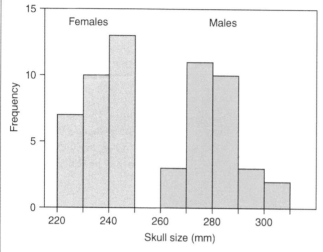

**Figure 4.6**  Skull sizes of 30 female and 29 male gorilla skulls. Data from O'Higgins (1989).

### 4.3.3  Mann–Whitney $U$ test for equality of position

The Mann–Whitney test is often referred to as a "non-parametric" alternative to the $t$ test. This means that it does not assume any particular distribution except that the two population distributions should ideally be equal under the null hypothesis. If your data set is highly skewed, multimodal (having several peaks in the distribution), or with strong outliers, it may violate the assumptions of the $t$ or Welch test too strongly for comfort. The Mann–Whitney $U$ test is then a possible alternative.

The null hypothesis of the test is

$H_0$: for randomly selected values $x$ and $y$ from two populations, the probability of $x > y$ is equal to the probability of $y > x$.

This definition is precise, although a bit awkward. If the data are on a continuous scale and we assume that the only difference between the two populations is one of location, not the shape of the distribution, the Mann–Whitney test reduces to a test of equality of medians, and this is often how the test is presented.

The Mann–Whitney test operates by pooling the two samples (putting them in one "hat") and then sorting the numbers in ascending order. The *ranks* are the positions in the sorted sequence such that the smallest number has rank 1, the second smallest has rank 2, etc. Equal numbers (ties) are assigned ranks according to special rules. Some ordinal variates may already be in a ranked form, such as a sample number in a stratigraphical column, where absolute levels in meters are not given. Other types of data, such as continuous measurements, can be converted to ranks by sorting and noting the positions in the sequence as just described.

If the medians of the two samples are almost equal, we would expect the sorted values from the two samples to be intermingled within the ranked sequence. If, on the other hand, the median of sample A is considerably smaller than that of sample B, we would expect the lower ranks to mainly represent values from A, while the higher ranks would mainly represent values from B. This is the basis of the Mann–Whitney U test.

If the Mann–Whitney test can cope with a wider range of data than the Student's t test, why not always use the former? One reason is that if your data are close to normally distributed, the parametric t test (or Welch test) will have slightly higher statistical power than the Mann–Whitney test and is therefore preferable.

There is a version of the Mann–Whitney test called the *Wilcoxon two-sample test*. The difference between the two is purely in the method of computation; they are otherwise equivalent. The equations for the Wilcoxon test are shown here (after Sokal and Rohlf 1995). Rank the pooled data set (both samples), taking account of ties by assigning mean ranks to tied points. Sum the ranks of the smaller sample, with sample size $n_y$. The Wilcoxon statistic C is then given by

$$C = n_x n_y + \frac{n_y(n_y+1)}{2} - \sum_{i=1}^{n_y} R(y_i)$$

The U statistic (equal to the Mann–Whitney U) is the smaller of C and $n_x n_y - C$:

$$U = \min\left(C, n_x n_y - C\right)$$

For large sample sizes ($n_x > 20$), the significance can be estimated using the t statistic:

$$t = \frac{U - \left(n_x n_y/2\right)}{\sqrt{n_x n_y\left(n_x + n_y + 1\right)/12}}$$

There is a correction term in the case of ties (Sokal and Rohlf 1995), but this is rarely of great importance. For smaller sample sizes and no ties, some programs will compute an exact probability using a look-up table or a permutation approach (see below).

The equations for the Mann–Whitney/Wilcoxon test differ somewhat between textbooks, but they should all be broadly equivalent.

---

**Example 4.4**

Returning to the *Stenopareia* trilobite data, the Mann–Whitney test in PAST reports $U = 121.5$, $p = 6.4685\text{E}{-}05$, meaning a statistically significant (at the $p < 0.05$ level) difference in size between the two species. This confirms the result of the t test.

In R, the Mann–Whitney test is called the Wilcoxon rank sum test, and the test statistic is called W instead of U, but otherwise we get the same result:

```
> wilcox.test(mydata$glaber, mydata$linnarssoni)
        Wilcoxon rank sum test with continuity correction
data:  mydata$glaber and mydata$linnarssoni
```

```
W = 121.5, p-value = 6.469e-05
alternative hypothesis: true location shift is not equal to 0
```

In Python, we use the *mannwhitneyu* function in the *scipy.stats* package. We also use the *dropna* method to remove the missing values in each of the data columns:

```
print(stats.mannwhitneyu(mydata['linnarssoni'].dropna(),mydata
['glaber'].dropna()))
MannwhitneyuResult(statistic=609.5, pvalue=
6.468544278861686e-05)
```

The test statistic is here reported as 609.5 rather than 121.5 as in PAST and R. This is not an error but is due to the fact that the Mann–Whitney test can report two different but equivalent test statistics resulting in the same *p* value.

### 4.3.4 Kolmogorov–Smirnov test for equality of distribution

While for example Student's *t* and the *F* tests compare certain parameters (mean and variance) of two normal distributions, the Kolmogorov–Smirnov (K–S) test is a non-parametric test that simply compares the complete shapes and positions of the distributions. In other words, the null hypothesis is

$H_0$: the two samples are taken from populations with equal distributions.

The Kolmogorov–Smirnov test is a "generalist" and has comparatively little power when it comes to testing specific parameters of the distribution. It may well happen that the K-S test finds no significant difference between distributions even if, for example, the Mann–Whitney test reports a significant difference in medians. It should also be noted that the K-S test is most sensitive to differences close to the median values and less sensitive out on the tails.

Another use for the K–S test is to compare the distribution in a sample with a theoretical distribution described by an equation. For example, we may wish to test whether the stratigraphic distances between consecutive fossil occurrences follow an exponential distribution with a given mean value, as they might be expected to do under a null hypothesis for random occurrence (but see Marshall 1990). In this case, parameters of the theoretical distribution should not be estimated from the given sample.

The Kolmogorov–Smirnov test is constructed around the cumulative distribution function (CDF) as estimated from the data (section 2.4). As shown in Fig. 4.7, the estimated (empirical) CDF(*x*) for each sample is the fraction of the data points smaller than *x*. It increases monotonically from zero to one.

The test statistic *D* for the two-sample K-S test is then simply the maximal difference between the two CDFs:

$$D = \max \left| \text{CDF}_1(x) - \text{CDF}_2(x) \right|$$

To estimate the probability *p* that the two samples have equal distributions (Press et al. 1992), first define the function $Q(\lambda)$:

$$Q(\lambda) = 2 \sum_{j=1}^{\infty} (-1)^{j-1} e^{j^2 \lambda^2}$$

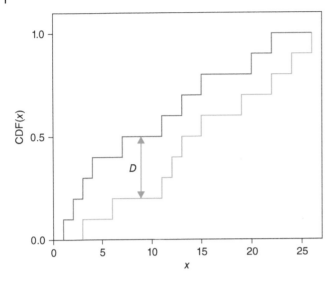

**Figure 4.7** The empirical cumulative distribution functions for two samples, and the maximal difference $D$.

(this sum is truncated when the terms get negligible).

Then, find the "effective" number of data points:

$$n_e = \frac{n_1 n_2}{n_1 + n_2}$$

Finally, we estimate $p$ as follows:

$$p(equal) = Q\left(\left[\sqrt{n_e} + 0.12 + \frac{0.11}{\sqrt{n_e}}\right]D\right)$$

Some alternative estimators of $p$ are given by Sokal and Rohlf (1995).

---

**Example 4.5**

From a multivariate morphometric study of Holocene living lingulide brachiopods by Kowalewski et al. (1997), we extract a data set consisting of two samples of measured shell lengths: *G. palmeri* from Vega Island, Mexico ($n = 30$) versus *G. audebarti* from Costa Rica ($n = 25$). The histograms are shown in Fig. 4.8. We here use a so-called bi-histogram (or back-to-back histogram), showing the second sample mirror-imaged below the horizontal axis, convenient for comparing two samples. We may feel that the distribution of *G. palmeri* is skewed to the right (right tailed), while *G. audebarti* is skewed to the left. However, using the Kolmogorov–Smirnov test, we arrive at the test statistic $D = 9.23$,

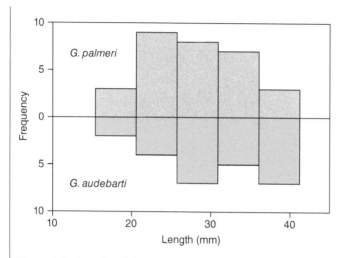

**Figure 4.8** Lengths of the valves of living lingulide brachiopods. Green bars: *G. palmeri* from Vega Island, Mexico (*n* = 30). Orange bars: *G. audebarti* from Costa Rica (*n* = 25). Data from Kowalewski et al. (1997).

with probability $p = 0.43$. Hence, there is no significant difference in distribution between the two samples.

### 4.3.5 Permutation tests

Consider the null hypothesis of equality of the means of two samples:

$H_0$: the samples are taken from populations with the same mean values.

The *t* test would be the standard parametric choice for testing this hypothesis. However, the *t* test makes a number of assumptions that may not hold, and we might like a more general approach that could work for any distribution and any parameter. Permutation tests (Good 1994) are often useful in such situations. Let us say that we have two values in sample *A* and two in sample *B*:

| A | B |
| --- | --- |
| 4 | 6 |
| 2 | 6 |

The observed difference in the means is $d\bar{x} = |(4+2)/2 - (6+6)/2| = 3$.

Let us further assume that the null hypothesis of equality is true. In this case, we should be able to put all the values in one "hat" and find that the observed difference in the means of the two original samples is no larger than could be expected if we took two samples of the same sizes from the pooled data set. Let us then consider all the possible ways of taking a sample of size two from the pool of size four, leaving the remaining two in the other sample (the order does not matter). There are six possibilities, as given below with the computed difference in the means:

| A | B | A | B | A | B | A | B | A | B | A | B |
|---|---|---|---|---|---|---|---|---|---|---|---|
| 4 | 6 | 4 | 2 | 4 | 2 | 2 | 4 | 2 | 4 | 6 | 4 |
| 2 | 6 | 6 | 6 | 6 | 6 | 6 | 6 | 6 | 6 | 6 | 2 |
| $\|3-6\| = \underline{3}$ | | $\|5-4\| = \underline{1}$ | | $\|5-4\| = \underline{1}$ | | $\|4-5\| = \underline{1}$ | | $\|4-5\| = \underline{1}$ | | $\|6-3\| = \underline{3}$ | |

In sorted order, the differences between the two samples from the pooled data set are 1, 1, 1, 1, 3, and 3.

In two out of six cases, the random samples from the pooled data set are as different as in the original samples. We therefore take the probability of equality to be $p(H_0) = 2/6 = 0.33$, and thus we do not have a significant difference. In fact, the smallest value of $p$ we could possibly get for such minute sample sizes is 1/6, implying that we could never obtain a significant difference (at $p < 0.05$ for example) even if the samples were taken from completely different populations. The test has very low power, because of the small sample sizes. This is another reminder not to interpret high $p$ values as proof of the null hypothesis!

The exact permutation test requires the investigation of all possible selections of items from the pool. For large sample sizes, this is impractical because of the amount of computation involved. Consider two samples of sizes 20 and 30 – there are more than 47 trillion possible selections to try out! In such cases, we must look at only a subset of the possible selections. The best approach is probably to pick selections at random, relying on the resulting distribution of test statistics to converge to the true distribution as the number of random selections grows. The resulting method is known as a *Monte Carlo permutation test*.

## 4.4 Multiple-sample tests

Often, we want to compare means, medians, or other parameters across more than two samples. The classic approach is the Analysis of Variance (ANOVA). There are different ways of grouping samples according to independent criteria, giving rise to a complicated family of ANOVA designs. Here, we will only discuss the basic one-way design.

### 4.4.1 One-way ANOVA

One-way ANOVA is a parametric test, assuming normal distribution, to investigate equality of the means of many groups: given a data set of measured dimensions of ostracods, with many specimens from each of many localities, is the mean size the same at all localities or do at least two samples have different means from each other? The null hypothesis is

$H_0$: all samples are taken from populations with equal mean values.

It is not statistically correct to test this separately for all pairs of samples using, for example, the *t* test. The reason for this is simple. Say that you have 20 samples that have all been taken from a single population. If you try to use the *t* test for each of the 190 pairwise comparisons, you will most probably find several pairs where there is a "significant" difference between the means at $p < 0.05$, because such an erroneous conclusion is indeed expected in 5% of the pairwise comparisons at this significance level even if all the samples are taken from the same population. We would then commit a type I error, concluding with a significant departure from the true null hypothesis of equality. ANOVA is an appropriate method in this case because it makes only one overall test for equality.

However, once overall inequality has been shown using ANOVA, it is possible to proceed with so-called "post hoc" tests in order to identify which particular pairs of samples are significantly different (in fact, it is valid to use these post hoc tests even without an initial ANOVA). Several such tests are in use; two of the common ones are *Bonferroni correction* and *Tukey's Honestly Significant Difference (HSD)*. With Bonferroni correction, the samples are subjected to pairwise comparison, using, for example, the *t* test, but the significance levels for these tests are set lower than the overall significance level in order to adjust for the problem mentioned earlier. Slightly more powerful versions of the Bonferroni correction are available, such as Holm-Bonferroni. Tukey's HSD is based on the so-called *Q* statistic. When sample sizes are unequal, it tends to become a little too conservative (reports too high *p* values) but is otherwise usually a good choice (see also Sokal and Rohlf 1995).

Traditional ANOVA assumes normal distributions and also the equality of variances, but these assumptions are less critical if the samples are of the same size. If the sample sizes are different, the variation in variance should not exceed say 50%. Several tests exist for the equality of variances, including the Fmax, Bartlett, Scheffé-Box, and Levene tests (you cannot use pairwise *F* tests, for the same reason that you cannot use pairwise *t* tests instead of ANOVA). These tests are, however, not much used, partly because we can accept significant but small differences in variance when we use ANOVA. If the assumption of normal distribution is completely violated, you may turn to a non-parametric alternative such as the Kruskal–Wallis test (see below).

Note that ANOVA for two groups is mathematically equivalent to the *t* test, so making a special case of this may seem unnecessary. Nevertheless, the use of the *t* test is an old habit among statisticians and is unlikely to disappear soon.

The idea behind ANOVA is to use an *F* test to compare the variance within the samples against the variance of all samples combined. If the variances of the individual samples are very small, but the variance of the whole data set is very large, it must be because at least two of the samples are quite different from each other. The resulting procedure is straightforward.

Let the number of samples be *m* and the total number of values *n*. First, we calculate the squared difference between each sample mean and the total mean of the pooled data set (all samples combined), and multiply each of these with the corresponding sample size. The sum of these values across all samples is called the *between-groups sum-of-squares* (BgSS):

$$BgSS = \sum n_i \left( \bar{x}_i - \bar{x} \right)^2$$

The BgSS divided by $m - 1$ (degrees of freedom) gives the *between-groups mean square* (BgMs). We now have a measure of between-groups variance.

Next, we compute the squared difference between each value and the mean of the sample it belongs to. All these values are summed, giving the *within-groups sum-of-squares* (WgSS):

$$WgSS = \sum \sum \left( x_{ij} - \bar{x}_i \right)^2$$

The WgSS divided by $n - m$ (degrees of freedom) gives the *within-groups mean square* (WgMs). All that is left is to compare the between-groups and within-groups variances using the $F$ test, with $F = BgMs/WgMs$. This finally gives a $p$ value for the equality of all group means.

---

**Example 4.6**

Cincotta et al. (2022) studied preserved melanosomes in an Early Cretaceous pterosaur from Brazil. Melanosomes are small organic structures responsible for the coloration of skin and feathers (Fig. 4.9). They reported that melanosomes in skin tissue from the crest, in monofilaments, and in branched feathers have different shapes, measured as their length-to-width ratios. Cincotta et al. interpreted this as evidence for different coloring in the three areas, which implies a function in display.

We will use a slightly reduced version of the data set, with measurements only from the ventral part of the crest. This gives a data set with a mean ratio of 2.37 in the crest

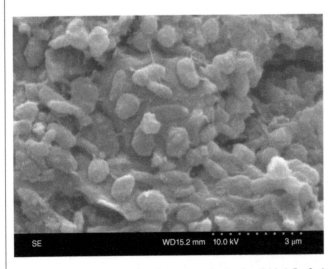

Figure 4.9  Morphology of melanosomes in the fossil bird *Confuciusornis* from the Lower Cretaceous of China. Courtesy of Stuart Kearns.

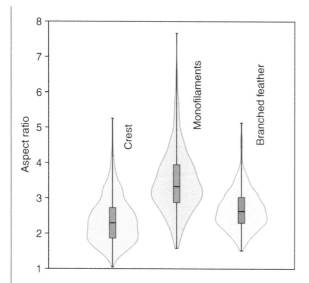

Figure 4.10 Box and violin plot of the length-to-width ratios of melanosomes in three different tissues in the pterosaur *Tupandactylus* cf. *imperator* from the Lower Cretaceous of Brazil. Data from Cincotta et al. (2022).

($n = 786$), 3.46 in the monofilaments ($n = 406$), and 2.68 in the branched feathers ($n = 878$). The data are summarized in Fig. 4.10.

It is clear from the plot that the distributions are asymmetric, with a tail towards higher values. In addition, the variances seem somewhat different in the three groups. Still, considering that ANOVA is fairly robust to violation of the assumptions, we proceed with some caution.

An alternative approach would be to apply some transformation to the data in order to force them into a distribution closer to normal. In this case, taking the logarithms (log-transform) would be suitable. However, such transformation makes the analysis more indirect and the results more difficult to interpret (e.g., Feng et al. 2014).

Table 4.1 shows the result of the one-way ANOVA in a standard table form.

The between-groups mean square is much larger than the WgMs, indicating a difference between groups. The $p$ value is so small that it is given as zero, so the null hypothesis of the equality of all the means is rejected at any reasonable significance level. We could report this as follows:

Table 4.1 ANOVA table for the melanosome data set.

|  | Sum of squares | Df | Mean square | F | p (same) |
| --- | --- | --- | --- | --- | --- |
| Between groups | 318.0 | 2 | 159.0 | 346.4 | 0 |
| Within groups | 949.0 | 2067 | 0.459 |  |  |
| Total | 1267.1 | 2069 |  |  |  |

**Table 4.2** Results of the Tukey's HSD test for all three pairwise comparisons in the melanosome data set.

| Comparison | $Q$ | $p$ | Conf. interval |
|---|---|---|---|
| Crest vs. monofilaments | 37.1 | <0.0001 | [0.99, 1.19] |
| Crest vs. branched | 13.2 | <0.0001 | [0.23, 0.39] |
| Monofilaments vs. branched | 27.0 | <0.0001 | [0.68, 0.87] |

The mean length/width ratio is 2.37 in the crest ($n = 786$), 3.46 in the monofilaments ($n = 406$), and 2.68 in the branched feathers ($n = 878$). The difference is statistically significant (one-way ANOVA, $F_{2,2067} = 346.4, p < 0.0001$).

Here, the degrees of freedom for the $F$ value are given in the subscript. Usually, we would report the actual $p$ value, but in this case, with $p$ vanishingly small, we report it as smaller than an arbitrarily small value, below any reasonable significance level. Also note that we always give information about the effect size, in this case simply the values of the means. We could also include the standard deviations or standard errors. When sample sizes are as large as these, the power of the statistical test is very high, and we would get significance even for scientifically insignificant differences.

We continue with the post-hoc test, to see which of the pairwise differences are significant. Table 4.2 lists the $Q$ statistics and $p$ values from Tukey's HSD test, together with a 95% confidence interval for the difference in means (computed with the multiple-$t$ procedure).

We see that all groups are significantly different from all others. This is also evident from the confidence intervals on differences, none of which contains zero.

### 4.4.2 Kruskal–Wallis test

The Kruskal–Wallis test is a non-parametric alternative to one-way ANOVA, just as Mann–Whitney's $U$ is a non-parametric alternative to the $t$ test. In other words, this test does not assume any particular distribution. For only two samples, the Kruskal–Wallis test will give similar results to Mann–Whitney's $U$, and it operates in a similar way, using not the original values but their ranks. The null hypothesis is that none of the samples "stochastically dominate" others. This can be understood as no difference in location between the distributions. In particular, if the distributions are assumed to be equivalent apart from position, then the null hypothesis of the Kruskal–Wallis test reduces to

$H_0$: all samples are taken from populations with equal median values.

The Kruskal–Wallis test uses a test statistic called $H$, based on the $k$ samples with a total sample size of $N$:

$$H = \frac{12}{N^2 + N} \sum_{i=1}^{k} \sum_{j} \frac{R(x_{ij})^2}{n_i} - 3(N+1)$$

where the $n_i$ are the individual sample sizes and $R(x_{i,j})$ is the pooled rank of object $j$ in sample $i$. In case of ties, $H$ needs to be divided by a correction factor given by Sokal and Rohlf (1995). For $k > 3$ or each of the $n_i$ larger than five, the significance can be approximated using a chi-squared distribution with $\nu = k - 1$ degrees of freedom.

---

**Example 4.7**

The lengths of the shells of the brachiopod genus *Dielasma* (Figs. 4.11 and 4.12) were measured in three samples $A$, $B$, and $C$ from different localities in the Permian of northern England (Hollingworth and Pettigrew, 1988). We wish to know if the length is significantly different in the three localities. First, we investigate whether we can use ANOVA. Looking at the histograms (Fig. 4.13), sample A seems bimodal, and sample B is highly skewed. Only sample C seems close to having a normal distribution. Remembering that ANOVA should be relatively robust to violations of the assumptions, we do not have to give up quite yet, but can look at the variances. Sample A has a variance of 30.8, sample B has 31.6, whereas sample C has only 3.9. If you insist on a statistical test, Levene's test reports a $p$ value of practically zero for equivalence of variance. As the final nail in the coffin, the sample sizes are very different ($n_A = 103, n_B = 79, n_C = 35$). The assumptions of ANOVA are obviously strongly violated, and we turn to the Kruskal–Wallis test instead (although its assumption of similar distributions does not seem to hold).

The Kruskal–Wallis test reports a highly significant difference ($H = 27.7, p < 0.0001$), so the equality of sizes can be rejected. Post-hoc tests for the Kruskal–Wallis test have been proposed, such as the Dunn's test, but are not very commonly used. Considering

(A)                                                          (B)

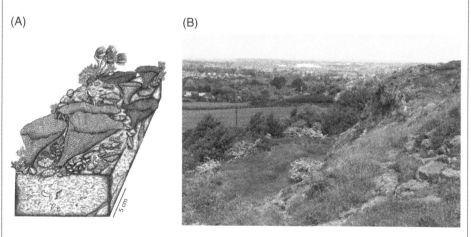

**Figure 4.11** (A) Reconstruction of the Permian seascape of the Tunstall Hills, northeast England. Nests of the brachiopod *Dielasma elongatum* occur on the lower right-hand part of the seafloor. A number of samples of this brachiopod have been collected from different parts of the reef complex. Redrawn from Hollingworth and Pettigrew (1988). (B) Permian exposures on the Tunstall Hills overlooking Sunderland. Brachiopods were collected from in and around the reef mounds, on the right.

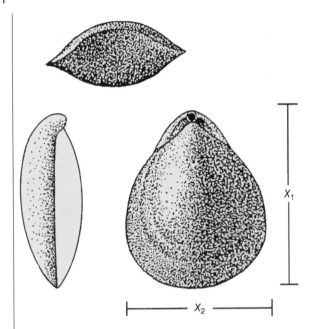

Figure 4.12   Measurement of sagittal length and width on the Permian brachiopod *Dielasma*.

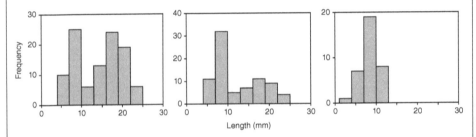

Figure 4.13   Histograms of lengths of shells of the brachiopod *Dielasma* from three Permian localities in NE England. Data from Hollingworth and Pettigrew (1988).

that sample A has a median of 15.2, sample B has a median of 9.0, while sample C has a median of 8.0, it seems that all the differences are important. This is confirmed by the Dunn's test with, e.g., Bonferroni correction, reporting that all three pairwise comparisons are significant.

## 4.5   Correlation

### 4.5.1   Linear correlation

Linear correlation between two variates $x$ and $y$ means that they show some degree of linear association, that is, they increase and decrease linearly together. Finding such correlations is of fundamental importance for many paleontological problems. Is the width of the pygidium correlated with the width of the cephalon across several trilobite species? Are the abundances of two species correlated across a number of samples? Note that in correlation analysis there is no concept of dependent and independent variable (as in linear regression, section 4.6), only of interdependency. It is important to remember that correlation does not imply causation in either direction but may be due to both variables being dependent on a common, third cause. For example, an observed correlation between the number of hostelries and the number of churches in cities is not likely to be due to either causing the other. Rather, both variables depend on the population size of the city.

Conventionally, the strength of linear correlation is measured using the linear correlation coefficient $r$, also known as Pearson's $r$ or the product-moment correlation coefficient (Fig. 4.14). In the following, we will assume that each of the two variates is normally distributed (binormal distribution, see section 5.1). The correlation coefficient can be calculated for samples from any distribution, but the usual parametric tests of significance

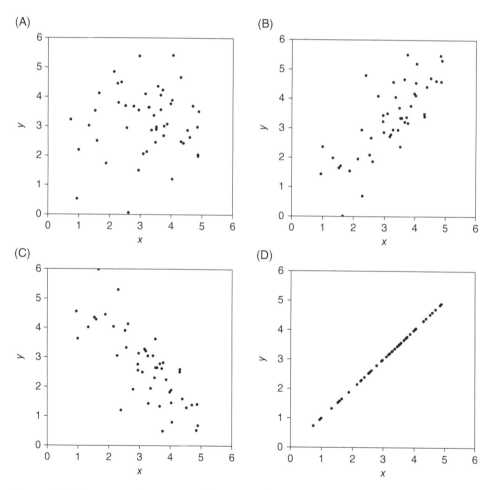

Figure 4.14 Some computer-generated data sets and their correlation coefficients. (A) $r = 0.068$, $p = 0.64$ (no significant correlation). (B) $r = 0.82, p < 0.001$ (positive correlation). (C) $r = -0.82$, $p < 0.001$ (negative correlation). (D) $r = 1, p < 0.001$ (complete positive correlation).

assume binormal distribution. Since this coefficient is based on our sample, it is only an estimate of the population correlation coefficient $\rho$.

If $x$ and $y$ lie on a perfectly straight line with positive slope, $r$ will take the value $+1$ (complete positive correlation). Conversely, we will have $r = -1$ if the points lie on a straight line with negative slope. Lack of linear correlation is indicated by $r$ being close to zero. Quite often the square of $r$ is given instead, with $r^2$ ranging from zero (no correlation) to one (complete correlation). This *coefficient of determination* is also useful but has a different interpretation: it indicates the proportion of variance that can be explained by the linear association. For a bivariate data set $(x_i, y_i)$, the linear correlation coefficient is defined as

$$r = \frac{\sum\left(x_i - \bar{x}\right)\left(y_i - \bar{y}\right)}{\sqrt{\sum\left(x_i - \bar{x}\right)^2}\sqrt{\sum\left(y_i - \bar{y}\right)^2}}$$

where $\bar{x}$ and $\bar{y}$ are the mean values of $x$ and $y$, respectively. We also define *covariance* for use in later chapters; this is similar to correlation but without the normalization for standard deviation:

$$\mathrm{COV} = \frac{\sum\left(x_i - \bar{x}\right)\left(y_i - \bar{y}\right)}{n - 1}$$

Although the linear correlation coefficient is a useful number, it does not directly indicate the statistical significance of correlation. Consider, for example, any two points. They will always lie in a straight line, and the correlation coefficient will indicate complete correlation. This does not mean that the two points are likely to have been taken from a population with high correlation! Assuming a normal distribution for each of the variables, we can test $r$ against the null hypothesis of zero correlation. If the resulting probability value $p$ is lower than the chosen significance level, we have shown a statistically significant correlation.

Assuming a binormal distribution, the null hypothesis of zero population correlation coefficient can be tested using Student's $t$ test, with

$$t = r\sqrt{\frac{n - 2}{1 - r^2}}$$

and $n - 2$ degrees of freedom (e.g., Press et al. 1992). As a non-parametric alternative, not assuming binormal distribution, we can do a permutation test by repeatedly re-ordering the pairs and calculating the proportion of re-orderings yielding a value of $r$ larger than or equal to the $r$ from the original sample (cf. section 4.3).

A Bayes factor (section 2.7) for Pearson's $r$ can be computed according to "Jeffrey's integrated Bayes factor" (Ly et al. 2016).

A *monotonic* function is one that is either consistently increasing (never decreasing) or consistently decreasing (never increasing). If the two variates are associated through a monotonic nonlinear relation (such as $y = x^2$ for positive $x$), the linear correlation coefficient will generally still not be zero, but harder to interpret. Such relationships are better investigated using non-parametric methods (see below).

A final, important warning concerns so-called *closed data*, consisting of relative values such as percentages. When one variate increases in such a data set, the other variate must by necessity decrease, leading to a phenomenon known as *spurious correlation*. Rather specialized methods must be used to study correlation in such cases (Aitchison 1986; Swan and Sandilands 1995; Reyment and Savazzi 1999).

---

**Example 4.8**

We measured whorl width and the strength (height) of the ribs on the sides of the shell in 14 specimens of the Triassic ammonite *Pseudodanubites halli*, all of similar diameter, and all taken from the same field sample (Fig. 4.15). We assume that both variates were taken from a normal distribution. Figure 4.16 shows the scatter plot of the values. The correlation coefficient is $r = 0.85$, significantly different from zero at

(A) (B)

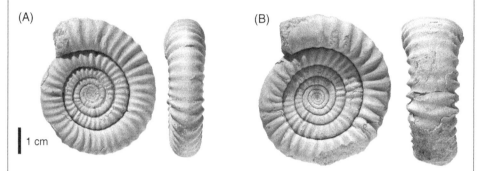

1 cm

Figure 4.15 A compressed (A) and a depressed (B) morph of the Triassic ammonite *Pseudodanubites halli* from Nevada. Hammer and Bucher (2005)/John Wiley & Sons, Inc.

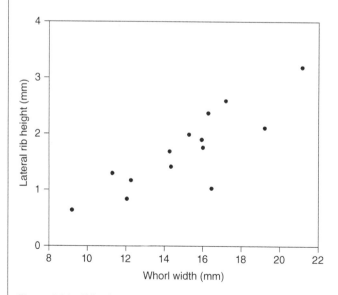

Figure 4.16 Whorl width and lateral rib height in 14 specimens of the Triassic ammonite *Pseudodanubites halli*. $r = 0.85$, $p < 0.001$.

$p = 0.0001$ ($t$ test). Likewise, a permutation test gives $p = 0.0003$. The proportion of variance in one variable "explained by" the other is given by the coefficient of variation, $r^2 = 0.72$. The Bayes factor is 235, which can be considered decisive evidence for correlation. It is of course normal for two size measures to correlate, but this particular correlation in ammonoids has received a special name and a considerable literature: *Buckman's law of covariation*.

## 4.5.2 Non-parametric correlation

The statistical procedure given earlier strictly tests for linear correlation, and it also assumes a binormal distribution. Non-parametric correlation does not make these assumptions. Like for the Mann–Whitney $U$ test, we proceed by ranking the data, that is, replacing the original values $(x_i, y_i)$ with their positions $(R_i, S_i)$ in the two sorted sequences, one for $x$ and one for $y$. Obviously, any perfectly monotonic relationship (linear or nonlinear) between the variates will then be transformed into a linear relationship with $R_i = S_i$.

Two coefficients with corresponding statistical tests are in common use for non-parametric bivariate correlation. The first one, known as the *Spearman rank-order correlation coefficient $r_s$*, is simply the linear correlation coefficient, as defined earlier, of the ranks $(R_i, S_i)$. Most of the properties of the linear correlation coefficient still hold when applied to ranks, including the range of −1 to +1, with zero indicating no correlation. For $n > 10$, the correlation can be tested statistically in the same way as described earlier.

While Spearman's coefficient is fairly easy to understand, the coefficient called Kendall's tau may appear slightly bizarre in comparison. All *pairs* of bivariate points $[(x_i, y_i), (x_j, y_j)]$ are considered – there are $n(n - 1)/2$ of them. Now if $x_i < x_j$ and $y_i < y_j$, or $x_i > x_j$ and $y_i > y_j$, the $x$ and $y$ values have the same "direction" within the pair, and the pair is called *concordant*. If $x$ and $y$ have opposite directions within the pair, it is called *discordant*. Ties are treated according to special rules. Kendall's tau ($\tau$) is then defined as the number of concordant pairs minus the number of discordant pairs, normalized to fall in the range −1 to +1. A simple statistical test is available for testing the value of Kendall's tau against the null hypothesis of zero association correlation (Press et al. 1992, see also Sokal and Rohlf 1995).

---

**Example 4.9**

Smith and Paul (1983) measured irregular echinoids of the genus *Discoides* through a stratigraphical section in the Upper Cretaceous of Wilmington, south Devon, England (Fig. 4.17). The maximum height and the maximum diameter of each specimen were recorded, together with the percentage of sediment with particle size smaller than 3.75 mm. Does the morphology of the echinoid correlate with grain size?

The height/diameter ratio and the fraction of fine particles are plotted against each other in Fig. 4.18. There seems to be some correlation between the two, but it does not seem very linear, flattening off to the right in the plot. We therefore decide to proceed using non-parametric correlation. The result is that Spearman's $r_s = 0.451$, and Kendall's

(A)                              (B)

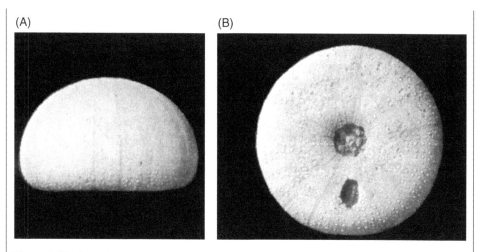

Figure 4.17    Lateral (A) and aboral (B) views of a test of the irregular echinoid *Discoides favrina* from Wilmington, England. From Smith and Paul (1983).

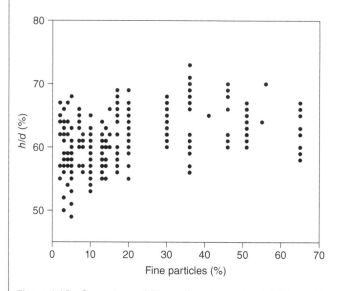

Figure 4.18    Percentage of fine sediment particles (<3.75 mm Ø) and height/diameter ratio for 270 specimens of the echinoid *Discoides*, Upper Cretaceous, Devon, England. Data from Smith and Paul (1983).

tau = 0.335. The $p$ value for the null hypothesis of no correlation between the two variates is very small for both these coefficients ($p < 0.0001$). We conclude that the echinoids tend to be higher and narrower in fine-sediment conditions, although a causal relationship cannot be inferred.

## 4.6 Bivariate linear regression

One of the most basic hypotheses we can have about a relationship between two variables $x$ and $y$ is that they are linearly related, that is, they follow the straight-line equation $y = ax + b$, where the slope $a$ and the intercept $b$ are constants (some authors interchange $a$ and $b$). However, due to measurement error and many other factors, a given set of $(x, y)$ pairs cannot be expected to fall precisely on a straight line. With linear regression, we attempt to fit a straight line to a bivariate sample, i.e., we are estimating the values of $a$ and $b$.

### 4.6.1 Ordinary least-squares linear regression

In this section, we will assume that there are no errors in the $x$ values (the independent variable) – they are precisely given. Our model consists therefore of a linear, "deterministic" component plus a "random" or "stochastic" error component (residual):

$$y_i = ax_i + b + e_i$$

The procedure of finding $a$ and $b$ such that the $e_i$ are minimized, given a set of $(x_i, y_i)$ values, is known as *linear regression* (the term linear regression is used also in a wider sense, meaning any fit to a model that is linearly dependent upon its parameters). There are several alternative ways of measuring the magnitude of the vector *e*, giving rise to different methods for linear regression. The most common approach is based on *least squares*, i.e., the minimization of the sum of squares of $e_i$. This optimization problem has a simple closed-form solution.

Although the least-squares solution of the regression problem is quite general and makes no assumptions about the nature of the data, we would like to be able to state in statistical terms our confidence in the result. For this purpose, we will make some assumptions about the nature of the residuals $e_i$, namely that they are taken independently from a normal distribution with zero mean and that the variance is independent of $x$. It is a good idea to plot the residuals ($y_i$ values minus the regression line) in order to informally check these assumptions.

Given that the assumptions hold approximately, we can compute standard errors on the slope and the intercept. We can also compute a "significance of regression" by testing whether the slope equals zero (using a $t$ test).

Nonlinear relationships between $x$ and $y$ can sometimes be "linearized" by *transforming* the data. This method makes it possible to use linear regression to fit data to a range of models. For example, we may want to fit our data points to an exponential function:

$$y = ce^{ax}.$$

Taking the logarithms of the $y$ values, we get the linear relationship

$$\ln y = \ln c + \ln e^{ax} = ax + \ln c.$$

We will see another example of a linearizing transformation in section 6.1.

The computation of an ordinary linear regression proceeds as follows. First, the slope $a$ is estimated, using

$$a = \frac{\sum\left(x_i - \bar{x}\right)\left(y_i - \bar{y}\right)}{\sum\left(x_i - \bar{x}\right)^2}$$

where $\bar{x}$ and $\bar{y}$ are the mean values of $x$ and $y$, respectively. The intercept is then given by

$$b = \bar{y} - a\bar{x}$$

With the assumptions given above, the standard error (or standard deviation) of the estimate of the slope is given by

$$s_a = \sqrt{\frac{\sum\left(y_i - ax_i - b\right)^2}{\left(n-2\right)\sum\left(x_i - \bar{x}\right)^2}}$$

The standard error of the intercept is

$$s_b = \sqrt{\frac{\sum\left(y_i - ax_i - b\right)^2 \left(1/n + \bar{x}^2 \Big/ \sum\left(x_i - \bar{x}^2\right)\right)}{\left(n-2\right)}}$$

Given the standard errors, we can compute 95% confidence intervals on the slope and intercept using the two-tailed 5% point of Student's $t$ distribution with $n - 2$ degrees of freedom:

$$a \pm t_{0.025, n-2}\, s_a$$

$$b \pm t_{0.025, n-2}\, s_b$$

These standard errors and confidence intervals are based on a number of assumptions, such as independence of points, normal distribution of residuals, and independence between variables and variance of residuals (homoskedasticity). A more robust estimate of the confidence intervals can be achieved by *bootstrapping* (section 4.1). A large number (e.g., 2000) of replicates are produced, each by random selection among the original data points, with replacement. Confidence intervals can then be estimated from the distribution of bootstrapped slopes and intercepts.

The significance of regression can be computed in several different ways: we can use Student's $t$ test on the slope to see if it is significantly different from zero, test for linear correlation as in section 4.5, or use a special form of ANOVA (Sokal and Rohlf 1995). These methods should give equivalent results.

---

**Example 4.10**

We extracted marine genera from the Paleobiology Database using shareholder quorum subsampling (section 9.2) with the R package divDyn (Kocsis et al. 2019). The data from the Jurassic to Quaternary are given in Fig. 4.19, showing a steady increase in diversity. Is a straight line a good model?

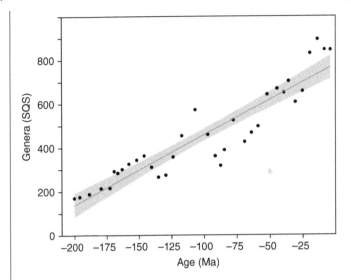

Figure 4.19    Counts of marine genera per stage from the Jurassic to Quaternary. Data from the Paleobiology Database resampled with the shareholder quorum subsampling (SQS) method. The fitted line is $y = 3.21t + 780$, and $r^2 = 0.85$. A 95% confidence interval for the regression line is also shown.

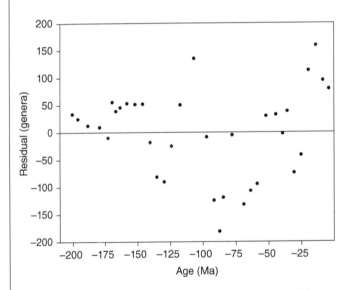

Figure 4.20    Residuals from the regression shown in Fig. 4.19.

The straight line fitted to the data is also shown in the figure, with a slope of $a = 3.21$ genera per million years, and an intercept at $b = 780$ genera at time zero (today). The significance of regression is reported as $p < 0.0001$.

The plotting of the residuals (Fig. 4.20) indicates some increase of variance with time, and also systematic trends over shorter intervals of time. A runs test

(section 12.5) confirms a structure in the residuals, reporting a significant departure from randomness at $p < 0.05$.

Because of this, and since the diversity at a given stage is obviously strongly dependent upon the previous stage, the assumptions underlying the calculation of error bars
do not really hold. Nevertheless, we note that the standard error on the slope is given
as 0.230, and the standard error on the intercept as 27.1. We can estimate a parametric
95% confidence interval on $a$ as [2.74, 3.68], in reasonable agreement with the bootstrapped confidence interval of [2.83, 3.64].

### 4.6.2   Reduced major axis regression

The linear regression method described earlier calculates the straight line that minimizes
the sum of squared distances from the data points to the line in the $y$ directions only. The $x$
values are assumed to be in exact positions, so that all errors are confined to the $y$ values. In
this sense, the $x$ and $y$ values are treated asymmetrically. This makes good sense if $x$ is supposed to be an independent variable and $y$ is dependent upon $x$. For example, we might want
to study shell thickness as a function of grain size. In this case, we would take grain size as
the independent variable, which is hopefully measured quite accurately. Shell thickness
would be a variable that might depend upon $x$ but also upon a host of "random" factors that
would be subsumed under the stochastic term $e$ in the regression equation. Regression with
minimization of residuals in the $y$ direction only is referred to as *Model I regression*.

Alternatively, we can minimize the residuals in both the $x$ and $y$ directions. This is called
*Model II regression*. Reduced major axis (RMA) regression (Miller and Kahn 1962; Sokal
and Rohlf 1995) is a Model II regression method that minimizes the sum of the product of
the distances from the data points to the line in the $x$ and $y$ directions. This is equivalent to
minimizing the areas of the triangles such as the one shown in Fig. 4.21. The two variables
are thus treated in an entirely symmetric fashion. Model II regression such as RMA is

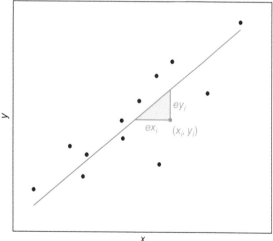

**Figure 4.21**   The RMA method for
linear regression takes into
consideration the errors in both the $x$
and the $y$ directions, by minimizing
the sum of the products of the errors:
$\Sigma\ ex_i \cdot ey_i$.

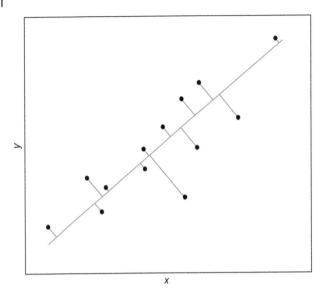

Figure 4.22   The MA method for linear regression minimizes the sum of the Euclidean distances from the points to the regression line.

usually the preferred method in morphometrics, where two measured distances are to be fitted to a straight line. When considering the length and thickness of a bone, there is no reason to treat the two variables differently. Nonetheless, ordinary least-squares (OLS) regression and RMA usually give similar, but not identical results.

It should be mentioned that several methods are available for Model II straight-line regression. The *major axis (MA)* method is particularly appealing because of its intuitive way of calculating the magnitude of the residual as the sum of Euclidean distances from each data point to the line (Fig. 4.22). The MA has the same slope as the first principal component axis and is computed as in PCA (section 6.2). However, the MA method has not been commonly used in paleontology. Bartlett (1949) suggested the "three-group method" which requires the data to be divided into three subsets. This method is, however, controversial (Kuhry and Marcus 1977; Sokal and Rohlf 1995).

The RMA method has been criticized by Kuhry and Marcus (1977) and other authors but remains the most commonly used Model II regression method.

---

**Example 4.11**

Figure 4.23 shows a scatter plot of valve length versus width in 51 specimens of a living lingulide brachiopod (*Glottidia palmeri* from the Gulf of California, Mexico). In this case, there is no reason to treat the two variables differently for the purposes of linear regression, so RMA is preferred over standard Model I linear regression. The standard linear regression line is given for comparison, showing a small but noticeable difference in the slope reported by the two different methods.

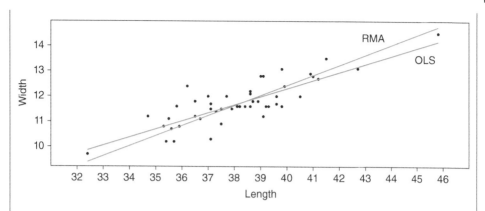

Figure 4.23  Length of the ventral valve plotted against valve width for a sample of the Recent lingulide brachiopod *Glottidia palmeri* from Campo Don Abel, Mexico (n = 51). The RMA regression line has slope 0.388 (bootstrapped 95% confidence interval [0.34, 0.45]), while the ordinary least squares linear regression (OLS) has slope 0.322 (confidence interval [0.25, 0.37]). Data from Kowalewski et al. (1997).

Equations for RMA regression were given by Warton et al. (2006). The slope $a$ is estimated by

$$a = \pm \sqrt{\frac{\sum(y_i - \bar{y})^2}{\sum(x_i - \bar{x})^2}}$$

where $\bar{x}$ and $\bar{y}$ are the mean values of $x$ and $y$, respectively. Note that the sign of the slope is left undetermined by this definition. The sign can be taken from that of the correlation coefficient (section 4.5). The intercept is then given by

$$b = \bar{y} - a\bar{x}$$

With the assumptions given above, the standard error (or standard deviation) of the estimate of the slope is estimated as

$$s_a = |a| \sqrt{\frac{1 - r^2}{n - 2}}$$

An estimate for the standard error of the intercept is

$$s_b = \sqrt{\frac{s_r^2}{n} + \bar{x}^2 s_a^2}$$

where $s_r$ is the estimate of standard deviation of residuals and $s_a$ is the standard error on slope.

For zero intercept ($b = 0$), set $\bar{x} = 0$ and $\bar{y} = 0$ for the calculation of slope and its standard error (including the calculation of $r$ therein), and use $n - 1$ instead of $n - 2$ for the calculation of standard error.

Confidence intervals for slope and intercept are more robustly estimated using a bootstrapping procedure (section 4.1).

## 4.7 Generalized linear models

The assumptions of OLS linear regression do not hold for several types of data commonly used in paleontology. For example, although we often use OLS for count data, a linear model is clearly not completely appropriate both because the linear model predicts negative counts for small values of the independent variable, and because the counts are discrete (not continuous) and therefore the residuals cannot be truly normally distributed. Similar problems arise for percentages of counts, which are restricted to the range 0–100. An even more extreme case is when the dependent variable is binary, such as presence or absence of a fossil.

These cases can be more properly handled with so-called generalized linear models (GLM). With GLM regression, we need to specify a *link function*, which provides the expected relationship between a simple linear model and the GLM model (or, more precisely, the mean of the model, which is allowed to vary with the independent variable). The link function is typically a nonlinear transformation. In addition, we need to specify a *distribution function* for the dependent variable. If we specify the "identity" link function (i.e., no transformation) and normal distribution, GLM reduces to OLS and will give the same results.

The technical implementation of GLM is somewhat complex because the least-squares solution must be found by numerical optimization, but its basic use is fairly straightforward. We will discuss the most common cases by example.

### 4.7.1 GLM regression of counts

Counts are always zero or positive integers. We therefore need a link function to produce a positive regression model. An exponential function is always positive and is a natural choice. The link function, which represents a transformation of the final model to the linear model, is then logarithmic. Moreover, the Poisson distribution is usually assumed for counts within fixed areas or time intervals. Thus, the standard choices for GLM regression of counts are the log link and Poisson distribution. It is usually assumed that the variance of the residuals is equal to the value of the regression model.

### 4.7.2 GLM regression of percentages or proportions

Like counts, percentages are always positive, but they are also limited upwards to 100 (or to 1 for a proportion). An appropriate model distribution for such data is the binomial distribution, and the most commonly used link function is the "logit":

$$g(\mu) = \ln \frac{\mu}{1-\mu}$$

where $\mu$ is the expected mean of the model, varying with $x$. For proportions, it is necessary to supply not only the proportions but also the sample size for each data point to the GLM algorithm.

**Example 4.12**

Rugstad (2022) noted the proportions of convex-up brachiopod shells through the Rytteråker Formation (Silurian) in Norway (Fig. 4.24). A total of 34 samples were collected through a 36 m thick succession, each containing from 1 to 34 shells. We will use GLM regression with the binomial distribution and the logit link for this data set, to see if there is a stratigraphic trend in the orientation of the shells. The data and the GLM regression line are shown in Fig. 4.25. The intercept is 0.32±0.31 (one standard error), and the slope is −0.031±0.015. The fitted function (inverse of the logit link) is thus

$$y = \frac{e^{-0.031x+0.32}}{1+e^{-0.031x+0.32}}$$

This is a nonlinear function but appears nearly linear in the plot. The slope is significantly different from zero, at $p = 0.041$, although the trend is weak.

Brachiopod shells that have been suspended in the water column tend to settle convex-down. The negative trend in convex-up shells could therefore imply more unsteady, high-energy conditions further up in the section, which fits with other information indicating a shallowing-up trend in the Rytteråker Formation, culminating in shallow reef facies near the top.

**Figure 4.24** Thin section of wackestone in the Rytteråker Formation (lower Silurian, Norway) showing mixed orientations of thick-shelled *Pentamerus* brachiopods. Scale bar 5 mm. Rugstad (2022)/University of Oslo.

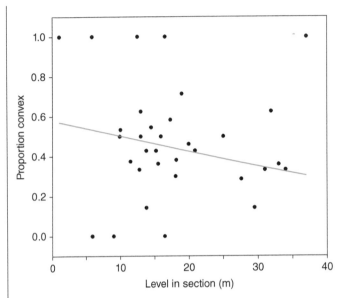

Figure 4.25   GLM regression of proportion of convex-up brachiopod shells in the Silurian of Norway. Data from Rugstad (2022).

### 4.7.3   GLM regression of binary data (logistic regression)

It is also possible to carry out a GLM regression on binary (presence-absence) data responding to a continuous independent variable. For this purpose, we typically use the Bernoulli distribution, which is a special case of the binomial distribution, and the logit link function. This results in a continuous, sigmoidal model curve varying from 0 to 1, which can be interpreted as an estimate for the probability of the event occurring, varying with the independent variable. This procedure is often called *logistic regression* (Fig. 4.26).

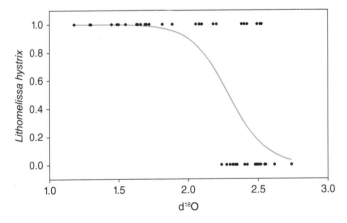

Figure 4.26   Logistic regression of presence-absence of the radiolarian *Lithomelissa hystrix* in a core through the base Holocene in Norway. The independent variable (oxygen isotopes) is a proxy for temperature. Data from Bjørklund et al. (2019).

## 4.8   Polynomial and nonlinear regression

There is little stopping us from fitting our bivariate data set to more complicated mathematical functions. Within the OLS framework, we can extend the linear model to a quadratic (second order) polynomial:

$$y_i = b_2 x_i^2 + b_1 x_i + b_0 + e_i$$

where the $b$s are the coefficients to be fitted, and $e$ is the residual which is to be minimized. Although this is a nonlinear function, it is a linear combination of terms with unknown coefficients. We can therefore find the least-squares solution with a closed-form equation, like when fitting a straight line. The same goes for higher-order polynomials (although we can run into numerical problems when fitting polynomials of very high order).

Such polynomial regression can serve two different purposes. First, we may have a theory that dictates, e.g., a cubic relationship between two variates, for example, that a volume scales as the cube of a length (an alternative approach would be to apply a third-root transformation to the volumes and then run a straight-line regression). The polynomial regression could then be used to assess the fit of the data to the theoretical model. Second (and this is perhaps the more common use of polynomial regression), we may want to simply fit any kind of curve to the data for aesthetical reasons. This is perhaps better done with one of the smoothing methods described in a later chapter.

The next step in complexity is to fit models that are nonlinear in the parameters, such as $y = e^{bx}$ or $y = \sin(bx)$. This is computationally more challenging, but nonlinear optimization methods such as the Levenberg-Marquardt method (Press et al. 1992) usually work well. In principle, any user-supplied function can be implemented (Fig. 4.27), but many programs (including PAST) include a number of predefined functional forms, each with a special code for initializing the optimization.

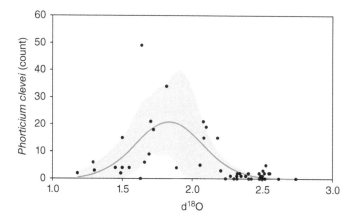

**Figure 4.27**   Nonlinear fit to a Gaussian function (red line) of the radiolarian *Phorticium clevei* in a core through the base Holocene in Norway, with 95% bootstrapped confidence interval. The independent variable (oxygen isotopes) is a proxy for temperature. The fitted Gaussian is $y = 20.8e^{(x-1.84)^2/(2 \cdot 0.24^2)}$. Data from Bjørklund et al. (2019).

### 4.8.1 Akaike information criterion

When faced with a choice between different models for a certain data set, it is often of interest to compare the relative success of the models in fitting the data. For this purpose, we can use some measure of goodness of fit, e.g., based on the sum of squared deviations from the data points to the model. The common regression methods work by maximizing the goodness of fit as represented by the likelihood, which is the probability of the data given the model (maximum likelihood).

Comparing goodness of fit works well when comparing models with the same number of parameters. However, our models often have different numbers of parameters, such as when comparing a linear model with two parameters to a quadratic model with three parameters. The goodness of fit will usually increase for more complex models with more parameters, eventually leading to "overfitting." An overfitted model appears to fit the data very well, but it has low predictive power: new data points that did not contribute to the model fitting will typically plot far outside the model prediction.

When comparing models with different numbers of parameters, we therefore need to take both the goodness of fit and the number of parameters into account. A popular method is the Akaike information criterion (Akaike 1974):

$$AIC = -2 \ln L + 2k$$

Here, $L$ is the maximized likelihood and $k$ is the number of parameters in the model.

## 4.9 Mixture analysis

We have discussed the normal distribution as a theoretical model for the distribution of a univariate population. The normal distribution has two parameters: the mean and the variance (or standard deviation). A more complex distribution model is the sum, or *mixture*, of two or more normal distributions, each with a vertical scaling (weight) as an additional parameter. Such a mixture will typically be bimodal or multimodal, with one peak for each normal component.

Mixture models are useful in paleontology, especially for size data. If a sample of fossils can be fitted by a mixture model, the components may represent annual age cohorts, arthropod instars (Hunt and Chapman 2001), or sexual dimorphism such as ammonite micro- and macroconchs.

The number of components in the model is chosen by the user, either based on a hypothesis (e.g., two components to investigate sexual dimorphism) or by minimizing the Akaike information criterion or other measure of fit.

The classical method for mixture analysis is the expectation maximization (EM) algorithm (Dempster et al. 1977). The model parameters are fitted by an optimization procedure that can get stuck on a local optimum, and the procedure is therefore typically run repeatedly with different starting conditions in order to ensure a good solution (PAST automatically executes 20 such iterations).

## Example 4.13

We return to the *Dielasma* data set from section 4.4. Sample 1 ($n = 103$) seems to show a bimodal distribution (Fig. 4.28). The mixture analysis program reports an Aikaike information criterion of 459.1 for a model with a single normal component (Table 4.3). For two and three components, the AIC reduces to 417.3, and 415.0, respectively, implying a better model fit. At four components, the AIC increases again. Although the AIC is slightly lower for three than for two components, we choose the two-component model for Sample 1 (Fig. 4.28). For the Gaussian on the left, the proportion is 0.346, the mean 7.66 mm, and the standard deviation 1.33 mm. For the Gaussian on the right, the mean is 17.7 mm and the standard deviation is 3.24 mm. Sample 3 ($n = 35$) gives a minimal AIC value for only one component, with mean 8.11 mm and standard deviation 1.94 mm (Fig. 4.28).

It is interesting that the mean and standard deviation for the single Gaussian in Sample 3 are similar to the mean and standard deviation for the left Gaussian in Sample 1. This could perhaps support a claim that the components represent first-year and second-year growth cohorts, with the second year missing from Sample 3.

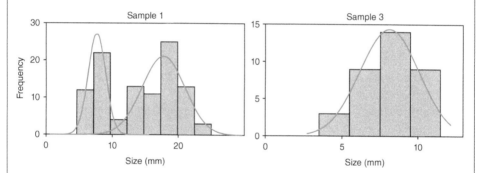

**Figure 4.28** Mixture modeling applied to two samples from the *Dielasma* data set. For Sample 1, we specify two components in the model. For Sample 3, we use a single component.

**Table 4.3** Results from the mixture analysis of the *Dielasma* data set, showing low Akaike values (i.e., good fit) for two and three components in Sample 1, and for one component in Sample 3.

| Number of components | Sample 1 AIC | Sample 3 AIC |
| --- | --- | --- |
| 1 | 459.1 | 85.89 |
| 2 | 417.3 | 87.84 |
| 3 | 415.0 | 95.28 |
| 4 | 420.1 | 100.9 |

## 4.10    Counts and contingency tables

The chi-squared test is a fundamental and versatile statistical test that is useful for many different paleontological problems, allowing the comparison of two sets of counted data in discrete categories. An obvious example is the comparison of two fossil samples, each with counts of fossils of different taxa. The null hypothesis to be tested is then

$H_0$: the two samples are drawn from populations with the same distributions

For comparing two samples, the chi-squared statistic is computed as

$$\chi^2 = \sum_{i,j} \frac{\left(n_{ij} - E_{ij}\right)^2}{E_{ij}}$$

where $n_{ij}$ is the count of category (taxon) number $i$ in sample $j = 1$ or $j = 2$. Moreover, $E_{ij}$ is the expected count if the null hypothesis were true:

$$E_{ij} = \frac{\sum_k n_{ik} \sum_k n_{kj}}{n}$$

where $n$ is the total count. The probability $p$ of equal distributions can be computed using the incomplete gamma function (Press et al. 1992) or merely looked up in a table. For this two-sample test, the number of degrees of freedom ($\nu$) equals the number of taxa minus one.

Continuing our taxa-in-samples example, a chi-squared test can also be used to test for equal compositions of more than two samples, simply by allowing $j > 2$ in the equations above. The data set will then be given in the form of a table, with taxa in rows and samples in columns or vice versa. Are the samples different? This is equivalent to asking whether the taxon categories show an association with the sample categories, that is, whether the two categorical (nominal) variables are coupled in some way. If this is not the case, then taxa seem to be similarly or randomly distributed across the samples. This type of investigation is known as *contingency table* analysis. The number of degrees of freedom is $(r-1)$ $(c-1)$ for $r$ rows and $c$ columns in the table.

The chi-squared test can also be used to compare a sample with a known, theoretical distribution. For example, we may have erected a null hypothesis that a certain fossil taxon occurs with equal frequency in a number of samples (categories) of equal sizes. If this were the case, we would expect $n/m$ fossils in each category, where $n$ is the total number of fossils and $m$ is the number of samples. This gives us our theoretical distribution. The number of degrees of freedom needs to be set according to whether the theoretical distribution was computed using any parameters taken from the data set. In the example above, the total number of fossils was used for normalizing the theoretical distribution, so we "lose" one degree of freedom, and $\nu$ equals the number of bins minus one. Any additional parameters that we use to produce the expected distribution and that are estimated from the data (such as mean and variance) must be further deducted from $\nu$.

The chi-squared test is not accurate if any of the cells contain fewer than five items (as a rule of thumb). This can be a serious problem for many paleontological data sets, where taxa are often rare and may be absent in one of the two samples to be compared. Apart from the hope that the test will be robust to mild violations of this assumption, we suggest increasing the sample size or removing rare taxa from the analysis altogether. For small contingency tables, a test has been made that does not have this requirement, Fisher's exact test. For larger numbers of categories, there are modern, computation-intensive methods available. One approach is to approximate the exact probability value using a "Monte Carlo" permutation procedure (section 4.3) where a large number of random samples are produced such that the sums of all objects in each of the two samples and the sums of all objects in each category are maintained. The chi-squared statistics of these random data sets are compared with the observed chi-squared statistics in order to estimate a probability value.

The *G* test is an alternative to the chi-squared test, generally giving similar results (Sokal and Rohlf 1995).

In addition to assigning a significance (*p* value) to the association, we need to quantify the strength. The $\chi^2$ statistic itself does increase with strength, but it also depends on the size of the contingency table – this is why we need to take the degrees of freedom into account when estimating *p* from $\chi^2$. Several coefficients have been proposed to compensate for this effect, a popular one is Cramér's *V*, varying from 0 to 1:

$$V = \sqrt{\frac{\chi^2}{n(q-1)}}$$

where *q* is the number of rows or the number of columns, whichever is smaller.

---

**Example 4.14**

We will compare the compositions of two fossiliferous samples, dominated by brachiopods (Fig. 4.29), in the Coed Duon section through the Upper Llanvirn (Darriwilian, Ordovician) of mid-Wales, a classic section studied by Williams et al. (1981). These authors collected rock samples of similar sizes (averaging 6.1 kg) throughout the section. The two samples (CD8 and CD9) are from successive levels within a uniform lithology, so we would assume that they are similar.

The two samples and the most abundant taxa they contain are given in Table 4.4.

The chi-squared statistic computes as $\chi^2 = 7.29$, with five degrees of freedom. The probability of equal distribution based on the chi-squared is given as $p = 0.20$, so we cannot reject the null hypothesis of equality of both distributions. Note that some of the cells contain fewer than five items, unfortunately mildly violating the assumptions of the test. However, Fisher's exact test and a Monte Carlo permutation test both give the same value ($p = 0.20$), giving further confidence in the result.

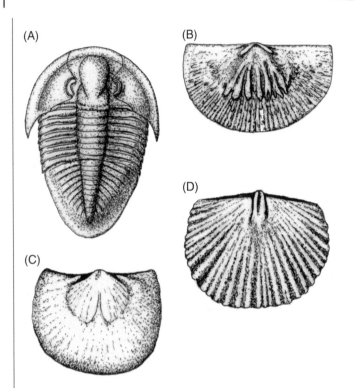

Figure 4.29   Commoner taxa from the mid-Ordovician Coed Duon section, mid-Wales: (A) The trilobite *Basilicus*; and brachiopods. (B) *Sowerbyella*. (C) *Macrocoelia*. (D) *Dalmanella*. These elements dominated the mid-Ordovician seafloor in mid-Wales. Drawn by Eamon N. Doyle.

Table 4.4   Counts of the most abundant taxa in two samples from the Coed Duon section.

|  | CD8 | CD9 |
| --- | --- | --- |
| *Macrocoelia* | 4 | 2 |
| *Sowerbyella* | 25 | 36 |
| *Dalmanella* | 15 | 8 |
| *Hesperorthis* | 3 | 2 |
| Crinoids | 8 | 3 |
| Ramose bryozoan | 4 | 5 |

# References

Aitchison, J. 1986. *The Statistical Analysis of Compositional Data*. Chapman & Hall, New York.

Akaike, H. 1974. A new look at the statistical model identification. *IEEE Transactions on Automatic Control* 19, 716–723,

Bartlett, M.S. 1949. Fitting a straight line when both variables are subject to error. *Biometrics* 5, 207–212.

Bjørklund, K.R., Kruglikova, S.B., Hammer, Ø. 2019. The radiolarian fauna during the Younger Dryas–Holocene transition in Andfjorden, northern Norway. *Polar Research* 38. https://doi.org/10.33265/polar.v38.3444.

Bruton, D.L., Owen, A.W. 1988. The Norwegian Upper Ordovician illaenid trilobites. *Norsk Geologisk Tidsskrift* 68, 241–258.

Cincotta, A., Nicolaï, M., Campos, H.B.N. et al. 2022. Pterosaur melanosomes support signalling functions for early feathers. *Nature* 604, 684–688.

Cohen, J. 1988. *Statistical power analysis for the behavioral sciences (2nd ed.)*. Academic Press, NY.

D'Agostino, R.B., Stephens, M.A. 1986. *Goodness-of-Fit Techniques*. Marcel Dekker, Inc., New York.

Davison, A.C., Hinkley, D.V. 1997. *Bootstrap Methods and Their Application*. Cambridge University Press, Cambridge, UK.

Dempster, A.P., Laird, N.M., Rubin, D.B. 1977. Maximum likelihood from incomplete data via the EM algorithm. *Journal of the Royal Statistical Society, Series B* 39, 1–38.

Feng, C., Wang, H., Lu, N., Chen, T., He, H., Lu, Y., Tu, X.M. 2014. Log-transformation and its implications for data analysis. *Shanghai Archives of Psychiatry* 26, 105–109.

Good, P. 1994. *Permutation Tests: A Practical Guide to Resampling Methods for Testing Hypotheses*. Springer Verlag.

Hammer, Ø., Bucher, H. 2005. Buckman's first law of covariation – a case of proportionality. *Lethaia* 38, 67–72.

Hollingworth, N., Pettigrew, T. 1988. Zechstein reef fossils and their palaeoecology. *Palaeontological Association Field Guides to Fossils* 3, 75p. Palaeontological Association, UK.

Hunt, G., Chapman, R.E. 2001. Evaluating hypotheses of instar-grouping in arthropods: a maximum likelihood approach. *Paleobiology* 27, 466–484.

Kocsis, A.T., Reddin, C.J., Alroy, J., Kiessling, W. 2019. The r package divDyn for quantifying diversity dynamics using fossil sampling data. *Methods in Ecology and Evolution* 10, 735–743.

Kowalewski, M., Dyreson, E., Marcot, J.D., Vargas, J.A., Flessa, K.W., Hallmann, D.P. 1997. Phenetic discrimination of biometric simpletons: paleobiological implications of morphospecies in the lingulide brachiopod *Glottidia*. *Paleobiology* 23, 444–469.

Kuhry, B., Marcus, L.F. 1977. Bivariate linear models in biometry. *Systematic Zoology* 26, 201–209.

Ly, A., Verhagen, J., Wagenmakers, E.-J. 2016. Harold Jeffreys's default Bayes factor hypothesis tests: explanation, extension, and application in psychology. *Journal of Mathematical Psychology* 72, 19–32.

Marshall, C.R. 1990. Confidence intervals on stratigraphic ranges. *Paleobiology* 16, 1–10.

Miller, R.L., Kahn, J.S. 1962. *Statistical Analysis in the Geological Sciences*. John Wiley & Sons, New York.

O'Higgins, P. 1989. A morphometric study of cranial shape in the Hominoidea. PhD thesis, University of Leeds.

Press, W.H., Teukolsky, S.A., Vetterling, W.T., Flannery, B.P. 1992. *Numerical Recipes in C*. Cambridge University Press, Cambridge, UK.

Razali, N., Yap, B.W. 2011. Power comparisons of Shapiro–Wilk, Kolmogorov–Smirnov, Lilliefors and Anderson–Darling tests. *Journal of Statistical Modeling and Analytics* 2, 21–33.

Reyment, R.A., Savazzi, E. 1999. *Aspects of Multivariate Statistical Analysis in Geology*. Elsevier, Amsterdam.

Royston, P. 1982. An extension of Shapiro and Wilk's W test for normality to large samples. *Applied Statistics* 31, 115–124.

Royston, P. 1995. A remark on algorithm AS 181: the W-test for normality. *Applied Statistics* 44, 547–551.

Rugstad, A. 2022. Quantitative paleoecology and gradient analysis of the Rytteråker Formation, Lower Silurian, Oslo Region. M.Sc. thesis, Institute for Bioscience, University of Oslo.

Shapiro, S.S., Wilk, M.B. 1965. An analysis of variance test for normality (complete samples). *Biometrika* 52, 591–611.

Smith, A.B., Paul, C.R.C. 1983. Variation in the irregular echinoid *Discoides* during the early Cenomanian. *Special Papers in Palaeontology* 33, 29–37.

Sokal, R.R., Rohlf, F.J. 1995. *Biometry: The Principles and Practice of Statistics in Biological Research*. 3rd edition. W.H. Freeman and Co., New York.

Swan, A.R.H., Sandilands, M. 1995. *Introduction to Geological Data Analysis*. Blackwell Science, Oxford, UK.

Warton, D.I., Wright, I.J., Falster, D.S., Westoby, M. 2006. Bivariate line-fitting methods for allometry. *Biological Review* 81, 259–291.

Williams, A., Lockley, M.G., Hurst, J.M. 1981. Benthic palaeocommunities represented in the Ffairfach Group and coeval Ordovician successions of Wales. *Palaeontology* 24, 661–694.

5

# Introduction to multivariate data analysis

In chapter 4, we studied data sets with one or two variates. These may be considered as special cases of more general situations, where an arbitrary number of variates are associated with each item. Such multivariate data are common in morphometrics, where we may have taken several measurements on each specimen; and in ecology (including biogeography), where several species have been counted at each locality. In this introductory chapter, we will discuss a few of the methods used to study multivariate data sets. Despite the differences in data types and investigative techniques, many multivariate algorithms can be easily adapted for both morphological and ecological investigations; thus, some of the most commonly used procedures are collected together here to avoid repetition. More specific methods will follow in later chapters.

In many respects, multivariate techniques are simply an extension of univariate techniques into multidimensional space. Thus, scalar parameters such as means and variances become vector quantities and coefficients of correlation are transformed from a single value for two variates into matrices for the pairwise relationships between variates. Multivariate distributions have similar properties as univariate ones but, of course, cannot be visualized in the same way. In addition, many of the univariate tests have multivariate analogs – Hotelling's $T^2$ test (section 5.2) can be thought of as a generalization of Student's $t$ test for example. Nevertheless, multivariate data analysis is invariably more complex than the univariate and bivariate cases, and an active area of research. Much multivariate data analysis is concerned with exploration and visualization of complex data rather than with hypothesis testing. Cluster analysis (automatic grouping of items; sections 5.4–5.5) is an example of this.

Multivariate data sets consist of *points in multidimensional space*. Just as a bivariate data set can be plotted in a two-dimensional scatterplot $(x, y)$, and a data set of three variates can be plotted as points in three-dimensional space $(x, y, z)$, a data set of seven variates can be thought of as a set of points in seven-dimensional space $(x_1, x_2, x_3, x_4, x_5, x_6, x_7)$. Position, variance, and distance can be defined in such a high-dimensional space, and statistical tests devised, but it is also often necessary to simplify multivariate data and present them in two or three dimensions to facilitate interpretation. Such reduction of dimensionality will invariably cause loss of information. Nonetheless, many methods are available to reduce the dimensionality of multivariate data sets while trying to preserve important information – these will be discussed in chapters 6 and 10.

*Paleontological Data Analysis*, Second Edition. Øyvind Hammer and David A.T. Harper.

## 5.1 Multivariate distributions

The concept of a univariate distribution as introduced in chapter 2 can be easily extended to the multivariate case. Figure 5.1 shows an example of a bivariate normal (binormal) distribution with two variates, each with univariate normal distribution.

In this case, the variances in the $x$ and $y$ directions are the same, and $x$ does not vary systematically with $y$ (i.e., there is no covariation, see section 4.5). Bivariate concentration regions around the mean value can then be represented by circles (Fig. 5.2A). If we let the variance in $y$ increase, the bivariate distribution will be stretched out in the $y$ direction and the concentration regions will turn into ellipses (Fig. 5.2B). If we introduce covariation between the variates, the bivariate distribution will be rotated and the concentration regions will consist of rotated ellipses (Figs. 5.2C, D).

In three dimensions, the multivariate normal distribution can be thought of as a "fuzzy ellipsoid" with density (frequency) increasing towards the center. The concentration regions will now be three-dimensional ellipsoids. In even higher dimensions, the concentration regions will be hyper-ellipsoids.

## 5.2 Parametric multivariate tests – Hotelling's $T^2$

By analogy with the parametric univariate tests of chapter 4 (Student's $t$ test, $F$ test, etc.), a number of multivariate versions have been devised. The multivariate analog of the $t$ test, comparing the multivariate means of two samples, is Hotelling's $T^2$ test (e.g., Davis 1986). The null hypothesis is thus

$H_0$: Data are drawn from populations with the same multivariate means.

Just as the $t$ test assumes a normal distribution and equal variances, Hotelling's $T^2$ test assumes a multivariate normal distribution and the equality of the variance-covariance matrices (Fig. 5.3). The latter can be tested with, for example, Box's $M$ test (an analog to the

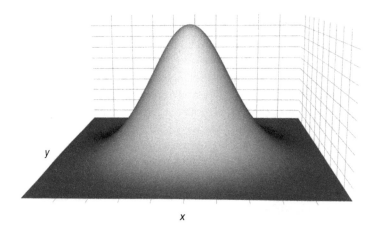

Figure 5.1   A bivariate normal distribution. Each variable, $x$ and $y$, has a univariate normal distribution as the other variable is kept constant.

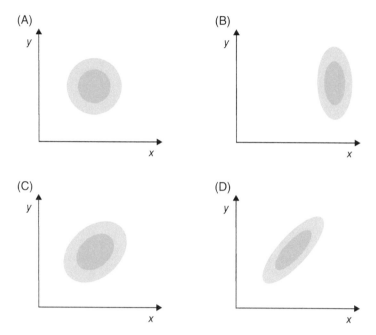

**Figure 5.2** 90% and 95% concentration ellipses for bivariate normal distributions. 90% of the values are expected to fall within the inner, 90% ellipse, while 95% fall within the outer, 95% ellipse. (A) Equal variance in x and y, no covariance. (B) Larger variance in y than in x, no covariance, mean displaced in the x direction. (C) Equal variance in x and y, weak positive covariance. (D) Equal variance in x and y, strong positive covariance.

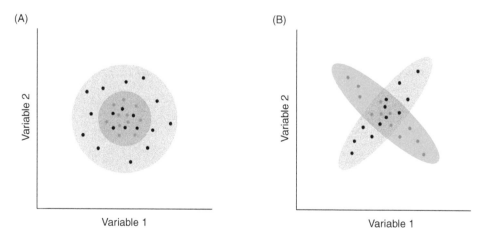

**Figure 5.3** Two examples where the assumptions of the Hotelling's $T^2$ test are not met. (A) The two samples (red and black) are from populations with the same means, but unequal variances. (B) The samples are from populations with the same means and variances, but unequal covariances. After Anderson (2001).

univariate $F$ test); but, this is not often done in practice. An alternative to Hotelling's $T^2$ has been devised for samples with unequal covariances (Anderson and Bahadur 1962), or a permutation test (section 4.3) can be used.

Hotelling's $T^2$ test for the equality of the multivariate means of two samples can be viewed as a special case of Multivariate ANalysis Of Variance (MANOVA) (section 6.5), which can be used also for more than two samples. It will fail for data sets with more variates than data points.

Hotelling's test is computed as follows. Let $\mathbf{A}$ be the matrix of $n_A$ observations on sample $A$ ($m$ variates in columns) and $\mathbf{B}$ the matrix of observations on sample B. Compute the $m \times m$ matrix $\mathbf{S^A}$ for $\mathbf{A}$:

$$S_{jk}^A = \sum_{i=1}^{n_A} A_{ij} A_{ik} - \frac{\sum_{i=1}^{n_A} A_{ij} \sum_{i=1}^{n_A} A_{ik}}{n_A}$$

and similarly for sample B. The pooled estimate $\mathbf{S}$ of the population variance-covariance matrix is then

$$S = \frac{S^A + S^B}{n_A + n_B - 2}$$

The *Mahalanobis distance squared* is given by

$$D^2 = \left(\bar{a} - \bar{b}\right)^T S^{-1} \left(\bar{a} - \bar{b}\right)$$

where $\bar{a}$ and $\bar{b}$ are the vectors of the means of each variate in the two samples. Incidentally, $D^2$ (or $D = \sqrt{D^2}$) is a useful general multivariate distance measure, taking into account the variances and covariances of the variates.

The $T^2$ test statistic is computed as

$$T^2 = \frac{n_A n_B}{n_A + n_B} D^2$$

The significance can be estimated by converting it to an $F$ statistic:

$$F = \frac{n_A + n_B - m - 1}{\left(n_A + n_B - 2\right)m} T^2$$

$F$ has $m$ and $(n_A + n_B - m - 1)$ degrees of freedom.

---

**Example 5.1**

O'Higgins (1989) measured the $(x, y)$ coordinates of eight landmarks in the midline of the skull (sagittal plane) in 30 female and 29 male gorillas. Using methods introduced in chapter 6 (section 6.11), the landmarks were standardized with respect to size, position, and rotation of the skull, to allow size-free comparison of landmark positions in males and females. The landmarks are shown in Fig. 6.28. Are males significantly different from females?

In this case, we have two samples of multivariate (16 variates) data, which are likely to be close to normally distributed (this can be informally checked by looking at the individual variates). Hotelling's $T^2$ test is therefore a natural choice. It reports a highly significant difference at $p < 0.0001$; so, the skulls of male gorillas are indeed differently shaped from those of the females. In this particular example, the statistics should be corrected for the fact that the standardization of the landmarks has reduced the number of degrees of freedom (see Dryden and Mardia 2016 for details).

## 5.3 Nonparametric multivariate tests – permutation test

The parametric Hotelling's $T^2$ test of section 5.2 assumes a multivariate normal distribution and the equality of variance–covariance matrices; it can be viewed as the multivariate version of the $t$ test. When these assumptions are strongly violated, we should turn to a nonparametric test. Again, by analogy with the univariate case (section 4.3), we can carry out a permutation test where we repeatedly reassign the specimens to the two samples (maintaining the two sample sizes) and count how often the distance between the permutated samples exceeds the distance between the two original samples. If the permutated distances are very rarely as large as or larger than the original distance, we have a significant difference between the means.

Again, the null hypothesis is

$H_0$: The data are drawn from populations with the same multivariate means.

To carry out this procedure, we need a way of measuring the distance between the two multivariate sample means. The Euclidean distance (section 5.4) between the multivariate means or the Mahalanobis distance is a possible choice. Such permutation tests were thoroughly treated by Good (1994). We will later (section 10.5) discuss a general framework for multivariate permutation tests, known as PerMANOVA, allowing several groups and different sampling designs.

An exhaustive permutation test (all possible permutations investigated) gives a probability $p$ of equality by

$$p = 1 - \frac{r-1}{P}$$

where $r$ is the rank position of the original distance among the $P$ permutations. For larger sample sizes, it is necessary to investigate only a smaller, random set of permutations.

---

**Example 5.2**

Returning to the example from section 4.3, we will consider the two samples of the illaenid trilobite *Stenopareia linnarssoni* – one from Norway ($n = 10$) and one from Sweden ($n = 7$). Four measurements were taken on the cephala of these 17 specimens (Fig. 6.8). We wish to carry out a multivariate comparison of the two samples.

In this case, the samples seem to follow multivariate normal distributions, and the variance–covariance matrices are not significantly dissimilar (Box's M, $p = 0.36$). We take this to imply that the assumptions of Hotelling's $T^2$ test are not strongly violated, and we can compare this procedure with the multivariate permutation test. Hotelling's $T^2$ test reports $p = 0.059$. A permutation test, using the Mahalanobis distance with 2000 random permutations, gives $p = 0.0565$, with a slight variation from run to run. The two tests are thus in good agreement.

In summary, there may be a difference between Norwegian and Swedish specimens, but it is not quite significant at $p < 0.05$. An inspection of the data with the aid of PCA (section 6.2) indicates that the Swedish specimens are generally larger and more antero-posteriorly compressed than the Norwegian ones. Perhaps a significant difference can be demonstrated if the sample sizes are increased, thus increasing the power of the test.

## 5.4   Hierarchical cluster analysis

Given a number of objects, such as individual fossils, or samples from different localities, we often want to investigate whether there are any groups of similar items that are well separated from other groups. In cluster analysis, such groups are searched for based on similarities in measured or counted data between the items (Sneath and Sokal 1973; Everitt 1980; Spath 1980; Gordon 1981). That is, we are not testing for separation between given groups that are known a priori from other information. Cluster analysis is, therefore, more typically a method for data exploration and visualization than a formal statistical technique.

All cluster analysis must start with the selection of a *distance or similarity measure*. How do you define the distance (and consequently also the similarity) between two items in multidimensional space? This will depend on the type of data and other considerations. In chapter 10, we will go through a number of distance measures for ecological data of the taxa-in-samples type. We have a choice of distance measure also for morphometric data and similar measurements.

Perhaps the most obvious option is Euclidean distance, which is simply the linear distance between the two points $x$ and $y$ in multidimensional space:

$$ED = \sqrt{\Sigma(x_i - y_i)^2}$$

Another possibility is to regard two items as similar if their variates correlate well. The correlation coefficient used can be parametric (Pearson's $r$) or nonparametric (Spearman's rank-order correlation $r_s$). This approach will, roughly speaking, compare the angles between the point vectors and disregard the length.

In addition to selecting a distance measure, we need to choose one of many available clustering algorithms. This is a somewhat inelegant aspect of cluster analysis: the clusters obtained will generally depend on technical details of the clustering algorithm, without any particular algorithm being universally accepted as "best." As a general recommendation, we suggest average linkage (UPGMA) for ecological data (taxa-in-samples) and Ward's method for morphometric data. These methods are described in the following paragraphs.

The general algorithm for agglomerative, hierarchical cluster analysis proceeds as follows. First, the two most similar objects are found and joined in a small cluster. Then, the second most similar objects (single items or clusters) are joined, and this proceeds until all objects are grouped in one "supercluster." The different algorithms differ in how they define the distance between two clusters. Since we proceed by joining items and clusters in progressively larger clusters, this approach is known as *agglomerative* clustering. It should be noted that a different class of clustering algorithms exists, known as *divisive* clustering, which works in the opposite way, starting with all items in one supercluster that is progressively subdivided. This approach is more rarely used.

As the items and clusters are being joined, a clustering diagram known as a *dendrogram* is gradually generated (Fig. 5.4). The dendrogram is a dichotomously branching hierarchical tree where each branching point corresponds to a joining event during the agglomeration. Furthermore, the branching point is drawn at a level (height above ground) corresponding

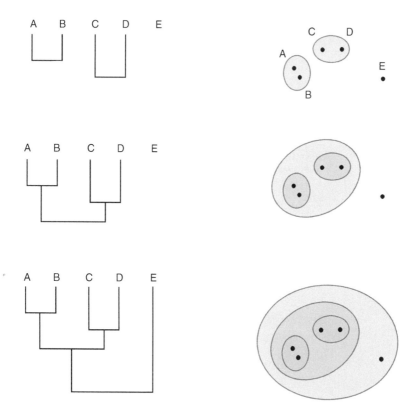

Figure 5.4  Agglomerative clustering of five two-dimensional data points A–E. Top row: Points A and B are most similar to each other and are joined in a cluster AB. Points C and D are the second most similar and are also clustered (at a lower similarity level) into CD. Middle row: The two clusters AB and CD are more similar to each other than either is to point E, so these two clusters are joined in a "supercluster." Bottom row: Finally, there are only two items left: cluster ABCD and point E. These are joined to complete the clustering.

to the similarity between the joined objects. This is a very important thing to note in the dendrogram because it indicates the degree of separation between clusters.

The clustering procedure will produce a dendrogram no matter how poorly the items are separated. Even a perfectly evenly spaced set of points will be successfully clustered – Fig. 5.5 shows the dendrogram resulting from a random data set. The "strength" of the clusters can only be assessed by investigating the clustering levels, as shown in the dendrogram. It must be noted again that clustering is not a formal statistical procedure, and it is difficult to assign any significance levels to the clusters. First of all, it is of course not a valid procedure to simply apply standard two-group tests such as the $t$ test to pairs of clusters, as the elements of the clusters have not been sampled randomly but have been selected to produce a difference between the clusters in the first place. A good review of possible approaches to the estimation of cluster significance and the associated problems is provided in the *SAS/STAT User Manual* (SAS Institute Inc. 1999).

Another point to remember is that there is no preferred orientation of dichotomies in the dendrogram. In other words, the clusters can be rotated freely around their roots without this having any significance.

Cluster analysis was of great importance in the school of *numerical taxonomy*, where the obtained clusters were used for the classification of species (Sneath and Sokal 1973). In modern paleontology, cluster analysis is perhaps less important, because it is perceived by many that it imposes a hierarchy even where it does not exist. Ordination methods such as PCA, PCoA, CA, and NMDS (chapters 6 and 10) will often be more instructive. Still, cluster analysis can be a useful, informal method for data exploration, such as in studies of intraspecific variation and particularly in ecology and community analysis (chapter 10).

The algorithms for agglomerative, hierarchical cluster analysis differ mainly in the way they calculate distances between clusters.

In *nearest-neighbor joining* (also known as *single linkage*) clustering, the distance between two clusters is defined as the distance between the closest members of the

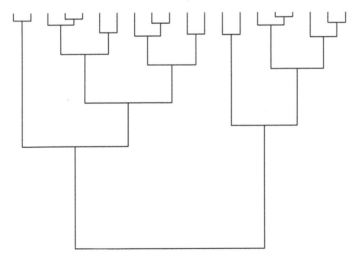

Figure 5.5   Dendrogram from a random data set (20 two-dimensional points with uniform distribution), clustered using Ward's method.

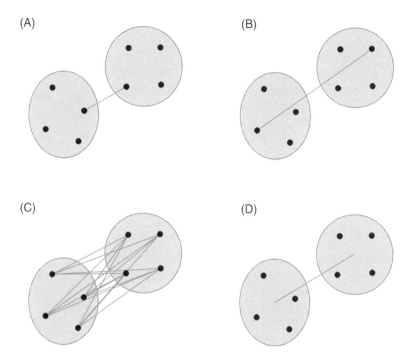

**Figure 5.6** Four ways of defining the distance between two clusters give rise to four different algorithms for agglomerative, hierarchical clustering. (A) Nearest neighbor (single linkage). (B) Furthest neighbor (complete linkage). (C) Average linkage (UPGMA). (D) Centroid clustering.

two groups (Fig. 5.6A). This method is simple, but tends to produce long, straggly clusters and unbalanced dendrograms (Milligan 1980). However, if the original groups really do occupy elongated, parallel regions in multivariate space, then nearest-neighbor joining will do a better job of recognizing them than most other methods.

In *furthest-neighbor joining* (also known as *complete linkage*; Sørensen 1948), the distance between two clusters is defined as the distance between the two most remote members of the two groups (Fig. 5.6B). This algorithm has precisely the opposite problem as nearest-neighbor joining: clusters are not at all allowed to elongate in multidimensional space. Furthest-neighbor joining is biased towards producing clusters of equal diameters and is also very sensitive to outliers (Milligan 1980; SAS Institute Inc. 1999).

In *average linkage clustering* (also known as *UPGMA* or *mean linkage*; Sokal and Michener 1958), the distance between two clusters is defined as the average of all possible distances between members of the two groups (Fig. 5.6C). This method is preferred by many workers, both for the philosophical appeal of all members of the two groups contributing equally to the distance measure and because it seems to work well in practice. Average linkage clustering is somewhat biased towards producing clusters of similar variances (SAS Institute Inc. 1999).

In *centroid clustering* (Sokal and Michener 1958), the distance between the two clusters is defined as the distance between the centroids (multidimensional means) of the two groups (Fig. 5.6D). Centroid clustering is not equivalent to average linkage clustering. When we calculate centroids, we implicitly treat the data points as if they reside in a

Euclidean space where only the Euclidean distance measure is appropriate. For example, if we have a presence-absence data set with values of zero and one, the centroid will not be constrained to a binary value or vector but can end up at any real-valued position. We cannot now use a binary distance measure for calculating the distance between the centroids. Centroid clustering is robust to outliers in the data, but otherwise does not seem to perform as well as average linkage or Ward's method (Milligan 1980; SAS Institute Inc. 1999).

*Ward's method* (Ward 1963) works in quite a different way. The idea here is to select clusters for fusion into larger clusters based on the criterion that we want within-group variance, summed over all clusters, to increase as little as possible. Again, the calculation of variances implicitly enforces the Euclidean distance measure. The method is biased towards producing clusters with similar numbers of items and is sensitive to outliers in the data (Milligan 1980), but otherwise seems to work well in practice.

---

**Example 5.3**

Crônier et al. (1998) studied the ontogeny of the Late Devonian phacopid trilobite *Trimerocephalus lelievrei* from Morocco. They recognized five growth stages, based on both morphological features and size. The centroid sizes (section 6.10) of 51 specimens are shown in Fig. 5.7, together with the classes numbered 1–5. We see that there are discontinuities between class 1 (with a single specimen) and class 2, and also between classes 3 and 4 and between 4 and 5. However, classes 2 and 3 are part of a size continuum, and class 5 seems to be split in two. Can the growth stages be recognized from sizes alone, using cluster analysis? In this case, there is only a single variate; so, the only sensible distance measure is the numerical difference in size. This example is so simple that cluster analysis may not help the interpretation of the data; but, its simplicity clearly highlights the comparison of results from different clustering algorithms.

Figure 5.8 shows the dendrograms resulting from three different algorithms, each of them having good and bad sides. Single linkage produces a somewhat badly resolved dendrogram; but, on the other hand, it looks quite "honest" and in good agreement with the visual impression from Fig. 5.7. Average linkage (UPGMA) gives a nicely balanced dendrogram, with two main clusters corresponding to size classes 1–3 and 4–5,

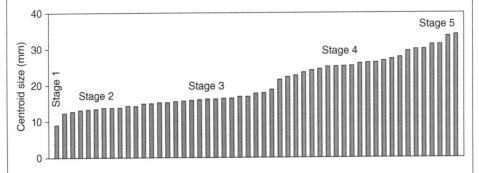

**Figure 5.7** Centroid sizes of 51 specimens of the Devonian trilobite *Trimerocephalus lelievrei.* Note the discontinuities between size classes, except between classes 2 and 3. Data from Crônier et al. (1998).

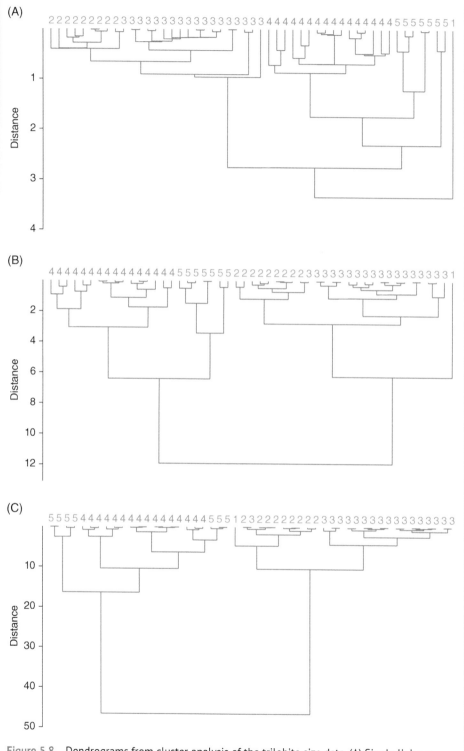

**Figure 5.8** Dendrograms from cluster analysis of the trilobite size data. (A) Single linkage. (B) Average linkage (UPGMA). (C) Ward's method.

a reasonable result. Ward's method gives even tighter clusters, overexaggerating the groupings in the data. For example, the implied close association between class 1 and class 2 seems unjustified, as does the linking of the three smallest class 5 specimens with class 4. Class 2 and class 3 seem to be "over-split," obscuring the continuity within these classes (note that one class 3 specimen is misclassified within class 2 in all three dendrograms).

Although this example is quite typical of the trade-off between loose and tight clustering, it is not always the case that average linkage performs "better" than other algorithms. This needs to be evaluated for each application of cluster analysis. An interesting simulation study comparing the performance of different hierarchical clustering algorithms is given in the *SAS/STAT User Manual* (SAS Institute Inc. 1999).

The example in this section illustrates the general problem of recognizing size classes due to, for example, an annual reproduction cycle, or instars. Using a mixture model (section 4.9) would probably be a better approach in this case.

## 5.5 K-means and k-medoids cluster analysis

Hierarchical cluster analysis generates a hierarchy of groups and subgroups based on similarities between the observations. The number of clusters found will depend on the data and on the interpretation of the dendrogram. In contrast, k-means clustering (Bow 1984) involves a priori specification of the number of groups. For example, we may have collected morphometric data on trilobites, and suspect that there may be three morphospecies. We want to see if the computer can group the specimens into three clusters, and if the assignments of individuals to clusters correspond with the taxonomic assignments made by an expert. We are not interested in any hierarchy of subgroups within the three main clusters.

The most common k-means clustering algorithms operate by the iterative reassignment of observations to clusters, attempting to produce as tight (low variance) clusters as possible. The clusters are initialized by assigning the $n$ points to the $k$ clusters at random or using some heuristic guesswork. Next, the centroids of the $k$ clusters are computed, and each point is reassigned to the cluster whose centroid is closest. The centroids of the clusters are computed again, and the procedure repeated until the points no longer move between clusters.

This procedure is not guaranteed to find the best solution, and the result will generally be different each time the program is run with the same input. It is recommended that k-means clustering is performed several times to get an idea about the range of results that can be achieved.

---

Example 5.4

Returning to the trilobite example from section 5.4, we can subject the data to k-means cluster analysis with five clusters. The results of four different runs, as shown in Table 5.1, are clearly not fully consistent. Only the boundary between size classes 4 and 5 is found in all four runs. The boundary between size classes 3 and 4 is detected in

Table 5.1 Four runs of k-means cluster analysis of the trilobite data from section 5.4. Horizontal lines show the cluster splits in each run.

| Size class | Cluster | Cluster | Cluster | Cluster |
|---|---|---|---|---|
| 1 | 1 | 1 | 1 | 1 |
| 2 | 1 | 1 | 2 | 1 |
| 2 | 1 | 1 | 2 | 1 |
| 2 | 1 | 1 | 2 | 1 |
| 2 | 1 | 1 | 2 | 1 |
| 2 | 1 | 1 | 2 | 1 |
| 2 | 1 | 1 | 2 | 1 |
| 2 | 1 | 1 | 2 | 1 |
| 2 | 1 | 1 | 2 | 1 |
| 2 | 1 | 1 | 2 | 1 |
| 3 | 1 | 1 | 2 | 1 |
| 3 | 2 | 2 | 3 | 1 |
| 3 | 2 | 2 | 3 | 1 |
| 3 | 2 | 2 | 3 | 1 |
| 3 | 2 | 2 | 3 | 1 |
| 3 | 2 | 2 | 3 | 1 |
| 3 | 2 | 2 | 3 | 1 |
| 3 | 2 | 2 | 3 | 1 |
| 3 | 2 | 2 | 3 | 1 |
| 3 | 2 | 2 | 3 | 1 |
| 3 | 2 | 2 | 3 | 1 |
| 3 | 2 | 2 | 3 | 1 |
| 3 | 2 | 2 | 3 | 1 |
| 3 | 2 | 3 | 3 | 1 |
| 3 | 2 | 3 | 3 | 1 |
| 3 | 2 | 3 | 3 | 2 |
| 3 | 2 | 3 | 3 | 2 |
| 3 | 2 | 3 | 3 | 2 |
| 4 | 3 | 4 | 4 | 2 |
| 4 | 3 | 4 | 4 | 2 |
| 4 | 3 | 4 | 4 | 2 |
| 4 | 3 | 4 | 4 | 3 |
| 4 | 3 | 4 | 4 | 3 |

(*Continued*)

Table 5.1  (Continued)

| Size class | Cluster | Cluster | Cluster | Cluster |
| --- | --- | --- | --- | --- |
| 4 | 3 | 4 | 4 | 3 |
| 4 | 4 | 4 | 4 | 3 |
| 4 | 4 | 4 | 4 | 3 |
| 4 | 4 | 4 | 4 | 3 |
| 4 | 4 | 4 | 4 | 3 |
| 4 | 4 | 4 | 4 | 3 |
| 4 | 4 | 4 | 4 | 3 |
| 4 | 4 | 4 | 4 | 3 |
| 4 | 4 | 4 | 4 | 3 |
| 4 | 4 | 4 | 4 | 3 |
| 4 | 4 | 4 | 4 | 3 |
| 5 | 5 | 5 | 5 | 4 |
| 5 | 5 | 5 | 5 | 4 |
| 5 | 5 | 5 | 5 | 5 |
| 5 | 5 | 5 | 5 | 5 |
| 5 | 5 | 5 | 5 | 5 |
| 5 | 5 | 5 | 5 | 5 |
| 5 | 5 | 5 | 5 | 5 |

three of four runs. The boundary between size classes 2 and 3 is also detected in three of the runs, although with one specimen misclassified into size class 2. These results are in general agreement with a visual inspection of Fig. 5.7.

K-means clustering implicitly uses Euclidean distance. Other distance measures are more appropriate for ecological (sections 10.3–10.4) or genetic data, however. K-medoids clustering (Kaufman and Rousseeuw 1990) can be compared to k-means clustering and requires the user to select the number of clusters. Unlike k-means, the clusters are centered on a point in the data set, rather than a cluster mean. Also, importantly, k-medoids allow any distance measure to be used.

## References

Anderson, T.W., Bahadur, R.R. 1962. Classification into two multivariate normal distributions with different covariance matrices. *Annals of Mathematical Statistics* 33, 420–431.

Anderson, M.J. 2001. A new method for non-parametric multivariate analysis of variance. *Austral Ecology* 26, 32–46.

Bow, S.-T. 1984. *Pattern Recognition*. Marcel Dekker, New York.

Crônier, C., Renaud, S., Feist, R., Auffray, J.-C. 1998. Ontogeny of *Trimerocephalus lelievrei* (Trilobita, Phacopida), a representative of the Late Devonian phacopine paedomorphocline: a morphometric approach. *Paleobiology* 24, 359–370.

Davis, J.C. 1986. *Statistics and Data Analysis in Geology*. John Wiley, Chichester, UK.

Dryden, I.L., Mardia, K.V. 2016. *Statistical Shape Analysis*. 2nd edition. Wiley, New Jersey.

Everitt, B.S. 1980. *Cluster Analysis*. 2nd edition. Heinemann Educational, London.

Good, P. 1994. *Permutation Tests: A Practical Guide to Resampling Methods for Testing Hypotheses*. Springer Verlag.

Gordon, A.D. 1981. *Classification: Methods for the Exploratory Analysis of Multivariate Data*. Chapman & Hall, New York.

Kaufman, L., Rousseeuw, P.J. 1990. Partitioning around medoids (program PAM). Ch. 2 in *Finding Groups in Data: An Introduction to Cluster Analysis*, pp. 68–125. John Wiley & Sons.

Milligan, G.W. 1980. An examination of the effect of six types of error perturbation on fifteen clustering algorithms. *Psychometrika* 45, 325–342.

O'Higgins, P. 1989. A morphometric study of cranial shape in the Hominoidea. PhD thesis, University of Leeds.

SAS Institute, Inc. 1999. *SAS/STAT User's Guide*, Version 8.

Sneath, P.H.A., Sokal, R.R. 1973. *Numerical Taxonomy*. Freeman, San Francisco.

Sokal, R.R., Michener, C.D. 1958. A statistical method for evaluating systematic relationships. *University of Kansas Science Bulletin* 38, 1409–1438.

Spath, H. 1980. *Cluster Analysis Algorithms*. Ellis Horwood, Chichester, UK.

Sørensen, T. 1948. A method of establishing groups of equal amplitude in plant sociology based on similarity of species content. *Det Kongelige Danske Videnskab. Selskab, Biologiske Skrifter* 5(4), 1–34.

Ward, J.H. 1963. Hierarchical grouping to optimize an objective function. *Journal of the American Statistical Association* 58, 236–244.

# 6

# Morphometrics

The term *morphometrics* refers to the measurement of the shape and size of organisms or their parts, and the analysis of such measurements. Morphometrics is important in the study of living organisms, but perhaps even more crucial to paleontology, because genetic sequence data are generally missing, and shape is all we have. There is a substantial literature associated with morphometrics, some of it influenced by classic works such as D'Arcy Thompson's (1917) *On Growth and Form* and Raup's (1966) elaboration of the concept of morphospace. Morphometrics is a rapidly evolving discipline with a range of available software. Morphometric techniques have enhanced and developed many areas of taxonomy and evolutionary biology and are crucial in the analysis of evolutionary-developmental problems (EvoDevo). Some example applications are:

- Taxonomy. In many cases, morphospecies can be successfully separated, defined, and compared using morphometric information.
- Microevolution. The study of how shape changes through time within a species.
- Ontogeny (growth and pattern formation in the individual organism) and its relationships to phylogeny through heterochrony.
- Ecophenotypic effects. How the environment influences shape and size within one species.
- Other modes of intraspecific variation, including polymorphism and sexual dimorphism.
- Asymmetry and its biological significance.

The measurement of fossil specimens is not trivial. First, if we are looking for subtle differences between specimens, we must be reasonably sure that the fossils have not been substantially deformed through post-mortem transport, burial, diagenesis, and tectonics (Webster and Hughes 1999). This may preclude the use of morphometrics in many cases, unless of course we are actually interested in studying compaction or microtectonics. Second, we must take all the usual precautions associated with capturing representative samples.

The methods of measurement will vary with the application and available equipment. Four main classes of measurement may be identified:

1) Univariate (single) measurements, such as length or width. Such data are treated with the usual univariate statistical methods outlined in chapter 4. The measurements are taken with calipers or from digital images using appropriate software.

*Paleontological Data Analysis*, Second Edition. Øyvind Hammer and David A.T. Harper.
© 2024 John Wiley & Sons Ltd. Published 2024 by John Wiley & Sons Ltd.

2) Multivariate distance measurements. A number of measurements are taken on each specimen, such as length, width, thickness, and distance between the eyes. The data are treated with the multivariate methods of this and the previous chapter. This approach is now known as "traditional" (or linear) multivariate morphometrics, in contrast with outline and landmark methods (geometric morphometrics).

3) Outlines. For some fossils, it can be difficult to define good points or distances to measure. To capture their shape, a large number of points can be digitized around their outlines. Special methods are available to analyze such data.

4) Landmarks. A number of homologous points are defined, such as the tip of the snout, the tips of the canines, and a few triple junctions between sutures connecting bones in the skull. The coordinates of these points are digitized. For two-dimensional coordinates $(x, y)$ this can be done easily on the computer screen from digital photographs (be aware of lens distortions), but three-dimensional measurement $(x, y, z)$ necessitates special equipment such as a coordinate measuring machine or CT scanner. Landmark analysis is a powerful technique and considered as the state of the art in morphometrics. One important advantage over traditional morphometrics is that the results of landmark analysis are easier to interpret in geometric terms.

The science of measuring (metrology) is a large field that is outside the scope of this book. One basic lesson from metrology is, however, that we should try to have an idea about our measuring errors. If possible, multiple measurements should be made on each specimen or at least on one test specimen, and the standard deviation be noted.

The order in which the specimens are measured should be randomized. Measuring all specimens from one sample first, and then all specimens from the next sample, can lead to systematic errors because of the skill of measurement improving through time, or conversely the quality deteriorating because of fatigue. Even the possibility of unconscious biasing of measurements to fit a theory must be eliminated.

## 6.1 The allometric equation

Already at this point, we must discuss one of the fundamental questions of morphometrics: to log or not to log? Many workers recommend taking the logarithms of all measurements (widths, lengths, etc.) before further analysis. Why might we want to do such a complicated thing? Our explanation of this will also introduce the fundamental concepts of allometry and the allometric equation. This is one of the few issues in our book that truly benefit from a mathematical first presentation, which we urge the reader to try to follow.

In the biological literature, the word "allometry" has taken two different meanings – one broad and one narrow. Allometry *sensu lato* simply means any change in proportion through ontogeny (or across species). Hence, the relative reduction of head size from newly born to adult human is an example of allometry (see also Fig. 6.1).

In the strict sense, the word allometry refers to a change in proportion through ontogeny that is well described by a specific mathematical model for differential growth, first formalized by Huxley (1932). The fundamental assumption of this model is that differential growth is regulated by maintaining a *constant ratio between relative growth rates*. Consider

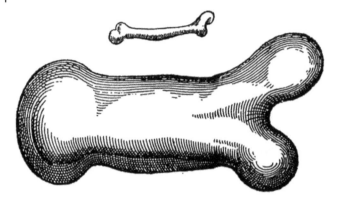

Figure 6.1   Galileo has an interesting discussion of biological scaling in his *"Dialogues Concerning Two New Sciences"* (1638). His illustration of bones in small and large animals is reproduced in this figure.

one measurement $x$, such as the length of the femur, and another measurement $y$, such as the width of the same bone. Using the notation of calculus, we write $dx/dt$ for the growth rate of $x$ (increase in length per time). The growth rate of $y$ is called $dy/dt$. Shape change through ontogeny is most likely achieved by regulating cell division rates, such that growth will be controlled in terms of relative rather than absolute growth rates. By this, we mean that growth rates are specified relative to the size of the organ, for example, as the percent size increase per unit time. The simplest hypothesis for the relationship between two relative growth rates is that one is a constant $a$ times the other:

$$\frac{dy/dt}{y} = a\frac{dx/dt}{x}$$

This is a differential equation, which is solved by a standard procedure. First, we multiply out $dt$:

$$\frac{dy}{y} = a\frac{dx}{x}$$

Integrating both sides (this step may seem mysterious to non-mathematicians), we get

$$\ln y = a \ln x + b$$

for an arbitrary constant of integration $b$. This shows that the logarithms of $x$ and $y$ are related by a linear equation. Now, take exponentials on both sides:

$$y = x^a e^b$$

As it is an arbitrary constant, we may just as well rename $e^b$ to $k$:

$$y = kx^a$$

We have now come to the remarkable conclusion that the length and width of the femur will be related through a power function, with a constant $a$ (the allometric coefficient) as the exponent. This holds at any point through ontogeny, and hence for any size, as time has been eliminated from the equation by mathematical manipulation. If $a = 1$, the ratio between $x$ and $y$ will be a constant ($k$), and we have *isometric* instead of allometric growth. If, for example, $a = 2$, we have *positive allometry* in the form of a parabolic relationship. If $a = 0.5$, we have negative allometry in the form of a square root relationship. Obviously, we may get positive or negative allometry simply by arbitrarily reversing the order of the variables in the equation.

Returning to the equations above, we noted that while $x$ and $y$ are related through a power function, their *logarithms* are linearly related. If we have collected a sample of femurs of different sizes, and we suspect allometry, we may therefore try to log-transform (take the logarithms of) our measurements. If the log-transformed values of $x$ and $y$ fall on a straight line with a slope ($a$) equal to one, we have isometric growth. If the slope is different from one, we have allometric growth according to the allometric equation (allometry *sensu stricto*). If the points do not fall on a straight line at all, we have allometry *sensu lato*.

The rationale for log-transforming morphometric data should now be clear: if the measurements adhere to the nonlinear allometric equation, the transformation will linearize the data, making it possible to use common analysis methods that assume linear relationships between the variables. Such methods include linear regression, principal components analysis (PCA), discriminant analysis, and many others. Similarly, it is not uncommon for morphometric data to have a distribution close to *log-normal*. This means that log-transformation will bring the data into a normal distribution, allowing the use of parametric tests that assume normality. On the other hand, log transformation adds an extra complicating step to the analysis, potentially making the results harder to interpret.

A further complication is that the allometric coefficient may change through ontogeny. If such a change happens abruptly, it gives rise to a sharp change in the slope of the log-transformed $x - y$ line, known to ammonite workers as a *knickpunkt*.

---

**Example 6.1**

Figure 6.2A shows a scatter plot of diameter versus umbilical width in 17 specimens of the ammonoid *Arctoceras blomstrandi* (see Fig. 6.3) from the Lower Triassic of Spitsbergen (Hansen et al. 2021).

In the linear scatter plot, there is a hint of disproportionally large umbilical width in the large specimens (the point cloud curves slightly upwards). The log-transformed data are well fitted by the RMA regression line ($R^2 = 0.97$, $p < 0.001$). The slope of the line is 1.47, significantly different from one (one-sample $t$ test; $t = 7.64$, $p < 0.001$). The bootstrapped 95% confidence interval for the slope is (1.38, 1.57). Hence, we have a statistically significant positive allometry in this case. The umbilical width is disproportionally larger in larger specimens, i.e., larger shells are more evolute. Hansen et al. (2021) also demonstrated a clear reduction in allometry upwards in the stratigraphic section.

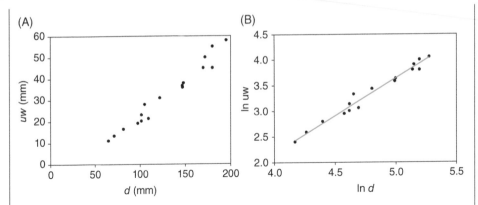

Figure 6.2   Shell diameter (*d*) versus umbilical width (*uw*) for 17 specimens of the ammonoid *Arctoceras blomstrandi*. (A) Original data. (B) Log-transformed data and RMA regression line with a slope of 1.47.

Figure 6.3   *Arctoceras blomstrandi* from the Lower Triassic of Spitsbergen. Scale bar 1 cm. Hansen et al. (2021)/John Wiley & Sons, Inc.

This procedure for detecting allometry should not be carried out uncritically. Even when a statistically significant allometry is found, the value of the slope can sometimes be "very close" to one, in which case the allometry is unlikely to be *biologically significant*. In this example, however, we consider the departure from isometry to be sufficiently large to accept it as a case of allometry.

Example 6.2

Manabe (1994) made morphometric measurements on a large number of *Ichthyosaurus* spp. specimens from the Lower Jurassic of England. The specimens cover a wide size range. Figure 6.4 shows allometric regressions of backbone length vs. skull length and skull length vs. eye size. Both pairs of measurements show negative allometry, with longer backbones corresponding to disproportionally smaller skulls, and longer skulls having disproportionally smaller eyes. In other words, the juveniles have large skulls and large eyes.

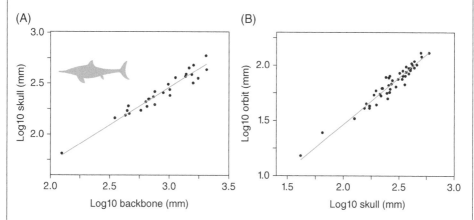

**Figure 6.4** Allometry in *Ichthyosaurus* spp. from England. (A) RMA regression of log-transformed backbone length vs. skull length, showing negative allometry with slope $0.739 \pm 0.037$ (std. err.), significantly different from 1 ($t = 7.02, p < 0.001; n = 28$). (B) Skull length vs. orbit (eye) size, also showing negative allometry (slope $0.845 \pm 0.032$; $t = 4.90, p < 0.001; n = 47$).

## 6.2 Principal components analysis

PCA is a procedure for finding hypothetical variables (components) that account for as much of the variance in your multidimensional data as possible (Hotelling 1933; Jolliffe 1986; Jackson 1991; Reyment and Jöreskog 1993). The components are orthogonal, linear combinations of the original variables (Figs. 6.5 and 6.6). This is a method of data reduction that in well-behaved cases makes it possible to present the most important aspects of a multivariate data set in a small number of dimensions, in a coordinate system with axes that correspond to the most important (principal) components. In practical terms, PCA is a way of projecting points from the original, high-dimensional variable space onto a two-dimensional plane, with a minimal loss of variance. In addition, these principal components may be interpreted (reified) as reflecting "underlying" variables with a biological significance, although this can be a somewhat speculative exercise. PCA is an extremely useful method with many applications, in particular within the field of morphometrics. The method does not make any statistical assumptions but will give best results for data with multivariate normal distribution.

(A)

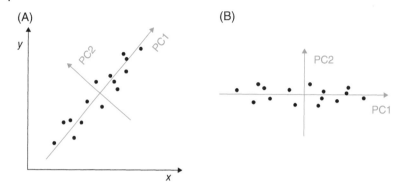

(B)

Figure 6.5   The principle of PCA in the trivial case of a bivariate (two-dimensional) data set, with variables $x$ and $y$. (A) The data points plotted in the coordinate system spanned by the original variables. PC1 is the direction along which variance is largest. PC2 is the direction along which variance is second to largest, and normal to PC1. (B) The data points are plotted in a coordinate system spanned by the principal components.

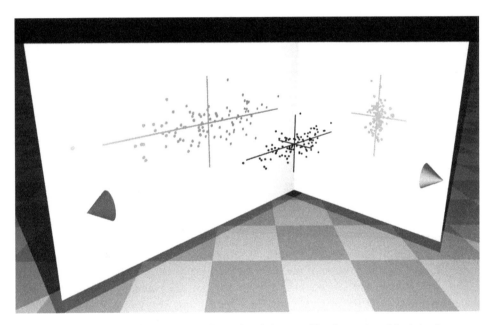

Figure 6.6   A cartoon of PCA for a three-dimensional data set. The data points (black dots) are originally given in the coordinate system shown as checkered floor plus the vertical. The first principal axis (longest black line) is the direction along which variance is maximal. The second principal axis (vertical black line) is normal to the first, with variance second to largest. The third principal axis defines the direction of the smallest variance. The right lamp projects the points onto the first and second principal axes. The left lamp projects the points onto the second and third axes.

PCA can be tricky to grasp initially. What is the meaning of those abstract components? Consider the following example. We have measured shell size $x$, shell thickness $y$, and length of ornamental spines $z$ on 1000 gastropods of the same species but from different latitudes. From these three variates, the PCA analysis produces three components. We are told that the first of these (component PC1) can explain 73% of the variation in the data, the second (PC2) explains 24%, while the last (PC3) explains 3%. Since PC1 explains so much variation, we assume that this component represents an important hypothetical variable that may be related to the environment.

The program also presents the vector elements or "loadings" of component PC1, that is, how much each original variable contributes to the component:

$$PC1 = -3.7x + 1.4y + 0.021z.$$

This tells us that PC1 is a derived variable that reduces sharply as $x$ (shell size) increases, but increases when $y$ (shell thickness) increases. The spine length $z$ has a very low loading on PC1. We guess that PC1 is an indicator of *temperature*. When temperature increases, shell size diminishes (organisms are often larger in colder water), but shell thickness increases (it is easier to precipitate carbonate in warm water). Spine length does not play an important part in PC1, and therefore seems to be controlled by other environmental parameters, perhaps wave energy or predation pressure. Plotting the individual specimens in a coordinate system spanned by the first two components supports this interpretation: we find specimens collected in cold water far to the left in the diagram (small values for PC1), while specimens from warm water are found to the right (large PC1 scores).

In the general case of $n$ variables $x_i$, there will be $n$ principal components, although some of these may be zero if there are fewer specimens than variables (see below). The principal component $j$ is given as the linear combination

$$PC_j = \sum_{i=1}^{n} a_{ij} x_i$$

where each $a_{ij}$ is the loading of the variable $x_i$ on the principal component $j$. An alternative way of presenting the loadings is as the linear correlation between the values of the variable and the principal component scores across all data points. This is equivalent to normalization of each loading with respect to the variance of the corresponding variable.

## 6.2.1 Transformation and normalization

In morphometrics, the variables are often log-transformed prior to PCA, converting ratios to differences that are more easily represented as linear combinations:

$$\log(a/b) = \log a - \log b$$

Another transformation that can sometimes be useful is to normalize all variables with respect to variance. This procedure will stop the first principal components from being dominated by variables with large variance. Such standardization would be justified when the variables are in completely different units, such as meters and degrees Celsius. PCA without standardization of variance is sometimes called *PCA on the variance-covariance matrix*, while PCA with standardization is called *PCA on the correlation matrix*.

Compositional data (relative proportions) must be analyzed with a special form of PCA (Aitchison 1986; Reyment and Savazzi 1999), because we need to correct for spurious correlations between the variables.

## 6.2.2   Relative importance of principal components

Each principal component has associated with it an *eigenvalue* (sometimes called latent root), which indicates the relative proportion of overall variance explained by that component. For convenience, the eigenvalues can be converted into percentages of their sum. The principal components are given in order of diminishing eigenvalues. The idea of PCA is to discover any tendency for the data to be concentrated in a low-dimensional space, meaning that there is some degree of correlation between variables. Consider the extreme case where the data points are concentrated along a straight line in $n$-space. The positions of the points can then be completely described by a single principal component, explaining 100% of the variance. The remaining $n - 1$ components will have eigenvalues equal to zero. If the data points are concentrated on a flat plane in $n$-space, the first *and* second eigenvalues will be non-zero, etc.

Incidentally, data sets with fewer data points than variables will always have fewer than $n$ non-zero eigenvalues. Consider an extreme example of only three data points in a high-dimensional $n$-space. Three points will always lie in a plane, and there will only be two non-zero eigenvalues.

It is rare for more than the first two or three principal components to be easily interpretable – the rest will often just represent "noise." It is difficult to give any strict rule for how many components should be regarded as important (but see e.g., Jolliffe 1986; Reyment and Savazzi 1999). It can sometimes be useful to show the eigenvalues as a descending curve called a "scree plot," to get an idea about where the eigenvalues start to flatten out (Fig. 6.7). Beyond that point, we have exhausted most of the correlation between variables, and the components are probably not very informative. On the other hand,

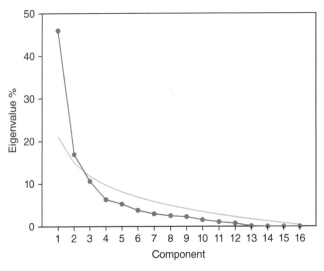

**Figure 6.7**   Scree plot from PCA, showing the relative importance of the principal components. Rather subjectively, we may choose to regard the first three principal components as informative – the curve flattens out after that point. The red line shows a "broken stick" model, which is the expected curve for a data set with no structure (spherical distribution). Only the two first eigenvalues exceed this null model.

it might be argued that the component corresponding to the smallest eigenvalue could sometimes indicate an interesting invariance – some direction in multivariate space along which there is nothing happening.

### 6.2.3   Algorithms for PCA

There are several algorithms available for PCA. We will assume that we have $m$ items and $n$ variates. The classical method (e.g., Davis 1986) is simply to produce an $n \times n$ symmetric matrix of variances (along the diagonal) and covariances of the variables (the variance-covariance matrix), or alternatively a similar matrix of correlation values, which will normalize all variables with respect to their variances. The $n$ eigenvalues $\lambda_j$ and the eigenvectors $\mathbf{v}_j$ of this matrix are then computed using a standard method (e.g., Press et al. 1992). The eigenvectors are the principal components, from which the loadings can be read directly. The PCA score of an item on axis $j$, which is plotted in the PCA scatter plots, is simply the vector inner product between $\mathbf{v}_j$ and the vector of the original data point.

A second method (ter Braak in Jongman et al. 1995) involves the so-called Singular Value Decomposition (SVD; Eckart and Young 1936; Golub and Reinsch 1970; Press et al. 1992) of the original $m \times n$ data matrix with subtracted means of each variable. SVD is considered a better approach because of improved numerical stability, though this is probably a minor concern in most cases.

Compositional data can be subjected to PCA by taking the eigenvalues $\lambda_j$ and eigenvectors $v_j$ of the *centered log-ratio covariance matrix* (Aitchison 1986; Reyment and Savazzi 1999). This is the covariance matrix on the transformed variables

$$y_{ij} = \log \frac{x_{ij}}{g(\mathbf{x}_i)}$$

where $x_{ij}$ is the original value (variables in columns) and $g(x_i)$ is the geometric mean along row $i$. There will be $n-1$ so-called logcontrast principal component scores for each item $i$, defined by the inner product

$$u_{ij} = v_j{}^T \log(x_i)$$

### 6.2.4   PCA is not hypothesis testing

It is sometimes claimed that PCA assumes some statistical properties of the data set such as multivariate normality. While it is true that violation of these properties may degrade the explanatory strength of the axes, this is not a major worry. PCA, like other indirect ordination methods, is a descriptive and explorative method without statistical significance anyway. There is no law against making any linear combination of your variables you want, regardless of the statistical properties of your data, if it is found to be useful in terms of data reduction or interpretation. However, keep in mind that the principal component vectors may be quite unstable with respect to sampling if the data set is not multivariate normal – the stability can be checked with resampling techniques such as bootstrapping or jackknifing (Reyment and Savazzi 1999). Finally, in the case of several independent samples, PCA of all samples simultaneously may not be the best approach. The technique of "Common Principal Component Analysis" (CPCA) was developed for such data sets (Flury 1988; Reyment and Savazzi 1999).

### 6.2.5 Factor analysis

Factor analysis is a term that has been used in different ways, sometimes as a synonym for PCA. In the strict sense, factor analysis covers methods that can be compared with PCA, but the number of components (factors) is chosen *a priori* and being lower than the number of variates (Davis 1986; Reyment and Jöreskog 1993). In one form of factor analysis, these factors (vectors) are allowed to rotate in order to maximize correlation between factors and variables. The orthogonality of the factors may be maintained, or, in a more aggressive approach, they are allowed to become oblique with respect to each other and thereby become correlated. Partly because of the complexity of carrying out and interpreting the results of factor analysis, and partly because it is felt by some that both the selection of number of factors and the rotation procedure introduce subjectivity, the method is now less commonly used than simple PCA.

---

**Example 6.3**

We will return to the example of section 4.3, concerning illaenid trilobites from the Ordovician of Norway and Sweden (Bruton and Owen 1988). Three width measurements and one length measurement (Fig. 6.8) were taken on the cephala of 43 specimens of *Stenopareia glaber* from Norway and 17 specimens of *Stenopareia linnarssoni* from Norway and Sweden.

Using methods described elsewhere in this chapter, we detect clear allometries in this data set, and it may therefore be appropriate to log-transform the measurements before further analysis. For simplicity, we will however use the non-transformed values in this example.

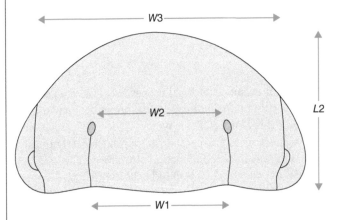

Figure 6.8    Measurements made on illaenid trilobites. Adapted from Bruton and Owen (1988).

Running PCA on the complete data set, we find that a staggering 98.8% of the total variance is explained by the first principal component. The loadings on this component are as follows:

$$PC1 = 0.59L2 + 0.39W1 + 0.35W2 + 0.61W3$$

Thus, all four variables have clear, positive loadings on PC1 – as we go along the direction of PC1 in the four-dimensional variable space, all four distance measurements increase. It is obvious that PC1 can be identified with general "size" in an informal sense. In morphometrics, it is typical that PC1 reflects size; the score on PC1 has even been used as a *definition* of size, as we will discuss in section 6.3. In our example, we are doing PCA directly on linear measurements. The first principal component, therefore, represents purely isometric growth. However, when variables are log-transformed, PC1 may involve an aspect of shape as well as size. This conflation of size and shape in the first principal component is, however, problematic (see section 6.3).

If we are interested in the variation of shape, we also need to consider the subtle variance explained by the next principal components. PC2 explains 0.73% of total variance. Figure 6.9 presents the scores of all specimens on PC1 and PC2. In addition, the loadings are presented as vectors. These vectors can be understood as the projection of the four-dimensional, orthogonal variable axes down along the two-dimensional plane. Their absolute lengths are not important – they can be scaled to produce a readable diagram – but their relative lengths and their directions indicate how the two principal components in the scatter plot should be interpreted. This type of combined scatter plot and variable vector plot is called a *biplot*. Returning first to PC1 (informally, the "size" axis), we see that all four variables tilt towards the right, showing how all four variables increase in value as we go towards higher scores on PC1. Concerning PC2, we see that all three width values increase, while length decreases as we increase the

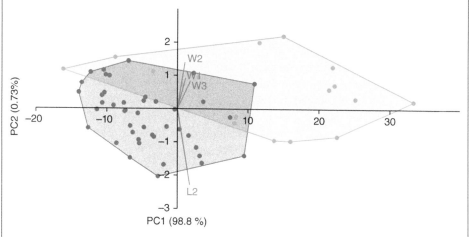

**Figure 6.9** Biplot of scores and loadings on PC1 and PC2 of the illaenid data of Bruton and Owen (1988). Blue dots: *S. glaber*; green squares: *S. linnarssoni*. PC1 is interpreted as a size axis, PC2 as a width/length difference axis. Adapted from Bruton and Owen (1988).

score on this component (W1, W2, and W3 point generally in the same direction as PC2, while L2 points in the opposite direction). We therefore interpret PC2 as a general width/length difference axis: specimens with high scores on this axis are wide and short, while specimens with low scores are long and narrow.

At this point, it should be noted that a width-length *difference* is a somewhat odd quantity, and it may seem more natural to discuss morphology in terms of a width/length *ratio*. However, a principal component is a linear combination of the original variables, which can only express sums and differences. If we had log-transformed the data before PCA, ratios would indeed have been converted into differences.

PC3 explains only 0.45 % of total variance. As the scores on PC3 increase, W3 increases while W1 and W2 decrease. It may be interpreted as an allometric axis concerning occipital width.

Concerning the taxonomy in this example, it seems that *S. linnarssoni* mainly occupies the upper right region of the PC1/PC2 scatter plot, although the separation from *S. glaber* is not good. This means that *S. linnarssoni* is generally larger (high scores on PC1), as we also saw in section 4.3, and also slightly wider relative to length (high scores on PC2). When it comes to the relative differences between W1, W2, and W3 (PC3), the two species do not seem to differ.

## 6.3 Multivariate allometry

The concepts of allometry and the allometric equation were introduced in section 6.1 for the bivariate case. An allometric growth trajectory was shown to produce a power-function curve, reducing to a straight line after log-transformation. In the multivariate case, with several distances involved, the allometric growth trajectory will similarly be a curved line in multidimensional space, reducing to a straight line after log-transformation (Fig. 6.10). Instead of looking at all possible pairs of variates in the search for allometry, it would be much easier and more satisfactory to calculate a single allometric coefficient for each variate with respect to a single size measure. The basic theory for such a multivariate extension of the allometric equation was developed by Jolicoeur (1963) and further discussed by Klingenberg (1996). The method was criticized by Reyment (1991). Kowalewski et al. (1997) presented a practical methodology including a bootstrap test, and we will follow their approach here.

As for the bivariate case, we start by log-transforming all values. The data set is then subjected to PCA. The first principal component (PC1) can be regarded as a multivariate linear regression line (using major axis regression). If PC1 captures a major part of the variance (say more than 80%), it may roughly represent a "size" axis, although this must be critically assessed in each case. The coefficients (loadings) of PC1 represent the multivariate slope of the straight line, and in analogy with the bivariate case we refer to them as allometric coefficients. To be precise, the allometric coefficient for a variate is estimated by dividing the PC1 loading for that variate by the mean PC1 loading over all variates. As for the bivariate case, a multivariate allometric coefficient significantly different from one indicates allometry.

The rationale for this procedure may get a little clearer by considering the equations. Say that we have three variates $x$, $y$, and $z$ (e.g., length, width, and thickness). These are

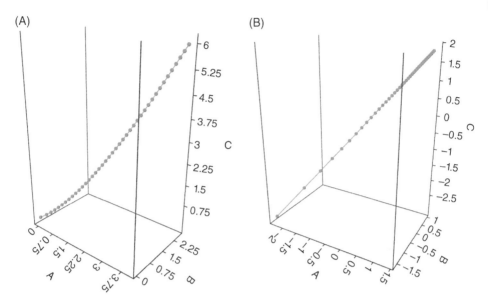

Figure 6.10  A computer-generated growth trajectory in three variables, with multivariate allometric coefficients set to 0.71, 1.00, and 1.29, respectively. (A) Original variables follow a curved trajectory in space. (B) Log-transforming the variables produces a straight line.

log-transformed and subjected to PCA, such that the first principal component score is given by the linear combination

$$u_1 = a \ln x + b \ln y + c \ln z,$$

where $a$, $b$, and $c$ are the loadings on PC1. But

$$u_1 = a \ln x + b \ln y + c \ln z$$

$$= \ln(x^a) + \ln(y^b) + \ln(z^c)$$

$$= \ln(x^a y^b z^c).$$

Hence, the loadings $a$, $b$, and $c$ can be considered allometric coefficients with respect to a common (logarithmic) size measure as given by the score on PC1.

The link with the bivariate case can be illustrated with a simple example. Consider the bivariate allometric relationship $y = x^2$. Log-transforming both variables gives $\ln y = 2 \ln x$, giving the familiar straight-line relationship where the allometric coefficient of $y$ with respect to $x$ is 2. What will this look like using the technique of multivariate allometry? The first principal component score will now have the form

$$u_1 = \ln x + 2 \ln y = \ln(xy^2).$$

The mean loading is $(1+2)/2 = 3/2$. The multivariate allometric coefficient on $x$ with respect to the overall size is then $1/(3/2) = 2/3$, and the coefficient on $y$ is $2/(3/2) = 4/3$. The coefficient on $y$ with respect to $x$ can be computed from these values – it is $(4/3)/(2/3) = 2$, precisely as when using the bivariate method.

The statistical testing of allometry can proceed by estimating a 95% confidence interval for each multivariate allometric coefficient. If this confidence interval does not include the value one, we claim significant allometry at the $p < 0.05$ level. Kowalewski et al. (1997) suggested the use of bootstrapping for this purpose.

The concept of multivariate allometric analysis may seem abstract at first sight, but the procedure is practical and the results easy to interpret. A contentious issue with Jolicoeur's method is the use of the first principal component score as a general size measure, which may not always be reasonable. If the method is to be used, the data and the principal components should be carefully inspected in order to informally check the assumption of size being contained in the first principal component score. Also, any conclusions about allometric effects should be cross-checked using bivariate analysis. In short, this method should only be used informally and with caution.

Hopkins (1966) suggested an alternative procedure based on factor analysis, and Klingenberg (1996) described the use of "CPCA", if several samples are available. We will not discuss CPCA in this book but refer to Reyment and Savazzi (1999).

---

**Example 6.4**

Benton and Kirkpatrick (1989) studied allometry and heterochrony in the rhynchosaur *Scaphonyx* from the Triassic of Brazil (Fig. 6.11).

We will concentrate on a subset of their data, with nine measurements taken on the skulls of 13 specimens of *S. fischeri* (Fig. 6.12). The data set includes small juveniles and covers a 5:1 size range.

The results of the multivariate allometry analysis are shown in Fig. 6.13. The 95% confidence intervals for all allometric coefficients but two include the value 1,

Figure 6.11 Two growth stages of the rhynchosaur *Scaphonyx*. Benton and Kirkpatrick (1989) Palaeontological Association/CC BY 4.0.

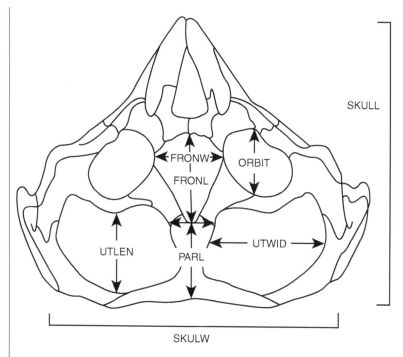

**Figure 6.12** Nine measurements taken on the skulls of the rhynchosaur *Scaphonyx* from the Triassic of Brazil. PARW (not labeled) is the width of the parietal bone. Benton and Kirkpatrick (1989)/Palaeontological Association/CC BY 4.0.

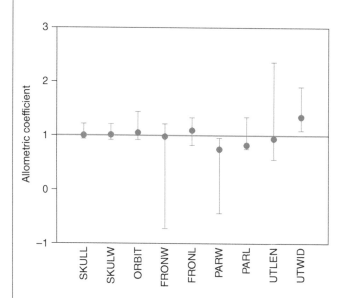

**Figure 6.13** Multivariate allometric coefficients for the *Scaphonyx* data set (dots). For each variate, the 95% confidence interval is indicated by a vertical line.

meaning that the variates do not show significant departure from isometry, using this method. The exceptions are PARW (parietal width) which shows negative allometry, and UTWID (width of upper temporal fenestra), which shows positive allometry. Some of the confidence intervals are very large, due to the relatively small sample size.

The negative allometry of PARW means that the parietal bone gets relatively narrower as the animal grows, and the positive allometry of UTWID implies wider temporal fenestra.

## 6.4   Linear discriminant analysis

Discriminant analysis (Fisher 1936; Anderson 1984; Davis 1986) projects multivariate data onto a coordinate system that will maximize *separation between given groups*. For example, we may have collected fossil shark teeth from two horizons, and we want to investigate whether they are morphologically separated, perhaps even belonging to different morphospecies. By plotting histograms of single variables such as tooth length or width, we do not find good separation of the two samples. However, we may suspect that they could be well separated along some linear combination of variables, for example, because of the ratio between length and width being different in the two samples. Discriminant analysis identifies the linear combination (discriminant axis or discriminant function) that gives maximal separation of the groups (Fig. 6.14).

It is important to note the difference between investigating the degree of separation and testing for difference in multivariate means using, for example, Hotelling's $T^2$ test. There may well be significantly different means without good separation, if the variances are large relative to the distance between the means.

The success of separation can be assessed by noting the percentage of specimens that can be correctly assigned to one of the two original groups based on their positions along the

(A)                                              (B)

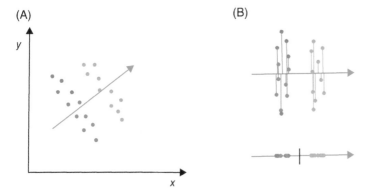

Figure 6.14   The principle of two-group discriminant analysis in the case of a bivariate (two-dimensional) data set, with variables *x* and *y*. (A) The data points plotted in the coordinate system spanned by the original variables. The discriminant axis is the direction along which separation between the two given groups is maximized. (B) The data points are projected onto the discriminant axis, and a cutoff point (vertical line) is chosen.

discriminant axis or their closest group means (typically using the Mahalanobis distance). Good discrimination has been used by some authors to assign the groups to different morphospecies, although others would argue that it is preferable to base a morphospecies definition on discrete characters. The degree of correct assignment necessary for taxonomical splitting can be debated, but it should definitely be better than 90% and perhaps even 100%. If there is a single member of group A within the morphological range of group B, it can be argued that there is no reason to assign them to different morphospecies.

If good separation is achieved, the linear combination, also known as a *predictor*, can later be used for classification, that is, to place new specimens of unknown affinity into one of the groups. It will usually turn out that the success of classification of the new specimens will be lower than that of the original training set. In other words, our original percentage of correct assignment was somewhat optimistic. This is not surprising, because the new specimens were not given the opportunity to contribute to the calculation of the discriminant function. A more realistic estimate of separation success may be achieved by removing one specimen at a time from the training set, recalculating the discriminant function based on all the other specimens, and classifying the one removed specimen accordingly. This is then repeated for all specimens, and the percentage of correctly classified specimens noted.

Once a discriminant function has been found, the coefficients of the linear combination indicate how strongly each variable is involved in it. It may then turn out that some variables are unimportant and may be discarded without significantly reducing separation. Swan and Sandilands (1995) described one of the several statistical tests available for this purpose.

For data sets with unequal variance-covariance matrices, methods other than classical discriminant analysis may give slightly better results, such as quadratic discrimination (Seber 1984). Other nonlinear classification methods are discussed in section 6.20. Discriminant analysis of compositional data was described by Aitchison (1986) and Reyment and Savazzi (1999).

---

**Example 6.5**

We will look at a data set containing three measurements (length $L$, width $W$, and depth $D$) of two species of terebratulide brachiopods from the Jurassic of Switzerland: *Argovithyris birmensdorfensis* ($n = 48$) and *A. stockari* ($n = 26$) (courtesy of Pia Spichiger).

A histogram of individuals projected onto the discriminate axis is shown in Fig. 6.15. We see that the two species occupy different regions, but the separation is small. Using a cut-off point at zero, the percentage of correct classification is 98.6. With some hesitation, we accept the discriminant function as a useful species discriminator. Perhaps we will go back to the specimens and check whether the ambiguous individuals were correctly classified prior to the analysis.

The discriminant function is $v = 0.28L - 1.11W + 1.01D$. Specimens with low $v$ are *A. stockari*, while *A. birmensdorfensis* receive high $v$. Since the coefficient on length is small, we conclude that the main difference between the two species is that *A. stockari* has large widths compared with depths (this will give small values for $v$).

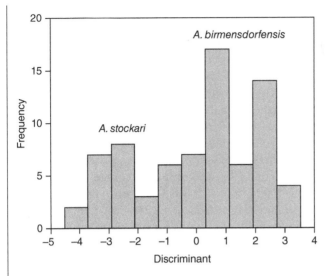

Figure 6.15 Histogram of discriminant projection values for two species of terebratulide brachiopod from the Jurassic of Switzerland. The cut-off point is set to zero.

### 6.4.1 Discriminant analysis for more than two groups

Discriminant analysis can be directly extended to more than two groups. This is often referred to as canonical variate analysis (CVA, e.g., Reyment and Savazzi 1999). For example, we may have made multivariate measurements on corals from four different localities and want to plot the specimens in the two-dimensional plane in a way that maximizes separation of the four groups. While discriminant analysis of two groups projects the specimens onto a single dimension (the discriminant axis), CVA of $N$ groups will produce $N-1$ axes (canonical vectors) of diminishing importance. As in PCA, we often retain only the two or three first CVA axes. Interpretation (reification) of the canonical vectors themselves must be done cautiously because of their lack of stability with respect to sampling (Reyment and Savazzi 1999). CVA of compositional data demands special procedures (Aitchison 1986).

---

**Example 6.6**

We will return to the example used in section 4.2. Kowalewski et al. (1997) measured living Holocene lingulide brachiopods on the Pacific coast of North and Central America, to investigate whether shell morphometrics could produce sufficient criteria for differentiating between species. Figure 6.16 shows the six measurements taken. In our example, we will concentrate on four of the seven groups studied by Kowalewski et al.: *Glottidia albida* from one locality in California, *G. palmeri* from two localities in Mexico, and *G. audebarti* from one locality in Costa Rica.

Figure 6.17 shows a CVA analysis of these four groups after log-transformation. The two samples of *G. palmeri* fall in an elongated group relatively well separated from *G. audebarti*. *G. albida* forms a somewhat more scattered group, showing some overlap with *G. audebarti*. Morphometric data may, in other words, be sufficient to delineate lingulide species, although the separation is not quite as clear as one could perhaps hope for. The biplot indicates that the lengths of the septae (Lmsep, Llvsep, and Lrvsep) may be a good species indicator, with the size of the shell (especially width) providing an additional character for separating *G. audebarti* from the other species.

It should be noted that Kowalewski et al. (1997) attempted to normalize away size by performing CVA on the residuals after regression onto the first principal component within each group. For details on such "size-free CVA," see their paper

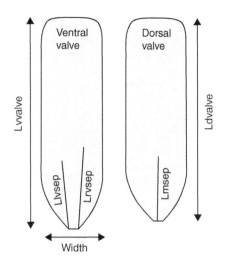

**Figure 6.16** Six measurements taken on the lingulide brachiopod *Glottidia*. Adapted from Kowalewski et al. (1997).

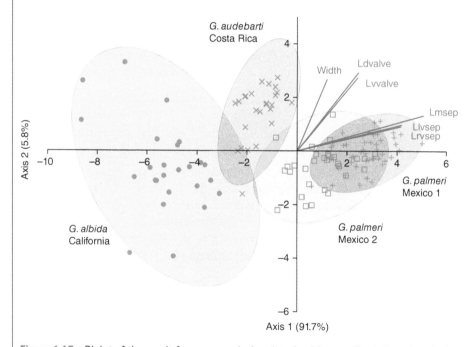

**Figure 6.17** Biplot of the result from a canonical variate (multigroup discriminant) analysis of four samples of living lingulide brachiopods from the Pacific coast of North and Central America. Data were log-transformed. Green dots: *Glottidia albida*, southern California. Brown crosses: *G. palmeri*, Campo Don Abel, Mexico. Blue squares: *G. palmeri*, Vega Island, Mexico. Red crosses: *G. audebarti*, Costa Rica. Variables are lengths of ventral and dorsal valves (Lvvalve and Ldvalve, lengths of left and right ventral septum (Llvsep and Lrvsep), length of median septum (Lmsep), and shell width (Width).

and references therein. A related approach is size removal by *Burnaby's method*, where all points are projected onto a plane normal to PC1 (or some other size-related vector) prior to analysis. As we noted in section 6.3, the use of PC1 as a size estimator is open to criticism.

## 6.5 Multivariate analysis of variance

One-way MANOVA (Multivariate ANalysis Of Variance) is the multivariate version of simple one-way ANOVA (section 4.4). It is a parametric test, assuming multivariate normal distribution, to investigate equality of the multivariate means of several groups. In a similar way as ANOVA assumes equal variances, MANOVA assumes equal variance-covariance matrices (cf. Fig. 5.3), although this requirement can be somewhat relaxed.

The null hypothesis is thus:

$H_0$: All samples are taken from populations with equal multivariate means.

In the section on ANOVA, it was explained why this cannot easily be tested by simple pairwise comparisons, using, for example, Hotelling's $T^2$ test (section 5.2). In paleontology, MANOVA is used mostly in morphometrics, to test the equality of several multivariate samples. This can be, for example, sets of distance measurements on fossils from different localities, or the coordinates of landmarks on fossils of different species.

Somewhat confusingly, several competing test statistics are in use as part of MANOVA. These include the popular *Wilk's lambda* and the possibly more robust *Pillai trace*.

---

**Example 6.7**

Bookstein (1991) presents a data set collected by Vilmann, with eight landmarks (section 6.10) from the skulls of rats at eight different ages, each age represented by 21 specimens. Is there a significant difference in shape? We prepare the data by Procrustes fitting (section 6.11), which normalizes for size. We can now regard the set of landmark coordinates as a multivariate data set, and test for difference between the groups using MANOVA. The "Pillai trace" test statistic equals 2.67, with a probability of multivariate equivalence between the groups of $p < 0.001$. We can therefore reject the null hypothesis of equal multivariate means, demonstrating allometry in this case.

---

## 6.6 Fourier shape analysis in polar coordinates

The digitized outline of a trilobite cephalon, a cross-section of an echinoid, or a bone in a fish skull represents a complex shape that cannot be handled directly by multivariate methods based only on linear measurements. We need some way of reducing the shape information to a relatively small set of numbers that can be compared from one specimen to the next. One way of doing so is to fit the outline to a mathematical function with a few adjustable model parameters. These parameters are then subjected to further statistical analysis.

As an introduction to outline analysis, we will start by explaining polar, also known as radial *Fourier shape analysis*. This method has largely been superseded by more modern approaches, but it provides a good introduction to the more advanced methods.

Any outline analysis starts with a large number of digitized $(x, y)$ coordinates of points around the outline. The coordinates may be acquired by tracing a photograph of the specimen on a digitizing tablet or using a mouse cursor on a computer screen. Automatic outline tracing using image-processing algorithms is available in some computer programs (e.g., F.J. Rohlf's freeware "tpsDig"), but this approach does not work for all types of images or shapes. For polar Fourier shape analysis, we also need to define a *central point* in the interior of the shape. The centroid of the digitized points is a natural choice. We then imagine vectors from the central point out to the outline at angles $\varphi$ from the horizontal (Fig. 6.18A). The lengths $r$ of these vectors describe the outline in *polar* coordinates, as a function $r(\varphi)$. Obviously, the function $r(\varphi)$ is periodic, repeating for every revolution of the vector. The next step in the procedure is to sample $r(\varphi)$ at equal angular increments. We will use eight points around the outline for illustration, but a larger number will normally be used. This resampling of the outline must proceed by interpolation between the original coordinates, using, for example, linear interpolation (Fig. 6.18B). The centroid of the new set of points is likely to be slightly displaced from its original position, and it is therefore necessary to re-compute the centroid and run the resampling again until its position has stabilized.

The $n = 8$ (in our example) evenly spaced samples of the polar representation of the outline (Fig. 6.18C) are then fitted to a periodic function with adjustable parameters. We use a *Fourier series* (see also section 12.1), which is a sum of sine and cosine functions with fixed frequencies but adjustable amplitudes. It can be shown that $n$ evenly spaced points can be fitted perfectly by $n/2$ sinusoids, each sinusoid being the sum of a sine and a cosine of the same frequency. The sinusoids are in *harmonic* relationship, meaning that their frequencies are multiples of the fundamental frequency of one period in the interval $[0, 2\pi]$. Thus, the first harmonic (fundamental) has one period per revolution; the second harmonic has

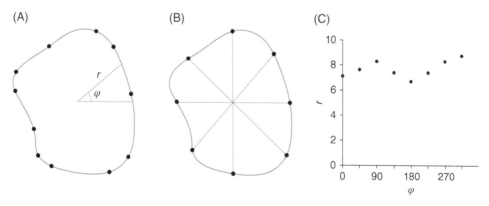

**Figure 6.18** The principle of Fourier shape analysis in polar coordinates. (A) Eleven points are digitized around the outline, at irregular intervals. The centroid of these points is computed and used as the origin for conversion to polar coordinates. (B) Using interpolation between the original points, eight new points are computed at constant angular increments. This will move the centroid slightly, so the procedure should be repeated until its position stabilizes. (C) Radius $r$ as a function of rotational angle $\varphi$ (polar coordinates). The eight points are fitted to a sum of sinusoids.

two periods per revolution; and so forth up to the highest frequency component with four periods per revolution in our example (see also section 12.1). The fitting of the amplitudes of both the sines and cosines to the given outline thus results in $n/2 + n/2 = n$ parameter values. So far, we have achieved only a 50% reduction in the amount of data necessary to specify the shape, from eight $(x, y)$ coordinates (16 numbers) to $4 + 4 = 8$ sine and cosine factors.

The next step is to realize that many of the sinusoids will usually have negligible amplitudes and can be discarded. First, if the centroid has been properly estimated, the fundamental (first harmonic) should be zero; a non-zero first harmonic would mean that the outline extended further out to one side of the center point, meaning that this center point could not be the centroid. Second, but more importantly, almost all natural objects are *smooth* in the sense that the amplitudes of the harmonics will generally drop with frequency. Usually, the first few harmonics will be sufficient to capture the main aspects of the shape, and the higher harmonics can be discarded. This leaves us with a relatively small set of parameters that can be subjected to further multivariate analysis.

Some of the cosine components are shown in polar representation in Fig. 6.19 (the corresponding sine components are similar but rotated). It can be clearly seen how higher order harmonics describe less smooth components of the outline.

There is a considerable amount of work involved in the digitization of outline data, although some semi-automatic methods have been developed. It is therefore practical to

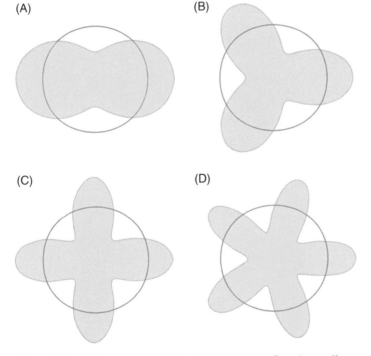

**Figure 6.19**  Cosine components in polar coordinates. Complex outlines can be decomposed into such partial outlines. A circle of unit diameter (zeroth harmonic) has been added to the second (A), third (B), fourth (C), and fifth (D) cosine components.

use only as many points as we need to get a good representation of the shape. As a general rule, we need at least a couple of points for each "bump" on the outline.

There are two serious problems connected with the use of polar coordinates in this method. First, the choice of center point (origin) is somewhat arbitrary, making comparison between specimens imprecise. Second, for many biological shapes, the radius vector crosses the outline more than once (Fig. 6.20), and the method therefore cannot be used at all. Both these problems are removed if we use elliptic or Hangle Fourier analysis, which are described in the next sections.

Cristopher and Waters (1974) applied polar Fourier shape analysis to the analysis of pollen shape.

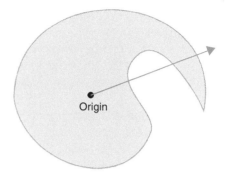

**Figure 6.20** For this complex shape, the radius vector from the centroid (origin) crosses the outline up to three times. Simple Fourier shape analysis cannot be used in such cases.

## 6.7 Elliptic Fourier analysis

The previous section introduced outline analysis by reference to the polar Fourier method. Two problems with this method are the arbitrary point of origin and the inability to handle complicated shapes. These problems are solved by the approach of elliptic Fourier analysis (EFA), as described by Kuhl and Giardina (1982) and Ferson et al. (1985).

EFA is not based on polar coordinates but on the "raw" Cartesian $(x, y)$ values (Fig. 6.21). The increments in the $x$ co-ordinate from point to point around the outline define a periodic

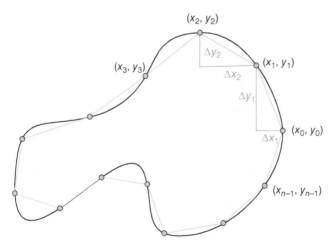

**Figure 6.21** The principle of elliptic Fourier shape analysis. The curved outline is the original shape. It is sampled at a number of discrete points, with coordinates $(x, y)$. The increments $\Delta x$ and $\Delta y$ are defined as $\Delta x_i = x_i - x_{i-1}$ and $\Delta y_i = y_i - y_{i-1}$. These increments are independently subjected to Fourier analysis with total chord length (length of straight lines) as the independent variable.

function, which can be subjected to Fourier decomposition into sinusoidal components. The increments in the $y$ direction are subjected to an independent decomposition. This results in a set of four coefficients for each harmonic, namely the sine and cosine amplitudes of the $x$ and the $y$ increments.

The original points do not have to be equally spaced, although the spacing should not be too irregular.

Most EFA routines standardize the size prior to analysis, by dividing all coordinates by a size measure. Some programs also attempt to rotate the specimens into a standardized orientation, but it is probably better to try to ensure this already under the process of digitization. Obviously, the procedure requires that each outline is digitized starting from the "same," homologous point.

---

**Example 6.8**

Crônier et al. (1998) studied the ontogeny of the Late Devonian phacopid trilobite *Trimerocephalus lelievrei* from Morocco (Fig. 6.22). Using the data of these authors, we will look at the outlines of 51 cephala from five well-separated size classes (growth stages), named A–E. Sixty-four equally spaced points were digitized around each outline. We are interested in whether the shape changes through ontogeny, and if so, how.

One of the cephala is shown in Fig. 6.23, as reconstructed from EFA coefficients. The first 10 harmonics (40 coefficients) seem to be sufficient to capture the main aspects of the shape.

PerMANOVA (section 10.5) of the sets of coefficients demonstrates a statistically significant difference between the outline shapes from different growth stages ($F = 4.25$; $p = 0.0011$). In other words, the shape of the cephalon does change through growth.

Figure 6.24 shows a principal components analysis of the sets of coefficients (since the groups are known *a priori*, CVA would also be appropriate). In this two-dimensional projection, the specimens are placed in order to maximize variance (section 6.2).

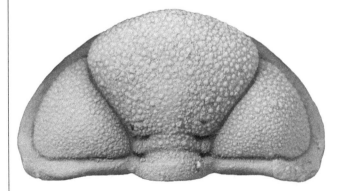

**Figure 6.22** Reconstruction of the cephalon of *Trimerocephalus lelievrei* from the Devonian of Morocco. Courtesy of Catherine Crônier.

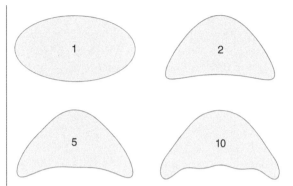

**Figure 6.23** One of the adult cephala from the data set of Crônier et al. (1998), reconstructed from 1, 2, 5, and 10 harmonics of the elliptic Fourier analysis. Crônier et al. (1998).

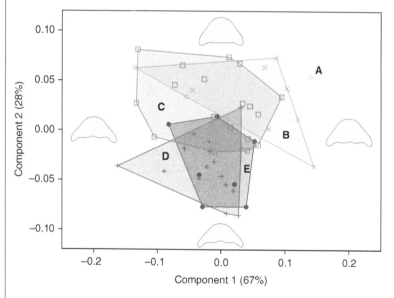

**Figure 6.24** Principal components analysis of EFA coefficients from five growth stages (A–E) of the trilobite *Trimerocephalus lelievrei* from the Devonian of Morocco. Within each growth stage, the specimens are enclosed by a convex hull polygon. Data from Crônier et al. (1998).

In order to assess the geometrical meaning of the two axes, we have also placed four synthetic outlines in their correct positions in the plot (PAST includes a function for this).

The first PCA axis (first principal component) explains 67% of the variance in the EFA coefficients. Specimens with negative scores on this axis are slender, while positive scores correspond with more robust forms. The growth stages do not separate clearly along this axis. The second axis is similar, but with a more pronounced gradient from wide forms with long genal spines, up to longer forms with short genal spines. Along this axis, there is a clear ontogenetic trend from growth stages A and B in the upper part of the diagram to stages D and E in the lower part. However, the last two growth stages (D–E) occupy similar regions. This can be interpreted as stabilization of cephalic shape at maturity.

The detailed algorithm is as follows. Let $k$ be the number of points $(x, y)$ in the outline, and $\Delta x_p$ the displacement in the $x$ direction between points $p-1$ and $p$ ($x_0 = x_k$ in a closed contour). Also, let $\Delta t_p$ be the length of the linear segment $p$:

$$\Delta t_p = \sqrt{\Delta x_p^2 + \Delta y_p^2}$$

Finally, let $t_p$ be the accumulated segment lengths and $T = t_k$ the total length of the contour, or more precisely, its approximation by the set of straight segments. We now define the cosine and sine coefficient for harmonic number $n$ ($n = 1$ for the fundamental or first harmonic), ($n = 2$ for the second harmonic, etc.) in the $x$ direction as follows:

$$A_n = \frac{T}{2n^2\pi^2} \sum_{p=1}^{k} \frac{\Delta x_p}{\Delta t_p} \left( \cos \frac{2\pi n t_p}{T} - \cos \frac{2\pi n t_{p-1}}{T} \right)$$

$$B_n = \frac{T}{2n^2\pi^2} \sum_{p=1}^{k} \frac{\Delta x_p}{\Delta t_p} \left( \sin \frac{2\pi n t_p}{T} - \sin \frac{2\pi n t_{p-1}}{T} \right)$$

The coefficients $C_n$ and $D_n$ for the cosine and sine parts of the increments in the $y$ direction are defined in the same way. From these equations, we see that the elliptic Fourier coefficients are computed by taking the inner products between the measured increments and the corresponding increments in a harmonic set of cosine and sine functions (cf. section 12.1).

For standardization of size, rotation, and starting point on the outline, see Ferson et al. (1985).

## 6.8 *Hangle* Fourier analysis

The "Hangle" method for analyzing closed outlines, proposed by Haines and Crampton (2000), is a competitor to EFA. Hangle has certain advantages over EFA, the most important being that fewer coefficients are needed to capture the outline to a given precision. This can be of importance for statistical testing (e.g., MANOVA) and discriminant analysis. The Hangle method is implemented in software by Haines and Crampton, and in PAST.

Instead of analyzing $x$ and $y$ increments along the outline, as in EFA, Hangle is based on spectral analysis of tangent angles around the outline, as described for eigenshape analysis below. As in EFA, the output consists of Fourier coefficients, which are the amplitudes of cosine and sine components of the harmonic series.

Haines and Crampton (2000) discussed thoroughly the problem of standardizing the starting point of the outlines and provided several alternative methods ("Hmatch," "Htree" and using the phase of the lowest harmonics).

---

**Example 6.9**

Figure 6.25 shows a principal components analysis of Hangle coefficients for the trilobite data set used in section 6.7. The ontogenetic sequence is clearest on the first PCA axis, in contrast with EFA, where ontogeny comes out on PC2. Otherwise, the result is similar to EFA. The differences are partly due to different estimation of the starting position.

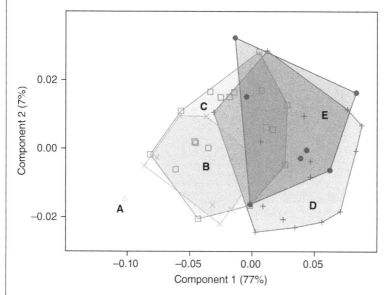

**Figure 6.25** Principal components analysis of Hangle coefficients from five growth stages (A–E) of the trilobite *Trimerocephalus lelievrei* from the Devonian of Morocco. Compare with Fig. 6.24. Data from Crônier et al. (1998).

---

## 6.9 Eigenshape analysis

Eigenshape analysis (Lohmann 1983; Rohlf 1986; MacLeod 1999) can, somewhat imprecisely, be thought of as PCA of the raw outlines, without going through a transformation stage such as Fourier analysis. This direct approach is appealing, because any additional transformation takes the data further away from the fossils, making interpretation of the result less intuitive. Eigenshape analysis can be carried out in different ways, but a specific protocol will be outlined below (see also MacLeod 1999):

1) Digitize each outline and produce an equally spaced set of points by interpolation between the original points. The number of interpolated points along the outline can either be a fixed number (such as 100), or it can be calculated automatically to obtain a good fit to the original points.

2) Going along the outline from a fixed point (homologous on all outlines), calculate the *tangent angle* from one point to the next. For *m* interpolated points on a closed outline,

there are $m$ tangent angles, constituting a vector $\varphi$ describing the shape (a similar shape vector was used by Foote (1989) and Haines and Crampton (2000) as input to Fourier analysis).

Given a set of $n$ equally spaced $(x, y)$-coordinates around a contour, the tangent angles are calculated in the following way (modified after Zahn and Roskies 1972). First, compute the angular orientations of the tangent to the curve:

$$\varphi_i = \tan^{-1} \frac{y_i - y_{i-1}}{x_i - x_{i-1}}$$

For a closed contour of $n$ points, we use

$$\varphi_1 = \tan^{-1} \frac{y_1 - y_n}{x_1 - x_n}$$

This vector is then normalized by subtracting the angular orientations that would be observed for a circle:

$$\varphi_i^* = \varphi_i - \varphi_1 - \frac{2\pi (i-1)}{n}$$

3) The shape vectors for the $n$ shapes are subjected to PCA, giving a number of principal components that are referred to as *eigenshapes*. The eigenshapes are themselves tangent angle vectors, given in decreasing order of amount of shape variation they explain. The first (most important) eigenshapes define a low-dimensional space into which the original specimens can be projected. The eigenshapes can be plotted and interpreted in terms of geometry or biology, indicating the main directions of shape variation within the sample.

MacLeod (1999) described some extensions to the eigenshape method. First, he allowed the outlines to be open contours, such as a longitudinal cross-section of a brachiopod valve or the midline of a conodont. Second, he addressed an important criticism of outline analysis, namely that the method does not necessarily compare homologous parts of the outline across individuals. In his extended eigenshape analysis, homologous points (landmarks) on the outlines are made to match up across individuals, thus reducing this problem.

---

**Example 6.10**

We return to the trilobite data of section 6.7, with 51 cephalic outlines from different ontogenetic stages of *Trimerocephalus lelievrei* from Morocco. The eigenshape analysis reports that the first eigenshape explains 33.5% of variance, while the second eigenshape explains 13.9%. Figure 6.26 shows a scatter plot of eigenshape scores on the two first axes. The latest ontogenetic stages (D and E) occupy the upper region of the plot, indicating an ontogenetic development along the second eigenshape axis from the bottom (negative scores) to the top.

The eigenshapes themselves are plotted at four locations in the figure. The first eigenshape is not so useful, but the second clearly corresponds to the degree of elongation of the cephalon and the length of the genal spines.

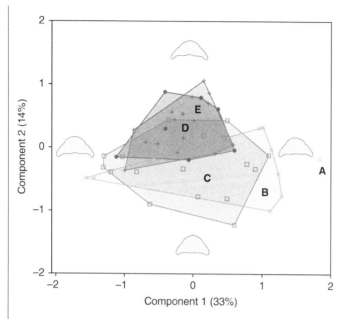

**Figure 6.26** Eigenshape analysis of five growth stages (A–E) of the trilobite *Trimerocephalus lelievrei* from the Devonian of Morocco. Compare with Fig. 6.24. Data from Crônier et al. (1998).

## 6.10   Landmarks and size measures

We have so far looked at three types of morphometric data, each with their own problems:

1) Univariate measurements, which are robust and easy to obtain and understand, but not sufficient to capture differences in shape.
2) Multiple distance measurements. These can be highly useful, but the results of the analysis can sometimes be difficult to interpret biologically, because the geometric relationships between the variables are not taken into account in the analysis. Also, the choice of measurements is subjective.
3) Outlines. Again, the results of the analysis can be difficult to interpret. Outline analysis can also be criticized for not comparing homologous elements (but see MacLeod 1999).

Currently, it is considered advantageous by many to base morphometric analysis on the positions of *landmarks*, if possible. A landmark can be defined as a point on each object, which can be recognized across all objects in the data set. Dryden and Mardia (2016) present two different classifications of landmarks:

Classification A:

1) *Anatomical landmarks* are well-defined points that are considered homologous from one specimen to the next. Typical examples are points where bone sutures meet, or small elements such as tubercles.

2) *Mathematical landmarks* are defined on the basis of some geometric property, for example, points of maximal curvature or extremal points (e.g., the end points of bones).
3) *Pseudo-landmarks* are constructed points, such as four equally spaced points between two anatomical or mathematical landmarks.

Classification B (cf. Bookstein 1991):

1) *Type I* landmarks occur where tissues or bones meet
2) *Type II* landmarks are defined by a local property such as maximal curvature
3) *Type III* landmarks occur at extremal points or at constructed points such as centroids.

Typically, the coordinates of a set of landmarks are measured on a number of specimens. For two-dimensional landmarks, this can be done easily from digital images on a computer screen. Three-dimensional landmarks are more complicated to collect, ideally using a coordinate measuring machine, or from digital 3D data obtained with CT scanning, laser scanning, or photogrammetry.

The analysis of landmark positions (including the pseudo-landmarks of outline analysis) is often referred to as *geometric morphometrics*. A large number of methods have been developed in this field over the last few decades, and it has become clear that geometric morphometrics has clear advantages over the "classic" multivariate methods based on measurements of distances and angles (Rohlf and Marcus 1993). One of the important strengths of geometric morphometrics is the ease with which the results can be visualized and interpreted. For example, we have seen that PCA of multiple distance measurements produces principal components that can be difficult to interpret. In contrast, the principal components of landmark configurations can be visualized directly as shape deformations.

### 6.10.1 Sliding landmarks

It is usually preferable to place landmarks on well-defined, homologous points. However, in some cases, it may be necessary to include additional pseudo-landmarks to capture important aspects of the shape. *Sliding landmarks* are constrained to lie on an outline (in 2D) or surface (in 3D) but are otherwise allowed to move in order to minimize bending energy (sections 6.13 and 6.15).

### 6.10.2 Size from landmarks

The size of a specimen can be defined in many ways, such as distance between two selected landmarks, the square root of the area within its outline, or the score on the first principal component. *Centroid size* is defined as the square root of the sum of squared distances from each landmark to the centroid (Dryden and Mardia 2016). In the two-dimensional case, we have

$$S = \sqrt{\Sigma\left(x_i - \bar{x}\right)^2 + \left(y_i - \bar{y}\right)^2}$$

For small variations in landmark positions, with circular distributions, centroid size is uncorrelated with shape (Bookstein 1991).

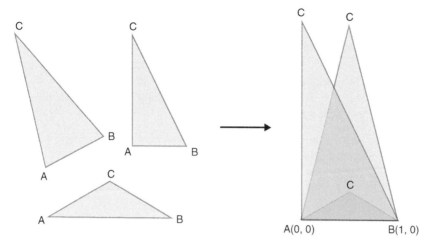

**Figure 6.27** The principle of Bookstein registration. Three triangles are registered to each other (translated, rotated, and scaled) using AB as the baseline. The coordinates of the remaining point C uniquely characterizes each shape and constitute the *Bookstein shape coordinates* of the triangles.

### 6.10.3 Landmark registration and shape coordinates

When digitizing landmarks from several specimens, it is generally impossible to ensure that each specimen is measured in the same position and orientation. If we want to compare shapes, it is therefore necessary to translate and rotate the specimens into a standardized position and orientation. In addition, we would like to scale the specimens to a standard size. Bookstein (1984) suggested doing this by selecting two landmarks A and B, forming a *baseline*. Each specimen is translated, rotated, and scaled so that landmark A is positioned at coordinates $(0, 0)$ and B is positioned at $(1, 0)$. Consider, for example, a set of triangles in 2D, with vertices A, B, and C. After the Bookstein registration, A and B are fixed and the shape of each triangle is uniquely described by the coordinates of the remaining free point C (Fig. 6.27). For any Bookstein-registered shape, the coordinates of the set of standardized landmarks except the baseline landmarks constitute the *Bookstein coordinates* of the shape.

One problem with this simple idea is that the baseline landmarks are forced to fixed positions, and landmarks closer to the baseline will therefore typically get smaller variances across specimens. This asymmetric treatment of landmarks is not entirely satisfactory, and the method of *Procrustes fitting* is therefore more commonly used.

## 6.11   Procrustes fitting

When digitizing landmarks, the specimens will usually have different sizes, and also different positions and rotations within the measuring equipment. It is therefore necessary to scale, rotate, and translate the sets of landmarks to bring them into a standardized size, orientation, and position before further analysis. We have seen one straightforward way of doing this, by selecting a baseline between two particular landmarks as a reference (Bookstein coordinates). An alternative approach is to minimize the total sum of the squared distances between corresponding landmarks. This is known as least-squares or Procrustes fitting. The resulting coordinates after fitting are called *Procrustes coordinates*.

Algorithms for Procrustes fitting in 2D and 3D are given by Dryden and Mardia (2016). For the 2D case, let $\mathbf{w}_1, \ldots \mathbf{w}_n$ be the $n$ landmark configurations to be fitted. Each $\mathbf{w}_i$ is a complex $k$-vector with $k$ landmarks, and the $x$- and $y$-coordinates are the real and imaginary parts, respectively. Assume that the configurations have been centered (location removed) by subtracting the centroid from each configuration.

1) Normalize the sizes by dividing by the centroid size of each configuration. The centered and scaled configurations are called *pre-shapes*, $\mathbf{z}_1, \ldots \mathbf{z}_n$.
2) Form the complex matrix

$$\mathbf{S} = \sum_{i=1}^{n} \mathbf{z}_i \mathbf{z}_i^*$$

where $\mathbf{z}_i^*$ is the complex conjugated transpose of $\mathbf{z}_i$. Calculate the first complex eigenvector $\hat{\mathbf{u}}$ (corresponding to the largest eigenvalue) of $\mathbf{S}$. This vector $\hat{\mathbf{u}}$ is the *full Procrustes mean shape*.
3) The fitted coordinates (full Procrustes fits, or shapes) are finally given by

$$\mathbf{w}_i^P = \frac{\mathbf{w}_i^* \hat{\mathbf{u}} \mathbf{w}_i}{\mathbf{w}_i^* \mathbf{w}_i}$$

where an asterisk again denotes the complex conjugated transpose.

Some comments on nomenclature are in place. The *full* Procrustes fit standardizes size, orientation, and position, leaving only shape. The *partial* Procrustes fit standardizes only orientation and position, leaving shape and size. *Ordinary* Procrustes analysis involves only two objects, where one is fitted to the other. *Generalized* Procrustes analysis (GPA) can involve any number of objects, which are fitted to a common consensus. The (full) Procrustes mean shape of a number of specimens is simply the average of their Procrustes-fitted shapes.

A large technical literature has grown up around the concept of a *shape space* (Kendall et al. 1999; Slice 2000; Dryden and Mardia 2016). Every shape, as described by its landmark coordinates, can be regarded as a point in a multidimensional shape space. Consider a triangle in 2D, with altogether six coordinate values. All possible triangles can be placed in a six-dimensional space. However, the process of Procrustean fitting reduces the dimensionality of the space. In the two-dimensional case, the scaling step reduces the dimensionality by one (removal of a scaling factor); the translation step reduces it by a further two (removal of an $x$ and a $y$ displacement); and rotation by one (removal of an angle). Altogether, Procrustes fitting in our triangle example involves a reduction from six to two dimensions (compare with the Bookstein coordinates of section 6.10).

In general, the shape space will be a curved hypersurface (a *manifold*) in the high-dimensional landmark coordinate space. It is therefore not entirely trivial to define a distance between two shapes – should it be a Euclidean distance, or taken along the curved surface (a great circle in the case of a hypersphere)? It has been found practical to use Euclidean distances within a *tangent space*, which is a flat hyperplane tangent to the shape space at a given point, usually the full Procrustean mean. An approximation to the tangent

(A)                                                          (B)

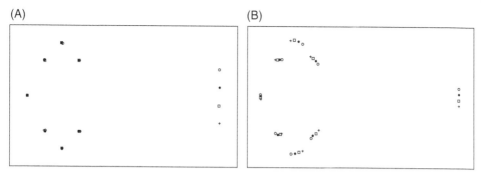

Figure 6.28   Least-squares (Procrustes) registration may not give satisfactory results when variation is concentrated to only a few landmarks. (A) Eight landmarks in four specimens (four different symbols). The shape variation is due almost exclusively to the distant landmark at the right. The illustrated registration solution may be regarded as satisfactory because it clarifies the localization of variance. (B) Procrustes registration of the same shapes. The shapes are rotated, scaled, and translated to minimize total squared differences between landmark positions, resulting in a "fair" distribution of error among all landmarks. This obscures the structure of the shape variation.

space coordinates for an individual shape is achieved simply by subtracting the full Procrustes mean from that shape, giving the *Procrustes residual* (Dryden and Mardia 2016). The smoothness of the shape space ensures that the local region around the tangent point (the pole) can be approximated by a plane, so all is fine as long as the spread in the data is not too great.

In spite of its popularity, least-squares Procrustes fitting is not the last word in landmark registration. Consider a situation where all landmarks but one are in almost identical positions in different shapes. The one diverging landmark may, for example, be at the end of a long, thin spike – minor differences along the spike may add up to a disproportionally large variance in the landmark coordinates of the tip. When such a set of shapes is Procrustes fitted, the differences in this one landmark will be evenly spread over all land-marks (Fig. 6.28), which may not be what we want (Siegel and Benson 1982). To address such problems, a plethora of so-called *robust* or *resistant* registration methods have been suggested – Dryden and Walker (1999) give a technical description of some of them. See also Slice (1996).

---

**Example 6.11**

We will look at a data set produced by O'Higgins (1989) and further analyzed by O'Higgins and Dryden (1993). The positions of eight landmarks were measured in the midline of the skull (sagittal plane) in 30 female and 29 male gorillas (Fig. 6.29). The raw data, as shown in Fig. 6.30A, are difficult to interpret and analyze directly, because the specimens are of different sizes and have been positioned and oriented differently in the measuring apparatus. After Procrustes fitting (Fig. 6.30B), the data are much clearer, and small but systematic differences between males and females are visible.

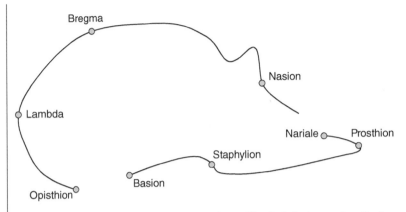

Figure 6.29   Landmarks in the midplane of a gorilla skull. Redrawn after Dryden and Mardia (2016).

(A)                                                          (B)

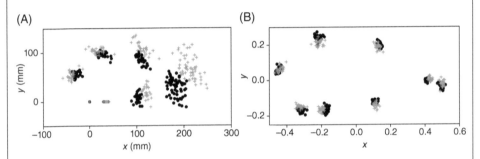

Figure 6.30   (A) Eight landmarks in 59 gorilla skulls. Red crosses are males and black dots females. (B) The same data set after Procrustes fitting, removing differences in size, translation, and rotation. A difference in mean position between males and females is apparent for some of the landmarks.

## 6.12   PCA of landmark data

Many of the standard multivariate analysis techniques, such as cluster analysis, Hotelling's $T^2$, discriminant analysis, and MANOVA, can be applied directly to Procrustes-fitted landmarks. A particularly useful method for direct analysis of landmarks is PCA, which will bring out directions of maximal shape variation. Since the principal components now correspond to displacement vectors for all the landmarks away from the mean shape, they can be easily visualized (Fig. 6.32).

---

**Example 6.12**

We continue the analysis of the gorilla data set of section 6.11. Figure 6.31 shows the scatter plot of PCA scores on PC1 (45.9% variance) and PC2 (16.9% variance). Males and females evidently occupy different regions of the plot, although with some overlap.

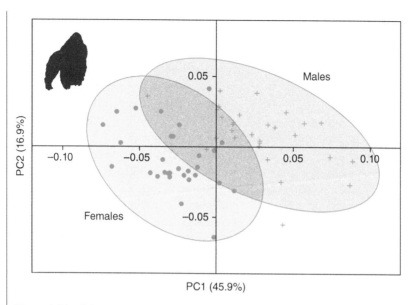

Figure 6.31  Principal component scores for the Procrustes-fitted gorilla landmark data (Procrustes residuals), with 95% concentration ellipses for females (dots) and males (crosses).

To interpret the meaning of the principal component axes, we need to look at the PC loadings. We could have presented them in a biplot, but a better way of visualizing them is as displacement vectors away from the mean shape. Figure 6.32 shows such a landmark displacement diagram for PC1. Evidently, the landmarks in the anterior and posterior regions move up while those in the middle of the skull go down, causing a concave-up bending of the skull as we go along the positive PC1 axis (compare with the relative warp deformation grids of section 6.15).

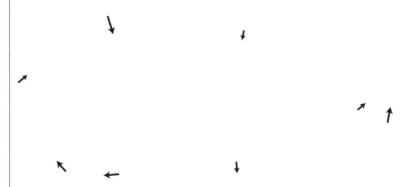

Figure 6.32  Gorilla landmark displacements corresponding to a PC1 score of 0.10, at the "male" end of PC1 (cf. Fig. 6.31). The arrows point from the mean landmark positions to the displaced positions.

## 6.13 Thin-plate spline deformations

Consider the shape of a fish drawn onto a rubber sheet with a square grid. The sheet can now be stretched and deformed in various ways, transforming the shapes of both the fish and the grid. Although such grid transformations have been known for hundreds of years, it was D'Arcy Wentworth Thompson who first realized that this method could be helpful for visualizing and understanding phylogenetic and ontogenetic shape change. The grid transformations in his famous book *On Growth and Form* (Thompson 1917) have become icons of modern morphometrics (Fig. 6.33). Thompson drew his grid transformations by hand, but today we can use more automatic, well-defined, and precise methods.

In the following, the word "shape" should be read as "set of Procrustes-fitted landmarks." A transformation from a source shape to a target shape involves the displacement of the source landmarks to the corresponding target landmarks. This is a simple vector addition. However, we also want arbitrary points in between the landmarks (including the grid nodes) to be displaced, in order to produce a full visualization of the deformed grid. This is an *interpolation* problem, because we need to interpolate between the known displacements of the landmarks. We want this interpolation to be as smooth as possible, meaning that we want to minimize any sharp, local bends and compressions. Luckily, mathematicians have found an interpolation method that fits our need perfectly, being *maximally*

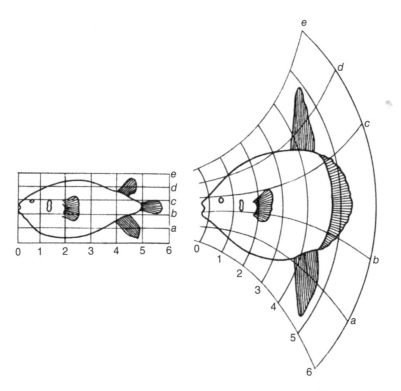

Figure 6.33 Grid transformation from one fish to another. From Thompson (1917).

*smooth* in a certain sense. This interpolator is known as the *thin-plate spline* (TPS), because of its mechanical interpretation (Bookstein 1991; Dryden and Mardia 2016).

The TPS operates on the displacements in the $x$ and $y$ directions independently. Considering first the former, each landmark $i$ is moved a known distance $\Delta x_i = x_i^T - x_i^S$, positive or negative, from the source to the target position. Now consider a thin, flat, stiff but flexible plate, suspended horizontally. At the position of each landmark in the source configuration, force the plate a distance $\Delta x_i$ vertically, for example by poking it with a pin (Fig. 6.34). The plate will then bend into a nice, smooth form, which has the smallest possible curvature under the constraints imposed by the point deflections (this is because the plate minimizes its bending energy, which is proportional to curvature). Relatively simple equations describing how a thin plate bends under these conditions are available, under certain model simplifications. At any point in between the landmarks, including the grid nodes, we can now read the interpolated deflections in the $x$ directions. This operation is then repeated for the $\Delta y_i$ deflections on an independent plate, producing a complete grid transformation. The TPS can be extended similarly to three dimensions.

The TPS method can produce beautiful figures that greatly aid interpretation of shape changes. Nonetheless, we may ask what the biological meaning of the grid transformation is supposed to be. Real organisms have practically nothing to do with thin plates. Ontogenetic and phylogenetic deformations result primarily from differential (allometric) growth, and the TPS does not explain the transformation in terms of such effects. The main attraction of the TPS is perhaps that it is geometrically parsimonious: in a certain mathematical sense, it is the simplest deformation that can bring the source shape onto the target. Still, we can hope that morphometricians will one day come up with a method for decomposing a shape change into a small set of allometries and allometric gradients that are more directly connected with real biology (Hammer 2004).

Methods have been developed for giving some biological meaning to the TPS transformation. For example, we can plot an *expansion map*, showing the local expansion or contraction of the grid with a gray scale, color scale, or contour map (Fig. 6.36). Such a

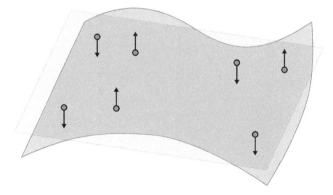

**Figure 6.34** Physical interpretation of the principle of thin-plate splines. Seven original landmarks (dots) are displaced vertically from their original positions on a thin, flat, stiff but flexible plate (pink). This forces the plate to bend into a smooth configuration (orange), interpolating the displacements of the landmarks. For two-dimensional landmarks, this operation is applied to the $x$ and the $y$ displacements independently.

map can give an idea about local growth rates and allometric gradients. Another important biological process is *anisometric growth*, meaning that growth is faster in one direction, causing elongation. This can be visualized by placing small circles onto the source shape and letting them be deformed by the transformation. The resulting *strain ellipses* have a major and a minor axis, showing the (principal) directions of maximal and minimal growth (Fig. 6.37).

---

**Example 6.13**

We continue the gorilla example from section 6.11. We want to visualize the transformation from the mean female skull to the mean male. After Procrustes fitting (section 6.11), the mean positions of the landmarks of the female and male skulls were computed, forming our source and target shapes. Figure 6.35 shows the resulting TPS transformation grid, clarifying the directions and degrees of compression and extension.

To make the expansions and compressions even clearer, we also compute the change in area over the transformation grid (Fig. 6.36).

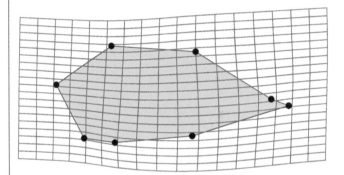

Figure 6.35  TPS transformation from the mean female gorilla skull, on an originally rectangular grid, to the mean male. The male landmarks are shown as points. Note the dorsoventral compression in the posterior part of the skull (left).

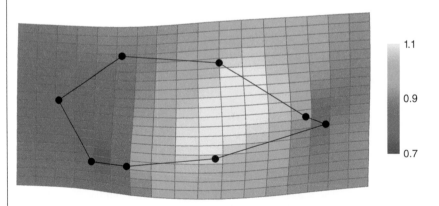

Figure 6.36  Color coding of the relative expansions (expansion factors larger than one) and contractions (smaller than one) of local regions of the transformation grid. Note the compression in the posterior region and expansion in the anterior.

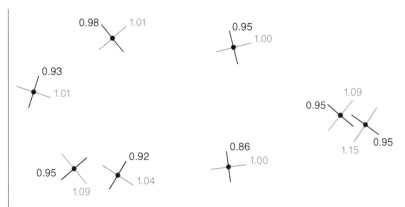

**Figure 6.37** Principal deformations (principal axes of the strain ellipses) of the transformation from mean female to mean male at each landmark, showing anisometric allometry. Major axes (directions of maximal extension) are in red, and minor axes (minimal extension) are in black.

Finally, the principal directions of expansion/contraction at the landmarks are shown in Fig. 6.37. Together with the "growth map" of Fig. 6.36, this gives a relatively complete picture of the local deformations around the landmarks.

The mathematical operations involved in the production of 2D TPSs are described for example by Dryden and Mardia (2016). They can be extended to 3D.

Let $\mathbf{t}_i$ be the position of landmark $i$ in the source shape (a two-vector), and $\mathbf{y}_i$ the position of the same landmark in the target shape. The full source and target shapes $\mathbf{T}$ and $\mathbf{Y}$ for $k$ landmarks are then $k \times 2$ matrices

$$\mathbf{T} = \begin{bmatrix} \mathbf{t}_1 \\ \mathbf{t}_2 \\ \vdots \\ \mathbf{t}_k \end{bmatrix}$$

$$\mathbf{Y} = \begin{bmatrix} \mathbf{y}_1 \\ \mathbf{y}_2 \\ \vdots \\ \mathbf{y}_k \end{bmatrix}$$

Define a univariate distance weighting $\sigma$ as a function of a two-vector $\mathbf{h}$:

$$\sigma(\mathbf{h}) = |\mathbf{h}|^2 \ln|\mathbf{h}| \text{ for } |\mathbf{h}| > 0$$

$$\sigma(h) = 0 \text{ for } |\mathbf{h}| = 0$$

where $|\cdot|$ is the Euclidean vector length. Let the $k$-vector $\mathbf{s(t)}$ be a function of any position $\mathbf{t}$:

$$\mathbf{s(t)} = \begin{bmatrix} \sigma(\mathbf{t} - \mathbf{t}_1) \\ \sigma(\mathbf{t} - \mathbf{t}_2) \\ \vdots \\ \sigma(\mathbf{t} - \mathbf{t}_k) \end{bmatrix}$$

Compute the $k \times k$ matrix **S** (which we will assume to be non-singular) as

$$S_{ij} = \sigma(\mathbf{t}_i - \mathbf{t}_j)$$

Construct a $(k + 3) \times (k + 3)$ matrix

$$\Gamma = \begin{bmatrix} \mathbf{S} & \mathbf{1}_k & \mathbf{T} \\ \mathbf{1}_k^T & 0 & 0 \\ \mathbf{T}^T & 0 & 0 \end{bmatrix}$$

where $\mathbf{1}_k$ is a $k$-vector of all ones. Invert $\Gamma$ and divide into block partitions

$$\Gamma^{-1} = \begin{bmatrix} \Gamma^{11} & \Gamma^{12} \\ \Gamma^{21} & \Gamma^{22} \end{bmatrix}$$

where $\Gamma^{11}$ is $k \times k$ and $\Gamma^{21}$ is $3 \times k$. Compute the $k \times k$ matrix

$$\mathbf{W} = \Gamma^{11}\mathbf{Y}$$

and the 2-vector **c** and the $2 \times 2$ matrix **A** such that

$$\begin{bmatrix} \mathbf{c}^T \\ \mathbf{A}^T \end{bmatrix} = \Gamma^{21}\mathbf{Y}$$

We now have all the building blocks necessary to set up the pair of TPSs, which sends a given point **t** onto a point **y**:

$$\mathbf{y} = \mathbf{c} + \mathbf{A}\mathbf{t} + \mathbf{W}^T\mathbf{s}(\mathbf{t})$$

In particular, this function will send the source landmark $\mathbf{t}_i$ onto the target position $\mathbf{y}_i$. The *expansion map* is most accurately computed using the determinant of the so-called *Jacobian matrix* of the TPS transformation. The Jacobian, as given in most textbooks on analytic geometry, is defined by certain partial derivatives of the transformation.

## 6.14  Principal and partial warps

A TPS deformation from one shape to another will normally involve some large-scale deformations such as expansion of the braincase relative to the face, and some local, small-scale deformations such as reduction of the distance between the eyes. It is possible that these components of the deformation have different biological explanations, and it is there-fore of interest to separate the full deformation into components at different scales. This idea is reminiscent of spectral analysis (section 12.1), where a univariate function is decomposed into sinusoids with different wavelengths.

The first step in this procedure is to look only at the source shape (also known as the *reference*). Movement of landmarks away from the reference configuration will bend the plate, involving global and local deformations, and our goal is to be able to re-express any movement away from the reference as a sum of such components. The *principal warps* (Bookstein 1991) are smooth, univariate functions of the two-dimensional positions in the grid plane, with large-scale (first principal warp) and shorter-range (higher principal warps) features. The principal warps are usually not of principal interest but constitute an important step in the decomposition of a deformation.

The principal warps are calculated using the reference only. The actual decomposed deformation to a given target shape is provided by the corresponding *partial warps* (Bookstein 1991). A partial warp is a mapping that directly represents a displacement of any given point from a reference to a target, and it is therefore a bivariate function of position (displacement along the two main directions of the grid plane).

Partial warps represent useful decompositions of shape change and can aid the interpretation of shape variation. However, it must be remembered that they are purely geometrical constructions that may or may not have direct biological relevance. To regard a partial warp as a biological homology is questionable.

To complete the confusion, the *partial warp scores* are numbers that indicate how strongly each principal warp contributes to a given transformation from source to target. There are two partial warp scores for each principal (or partial) warp, one for each axis of the grid. When studying a collection of specimens, the mean shape is often taken as the reference, and all the specimens can be regarded as deformations from this reference. The partial warp scores of all specimens can then be presented in scatter plots, one for each partial warp.

## 6.14.1 The affine (uniform) component

All the partial warps represent bending and/or local compression or expansion. Put together, they do not completely specify the deformation, because they miss the operations of uniform scaling along each of the two axes, and uniform shearing. Together, these linear operations constitute the *affine* component of the deformation, sometimes referred to as the zeroth partial warp. The affine component will deform a square into a parallelogram. Just as each partial warp has an associated partial warp score, it is possible to define a bivariate affine score.

## 6.14.2 Partial warp scores as shape coordinates

A nice property of the partial warp scores (including the affine component) is that they are coordinates of the space spanned by the partial warps, and this space is a tangent plane to shape space with the reference shape as a pole. While the Procrustes residuals in 2D have $2k$ variables for $k$ landmarks, these numbers lie in a subspace of $2k - 4$ dimensions, because four dimensions were "lost" by the Procrustes fitting: two degrees of freedom disappeared in the removal of position, one in the removal of size, and one in the removal of rotational orientation. For statistical purposes, partial warp scores may be better to use than Procrustes residuals, because the "unnecessary" extra dimensions have been eliminated to produce a $(2k - 4)$-dimensional tangent space. This happens because there are $k - 3$ non-affine partial warps for $k$ landmarks (ordered from a global deformation represented by the first partial warp to a local deformation represented by the last partial warp). When we include the affine component we have $k - 2$ partial warps, each associated with a bivariate partial warp score. The total number of variates in partial warp score space is therefore $2(k - 2) = 2k - 4$ as required.

---

**Example 6.14**

Returning to the gorilla data set of section 6.11, we will focus on the deformation from the mean female to the mean male. Figure 6.38 shows the affine component (zeroth partial warp) and the first, second, and fifth partial warps. The lower partial warps are large-scale (global) deformations, getting progressively more small scale (local) for the higher partial warps.

The partial warp scores for the first five partial warps are presented in Fig. 6.39. Note that each partial warp corresponds to a bivariate partial warp score, which is conveniently plotted in a scatter diagram. It seems that the males and females are separated best on the first and fourth partial warps and also on the affine component.

The equations for computing the principal and partial warps and the affine component were given by Dryden and Mardia (2016). General methods for the computation of the affine component in 2D and 3D were discussed by Rohlf and Bookstein (2003).

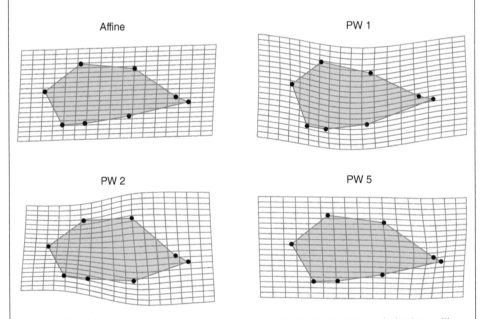

Figure 6.38   Partial warp deformations from mean female shape to mean male in the gorilla skull data set. The deformations are amplified by a factor of three for clarity. The first partial warp is a large-scale bending of the whole shape. The second partial warp represents more localized deformations. The fifth partial warp involves mainly the displacement of a single landmark in the snout.

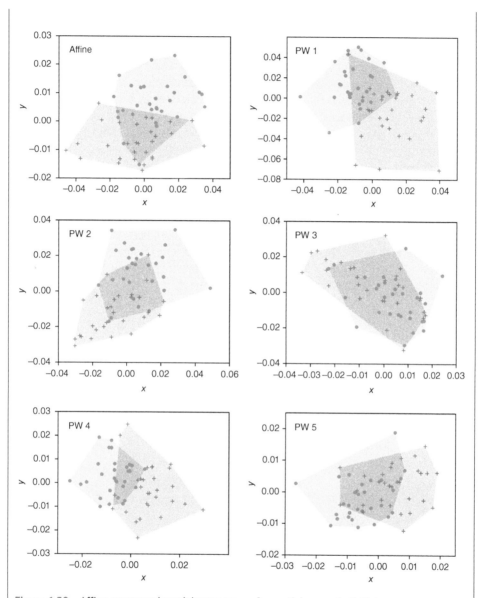

Figure 6.39 Affine scores and partial warp scores for partial warps 1–5. Males are crosses, females are dots. The sexes seem to differ more clearly in their first and fourth partial warp scores.

## 6.15 Relative warps

In section 6.12, we discussed the use of Principal Component Analysis for a set of shapes (landmark configurations) in order to analyze the structure of variability within the sample. This standard version of PCA maximizes variance with respect to a Euclidean distance measure between landmarks. Alternatively, the distance between two shapes can be measured with respect to the "bending energy" involved in the TPS transformation from one shape to the other (the bending energy metric is described by Dryden and Mardia 2016). PCA, with respect to bending energy, can be regarded as PCA of the TPS deformations from the mean shape to each individual shape. Each deformation can then be expressed in terms of a set of principal components, which are themselves deformations known as *relative warps*. As we are used to from standard PCA, the first relative warp explains the largest variance in the data set. The principal component scores are called *relative warp scores*.

Relative warp analysis with respect to the bending energy tends to put emphasis on the variation in large-scale deformations. It is also possible to carry out the analysis with respect to inverse bending energy, which tends to put more emphasis on small-scale deformations. Standard PCA of the landmarks, as described in section 6.12, is more neutral with respect to deformation scale. It seems that this has become the standard method in the literature, in which case a "relative warp" basically means a TPS visualization of a principal component of a set of landmark configurations. This combination of PCA and TPS is a cornerstone of geometric morphometrics.

---

**Example 6.15**

Figure 6.40 shows the relative warp deformation for a PC1 score of 0.10, based on the standard PCA of the gorilla landmarks. Referring to Fig. 6.31, this represents an "extreme male" at the right of the PCA scatter plot.

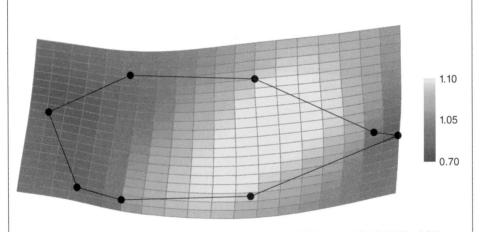

1.10

1.05

0.70

Figure 6.40 Relative warp deformation corresponding to a PC1 score of 0.10 (cf. Fig. 6.31), with expansion map.

## 6.16    Regression of warp scores

The partial or relative warp scores for a number of specimens may be subjected to linear regression (section 4.6) upon a univariate variable such as size or position along an environmental gradient. This may show whether shape varies systematically with e.g., size (allometry), along an environmental gradient (ecophenotypic effects) or with stratigraphic position (ecophenotypic effects and/or microevolution). Moreover, since the regression results in a model for how shape changes with the independent variable, shape can be interpolated within the original range and extrapolated beyond it to visualize the precise nature of shape change.

   Although it seems likely that such a linear model might be insufficient in some cases, it has turned out to be an adequate first-order approximation for many problems.

---

**Example 6.16**

We will illustrate regression of warp scores using the "Vilmann" data set from Bookstein (1991). Eight landmarks have been digitized in the midplane of the skull (without the jaw) of 164 rats, aged from 7 to 150 days. After Procrustes fitting, we carry out a relative warp analysis (PCA). PC1 explains 81.9% of shape variation (Fig. 6.41), while PC2 explains 7.9%.

   Figure 6.42 shows the PC1 scores plotted against centroid size. Since PC1 explains so much of the shape variation, we may regard this as a plot of shape against size. There is obviously a systematic shape change through growth, and it can be modeled well as a straight line ($R^2 = 0.92, p < 0.0001$). Similar regressions of PC2 and PC3 scores against size do not show significant correlations.

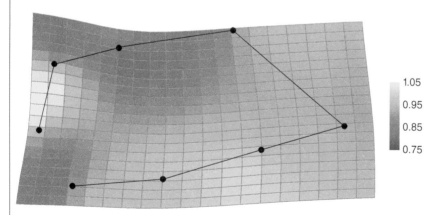

**Figure 6.41**    The deformation corresponding to PC1 (the first relative warp) of the rat skull data set, for a PC1 score of −0.10. This deformation explains 81.9% of the shape variation in the data.

---

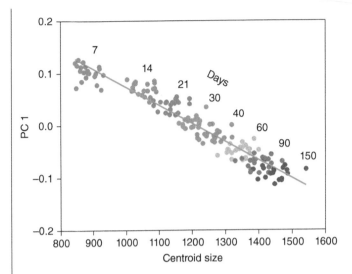

Figure 6.42 Centroid size versus first relative warp scores for the rat skull data. The data points are shown with different colors for the eight age groups (7, 14, 21, 30, 40, 60, 90, and 150 days). OLS regression, $R^2 = 0.92, p < 0.0001$.

## 6.17 Common allometric component analysis

Mitteroecker et al. (2004) introduced a procedure called common allometric component (CAC) analysis for studying size allometry in the framework of geometric morphometrics. This method can be applied to outline or landmark data in 2D or 3D. The idea is simply to carry out a linear regression of the Procrustes coordinates on log-transformed centroid sizes, giving a linear model for the allometric shape deformation (the CAC). Finally, the residuals from this regression are subjected to standard PCA, resulting in a set of *residual shape components* (RSCs).

Let **X** be the $n \times m$ matrix of Procrustes-fitted shape coordinates. In the implementation of CAC by Mitteroecker et al. (2004), these coordinates are centered on group means (if groups are specified). Further, **s** is the $n$-vector containing the log-transformed centroid sizes. The CAC is then given by the $m$-vector

$$\mathbf{a} = \frac{\mathbf{X}'\mathbf{s}}{\mathbf{s}'\mathbf{s}}$$

normalized as $\mathbf{a}' = \mathbf{a}/\sqrt{\mathbf{a}'\mathbf{a}}$ (unit Euclidean distance). The CAC can be visualized with landmark displacement vectors or a TPS deformation as usual. This CAC is projected out to produce a reduced data matrix

$$\mathbf{W} = \mathbf{X}\left(\mathbf{I} - \mathbf{a}'\left(\mathbf{a}'\right)'\right)$$

The principal components of **W** constitute the RSCs.

Bookstein (2021) argued against using Procrustes-fitted landmarks, where centroid size has been removed, as a basis for studies of allometry. He suggested instead to multiply back the centroid sizes to the landmark coordinates after Procrustes fitting, thus restoring the size component to the landmarks, and continue the analysis on these so-called *Boas coordinates*. This is a somewhat radical departure from current practice in geometric morphometrics but has intuitive appeal.

## 6.18 Landmarks in 3D

The shapes of most organisms are more accurately described in 3D than in 2D. With CT scanning (chapter 8), laser or structured light scanning, photogrammetry, or a coordinate measuring machine, it is possible to digitize 3D coordinates of landmarks. Although the algorithms and interpretations become a little more complex, most of the methods for the analysis of 2D landmarks can be extended to 3D. Thus, there are 3D versions of e.g., Procrustes analysis, PCA of landmarks, deformation grids, and CAC analysis. Although 3D morphometrics is not always possible for fossil specimens because of flattening, these techniques are likely to become increasingly common in paleontology.

---

**Example 6.17**

Hopkins and Pearson (2016) CT scanned 27 silicified cephala of the trinucleid trilobite *Cryptolithus tesselatus* (Fig. 6.43) from the Upper Ordovician of Virginia. They defined 24 fixed landmarks, 18 semi-landmarks (section 6.10) along the anterior edge of the cephalon, and 21 surface semi-landmarks on the glabella. The CAC (section 6.17) is shown in Fig. 6.44. The displacement vectors show that the growth allometry mainly consists of a narrowing of the posterior part of the cephalic margin, and also an upwards doming of the glabella. If we plot the CAC scores as a function of size (Fig. 6.45), a break in the linear relationship (a "knickpunkt") is indicated at a log centroid size of ca. 1.1, i.e., a linear size of ca. 13 mm. Hopkins and Pearson (2016) concluded that this rate shift occurred roughly at the same time as the cessation of the addition of fringe pits. Such studies are important to understand ontogenetic trajectories, and thus to disentangle evolution by heterochrony.

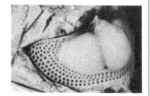

Figure 6.43   The morphological features of the *Cryptolithus* cephalon. Stereopair on left, oblique view on right. Hughes et al. (1975)/The Royal Society.

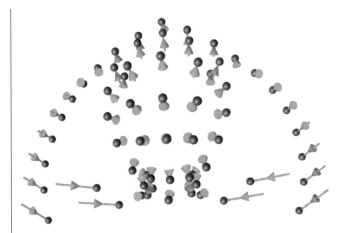

**Figure 6.44** Three-dimensional common allometric component for the *Cryptolithus* data set, visualized with PAST. Landmark displacements from the mean shape are shown as arrows pointing towards the spheres at positions corresponding to a CAC score of 0.20, i.e., a large specimen. Data from Hopkins and Pearson (2016).

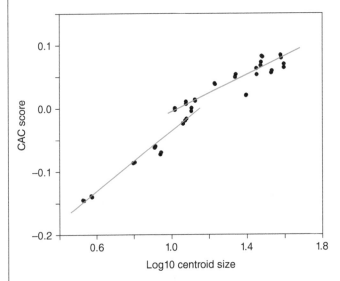

**Figure 6.45** Log-transformed size versus common allometric component score for the *Cryptolithus* data set. Informal linear growth trajectories (red) indicate a shift in developmental trajectory.

## 6.19 Disparity measures

The variance or "spread" of form within a group of organisms (the group may be defined taxonomically, temporally, or geographically) is a quantity of considerable interest. We will use the term "disparity" in a broad sense, although Foote (1992) reserved this term for morphological diversity across body plans. If disparity could be sensibly defined and measured, we could answer questions such as these: Were the Cambrian arthropods more diverse morphologically than recent ones (see Wills et al. 1994 for a review)? Does the idea of initial morphological diversification and subsequent effective stasis hold for particular taxonomic groups (e.g., Foote 1994)? Does disparity decrease in times of environmental stress, because of increased selection pressure?

Disparity can be measured in many ways. To some extent, the choice of disparity measure will be dictated by the taxonomic level of the study (intraspecific or at the level of classes or phyla). For lower taxonomic levels, a morphometric approach using continuous measurements may be the most appropriate, whereas at higher levels, a common set of measurements may be hard to define, and disparity must be measured using discrete characters as described in chapter 14.

## 6.19.1 Morphometric disparity measures

For continuous measurements, disparity within a single continuous character such as the length of the skull can be measured either using the variance (or standard deviation) or the range, i.e., the largest minus the smallest value. Although the variance seems a natural choice, it may not capture our intuitive sense of disparity. Foote (1992) gives the example of mammals, where the morphological range "from bats to whales" is very large. However, since most mammals are rat-like, the variance is small. The range is therefore perhaps a better choice of disparity measure.

For multivariate continuous data sets, the situation gets more complex. How can we combine the individual univariate disparities into a single disparity measure? One solution is to use the product of the individual disparities, which is equivalent to taking the volume of a hypercube in multivariate space. A more refined approach would be to take the volume of a hyperellipsoid (Wills et al. 1994). To limit the inflation of these values for large dimensionalities $k$, the $k$th root could be taken. One possible problem with the product is that it tends to vanish if only one of the characters has small disparity. A partial solution would be to use the scores on the first few axes of PCA. Alternatively, we could use the sum rather than the product of the individual disparities (Foote 1992). Ciampaglio et al. (2001) compared these and other disparity measures and concluded that they can all be useful depending on the purpose of the study.

## 6.19.2 Disparity measures from discrete characters

Discrete characters are treated in chapter 14. Foote (1992) and Wills et al. (1994) discussed their use for disparity measurement in some detail. We will consider two approaches. First, the *mean* (or median) *phenetic dissimilarity* within a group can be calculated as the mean or median of all possible pairwise dissimilarities between two members (often species) of the group. The dissimilarity between two members is the character difference summed over all characters, divided by the number of characters. If the characters are unordered, we simply count the number of characters where the states differ in the two species and divide by the number of characters (this is known as the simple matching coefficient or Hamming distance). Other distance measures can also be used. A variant on this is the weighted mean pairwise disparity (WMPD), which is the sum of all pairwise distances weighted by the numbers of comparable characters in the two taxa (Ciampaglio et al. 2001; Close et al. 2015; Reeves et al. 2021).

A second method attempts to use the discrete character states to place the individual species as points in a continuous multivariate space. The morphometric disparity measures described earlier can then be applied (Wills et al. 1994). This method, based on principal coordinates analysis (to be described in chapter 10) has the added advantage of visualizing the structure of disparity within and across the different groups.

These methods have other applications than estimating morphological disparity. Reeves et al. (2021) defined a number of ecological characters (e.g., diet and habitat) for Mesozoic marine tetrapods and used these to analyze the change in "ecospace" (rather than morphospace) occupancy through time.

### 6.19.3 Sampling effects and rarefaction

Disparity is expected to increase with larger sample size. As more new species are included in the study, a larger range of morphologies will be covered, and disparity goes up. Foote (1997) demonstrated that the range is more sensitive to increased sample size than the variance. In order to correct for sample size, it is possible to use a type of rarefaction analysis analogous to the technique used for ecological samples, as described in chapter 9 (Foote 1992).

### 6.19.4 Morphospaces

A multivariate space defined by morphological parameters constitutes a *morphospace* (McGhee 1999). Disparity may be described as the size of the region in morphospace occupied by the given group of organisms or taxa. We may discern between two types of morphospaces: empirical and theoretical (McGhee 1999). An empirical morphospace is a space spanned by measured quantities on sampled specimens, while a theoretical morphospace is a parameter space for a morphological or morphogenetic model. Combinations of the two are possible.

The most famous theoretical morphospace is undoubtedly that of Raup (1966, 1967), who constructed a simple geometric model of coiling shells, using three parameters: $W$ (whorl expansion ratio); $D$ (distance from coiling axis to generating curve); and $T$ (translation rate along the coiling axis). To a first approximation, most shells of gastropods, ammonoids, nautiloids, bivalves, and brachiopods can be described by such a model, and placed in the three-dimensional theoretical morphospace spanned by these three parameters (Fig. 6.46). It can then be shown how different groups tend to occupy different regions of morphospace, while other regions tend not to be occupied by real organisms. Raup's model and later amendments, alternatives, and applications were thoroughly reviewed by McGhee (1999).

## 6.20 Morphogroup identification with machine learning

Machine learning (ML) is a general and somewhat ill-defined term for computer methods that try to make models for complicated data sets, based on sample data (the "training set"). ML often (but not always) attempts to classify items into groups. Although ML is

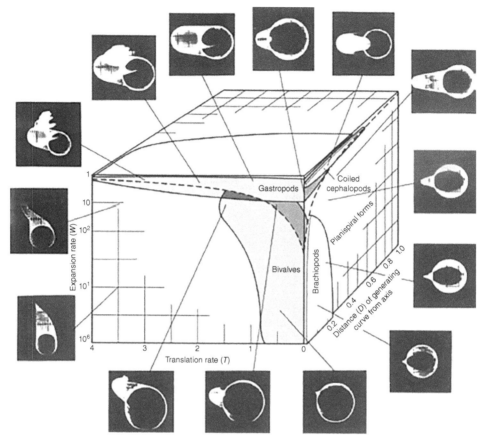

**Figure 6.46** The classical theoretical morphospace of Raup (1966), with simulated shells corresponding to particular values of the coiling parameters $W$, $D$, and $T$. Amazingly, the visualizations were produced with an analog computer connected to an oscilloscope. Some shapes occur in nature, others do not. Raup (1966)/SEPM Society for Sedimentary Geology.

often described as a subfield within, or even a synonym for, artificial intelligence (AI), many ML methods are historically more related to classical statistics. The field is currently in a phase of exponential growth, and we can only briefly mention a few important methods in this book.

We can roughly divide ML into two different approaches: supervised and unsupervised learning. In supervised learning, the computer is presented with a training set where the items are already assigned to known groups. The computer then builds a model (a classifier) that will correctly assign the training items to their groups. Finally, new items without known group can be classified using the model. Linear discriminant analysis (section 6.4) is an example of a supervised learning algorithm. More advanced algorithms for supervised learning have recently become extremely powerful in image classification, with popular apps identifying species of e.g., plants and insects from photos.

In unsupervised learning, the groups are not known *a priori*. Instead, the algorithm generates the groups during the learning phase. The clustering methods discussed in chapter 5 are examples of unsupervised learning.

ML methods can be applied in several fields of paleontology. We will later (section 10.11) discuss the use of advanced ordination methods, also considered part of the ML field, in paleoecology and paleobiogeography. However, most applications of ML in paleontology so far have been for the identification of morphospecies or other taxonomic groups based on fossil images or morphometric data. This can have at least three different purposes: (1) speeding up the identification of fossils, allowing, for example, automated counting of microfossils in slides, (2) making species identification more reproducible, and accessible to non-experts, and (3) making morphological description and selection of characters more objective, possibly generating unexpected hypotheses for evolutionary relationships.

### 6.20.1 K-nearest-neighbor classification

K-nearest-neighbor classification (KNN; Fix and Hodges 1951) is a simple but effective supervised ML algorithm. Based on a training set, classified *a priori*, it can assign new specimens to the given classes. The algorithm simply identifies the $k$-nearest neighbors in the training set, based on any distance measure (often Euclidean), and then selects the class based on a majority vote among the neighbors.

The principle is shown in Fig. 6.47. The training set contains two classes: blue ($n = 9$) and green ($n = 10$). The task is to classify the red point. With $k = 11$, the circle shows the set of nearest neighbors. There are four blue points in this set and seven green points. By the majority vote, the red point is assigned to the green group. However, this classification depends on the user-selected parameter $k$. With $k = 2$, only the two nearest blue points will be included in the set of nearest neighbors, and the red point will be assigned to the blue group instead. Smaller values of $k$ are more sensitive to noise and outliers, while larger values can cause assignment to remote, irrelevant classes.

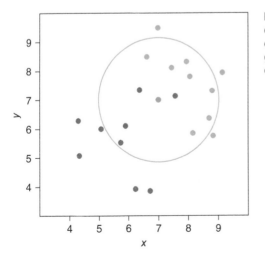

**Figure 6.47** Principle of k-nearest-neighbor classification with $k = 11$. The training set consists of points in two classes, blue and green. The red point is to be assigned to one of the classes.

The classification success can be considerably improved by weighting the points according to distance, such that points further away from the test point count less in the vote. One of the most effective weightings is the linear function (Dudani 1976)

$$w = \frac{d_{max} - d}{d_{max} - d_{min}}$$

where $d_{max}$ and $d_{min}$ are the largest and the smallest distance within the nearest neighbors. With this weighting, and $k = 11$, the test point in Fig. 6.47 is classified as blue instead of green.

Classification success can be evaluated by "jackknifing," i.e., cross-validation within the training set. Each point in the training set is removed in turn, each time classifying it based on the remaining data. The result is compared with the known classification, and the percentage of correct classifications is reported. For the example above, 17 out of 19 points are correctly classified, giving a percentage of 89.5%.

Like most modern supervised ML methods, KNN has the advantage over LDA (section 6.4) in that it can work also when the classes cannot be delineated by flat planes in the data hyperspace. It is therefore sometimes called a nonlinear classification method. In paleontology and archeology, Domínguez-Rodrigo and Baquedano (2018) found KNN to be among the most accurate ML methods for distinguishing butchery marks from crocodile bite marks in bones. Foxon (2021) also found KNN to perform well for a large ammonoid data set with shell coiling parameters, although other methods did slightly better.

## 6.20.2  Naïve Bayes

"Naïve Bayes" is a simple, fast, and flexible classification method. Although it makes some relatively unrealistic statistical assumptions (hence "naïve"), it can perform extremely well in many cases (Hand and Yu 2001). To explain this algorithm, we start with Bayes' theorem from section 2.7:

$$p\big(\theta \,|\, \text{data}\big) = \frac{p\big(\text{data}\,|\,\theta\big)p\big(\theta\big)}{p\big(\text{data}\big)}$$

where $\vartheta$ is the model parameter of interest; $p(\vartheta|\text{data})$ is the probability of the parameter given the data (posterior); $p(\text{data}|\vartheta)$ is the probability of the data given a parameter value (likelihood); $p(\vartheta)$ is the probability of the parameter before the data are collected (prior); and the denominator $p(\text{data})$ is the probability of the data independent of the parameter (evidence).

For the classification problem, the parameter of interest is the class $k$, given the data vector $\mathbf{x}$. Thus, Bayes' theorem takes the form

$$p\big(k \,|\, \mathbf{x}\big) = \frac{p\big(\mathbf{x}\,|\,k\big)p\big(k\big)}{p\big(\mathbf{x}\big)}$$

The class $k$ is usually chosen to maximize the posterior. Since the evidence $p(\mathbf{x})$ does not depend on $k$, it reduces to a constant that we do not need to evaluate. We can also use a "flat" prior $p(k) = 1/C$, where $C$ is the number of classes, or we can use the number of points in each class in the training set to weight the prior. The problem therefore reduces to the estimation of the likelihood $p(x|k)$. To make this estimation tractable, Naïve Bayes makes a bold assumption, namely that the variables are *independent*. This is entirely unrealistic, as all real data sets show correlations between variables, but as mentioned earlier, this is not as detrimental as it may sound. The independence assumption allows us to express the likelihood as a simple product of the likelihoods of the individual variables:

$$p(\mathbf{x}|k) = p(x_1|k) \cdot p(x_2|k) \cdot \ldots \cdot p(x_n|k).$$

Now, each of these likelihoods must be estimated by reference to the probability distribution of the variable in question, within each class $k$. For continuous variables, we can, for example, assume a normal distribution with mean and variance estimated within each class in the training set.

With a normal distribution model, the red test point in Fig. 6.47 is assigned to the green class using Naïve Bayes. The cross-validated classification success for the training set is 95%, as only a single point is misclassified (the blue point to the right of the red in Fig. 6.47 was classified as green).

### 6.20.3   Decision trees and random forests

Decision tree classification (e.g., Rokach and Maimon 2014) is an intuitive and transparent classification method, although its inner workings can be quite complex. In its basic form, a decision tree is a dichotomously branching "flow chart" where each node represents a choice of path depending on the value of a single variable.

For the example from Fig. 6.47, we can construct a decision tree as in Fig. 6.48. The algorithm first selects a single variable (feature) and a splitting point that produces a good classification for the training set. In this case, it selects the variable $x$, with a splitting

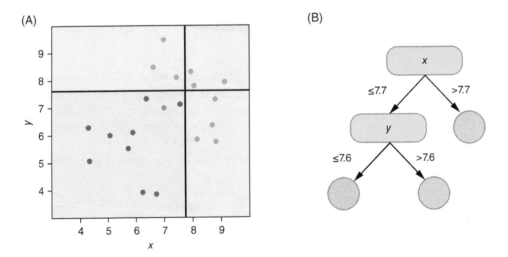

**Figure 6.48**   (A) The data set from Fig. 6.45, as partitioned by a decision tree (B).

point of 7.7. If we stop the construction here, the classifier will assign all points with $x \leq 7.7$ to the blue class, and all points with $x > 7.7$ to the green class. However, this would misclassify the three green points to the upper left. We therefore let the algorithm proceed with the construction of the decision tree, by adding a new node to the $x \leq 7.7$ branch. In this case, the variable $y$ is selected, with a splitting point of 7.6. Within this path, all points with $y \leq 7.6$ are assigned to the blue class, and all points with $x > 7.6$ to the green class. This decision tree will successfully classify all the points in the training set.

The red test point is classified as follows. At the root of the tree (Fig. 6.48B), we inspect the value of $x$, which is 7. This brings us down to the left branch, where we next inspect the value of $y$, which is also 7. This brings us to the node at the lower left of the tree, which selects the blue class.

A decision tree has several attractive properties. First, it is transparent and easy to understand. The decision tree can be inspected to reveal the details of the classifier. This contrasts with "black box" methods such as neural networks (below). Second, the decision tree is robust to outliers and rescaling of the input data, because it is simply based on a dichotomy between "small" and "large" values. Finally, it works well with mixed continuous and nominal data (e.g., sex or species).

One problem with decision trees is that as we increase the depth of the tree to make more detailed classifications, we will start to overfit the class boundaries. This increases the sensitivity to noise. The robustness of the classifier can be greatly improved by bootstrapping. Instead of generating a single decision tree, we resample the training set with replacement and generate a new tree for each bootstrap replicate. This ensemble of trees, poetically called a *random forest*, can smooth out the variance and provide a better classification by majority vote among the trees.

The random forest has become one of the most popular classification techniques. Modern implementations are complex, often combining the simple method described earlier with random resampling of variables (features), random selection of features at each node in the tree, etc. Breiman (2001) is a milestone paper in this field.

Domínguez-Rodrigo and Baquedano (2018) used the random forest successfully for classifying marks on fossil bones, and Foxon (2021) reported good results for their ammonoid data set.

## 6.20.4 Neural networks

Artificial neural networks consist of interconnected nodes called neurons, originally modeled on the presumed function of neurons in the human brain. The topology of these networks, the implementation of the neurons, and the way the networks are trained, vary widely. Here, we will briefly describe a very simple "feedforward" network that can be used for the classification of multivariate data (Fig. 6.49). In contrast with e.g., LDA, such a network can handle nonlinear class boundaries.

In this example, an *input layer* receives the values of the three variables $x$, $y$, and $z$ and distributes them to the next layer in the network, called a *hidden layer* because of its position in the interior of the network. In this case, all inputs are distributed to all the nodes in the hidden layer. Each input to each node in the hidden layer is multiplied by a certain weight value, before all these weighted inputs are added together in each node.

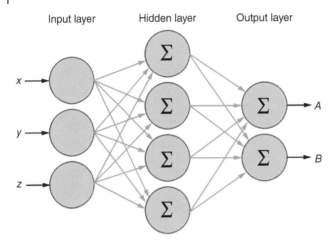

**Input layer**   **Hidden layer**   **Output layer**

$x$

$y$

$z$

$A$

$B$

**Figure 6.49** A simple neural network that can classify continuous data with three variables $(x, y, z)$ into two categories $A$ and $B$. The input layer simply distributes the input data to the hidden layer. The hidden layer and the output layer consist of neurons that sum their inputs, and usually apply a nonlinear threshold function to the sum. A number passing through a red arrow is multiplied with a weight value that is specific to each arrow. These weights are optimized during the training phase.

Moreover, this sum is often passed through a nonlinear function, classically some sigmoidal threshold function, but other functional forms are now also in common use.

In supervised learning, the network is trained with a large data set of $(x, y, z)$ vectors with known class memberships. For each data point in the training set, the desired output ($A = 1, B = 0$ for class $A$ data; $A = 0, B = 1$ for class $B$ data) is compared with the network output, and the weights on all node inputs are updated according to a special algorithm such as "backpropagation." After training, the network can be presented with new data to be classified.

This simple network consists of a single hidden layer. By adding more layers, neural networks can perform better and be used for more complex tasks. The use of multiple layers is referred to as "deep learning."

In the study of Domínguez-Rodrigo and Baquedano (2018), the neural network was one of the best performers for classifying cut and bite marks on fossil bones. In contrast, the neural network was a relatively poor classifier for the ammonoid data set of Foxon (2021).

### 6.20.5 Image classification and convolutional neural networks

Most modern image classification systems use neural networks. For example, free apps such as PlantNet (Affouard et al. 2017) are astonishingly successful in classifying plants to species level, based on mobile phone images taken in the field. Such systems can be trained on millions of images covering thousands of species. So-called convolutional neural networks (CNN) are especially well suited to this task, allowing images to be taken at various angles, scales, and translations, with varying lighting and backgrounds, and generalizing across morphological variation.

Given the success of image classification for living plants and animals, it should be applicable also to fossils. This could provide automatic identification and counting of microfossils and palynomorphs, and also has the potential to generate new taxonomic and phylogenetic hypotheses. So far, the many attempts in this area have been mostly experimental but show considerable promise. For example, Lallensack et al. (2022) trained a CNN on more than 1600 digitized outlines of theropodan and ornithischian dinosaur footprints, producing a classifier that could successfully assign new tracks to the two classes. This can be a difficult task even for human experts. Liu et al. (2022) classified fossil images to one of 50 major animal clades using a training set of more than 415,000 images automatically extracted from the Internet. Hsiang et al. (2019) and Marchant et al. (2020) classified planktonic and benthic foraminiferans to species.

## 6.21  Case study: the ontogeny of a Silurian trilobite

We will end this chapter with a case study concerning the ontogeny of a Silurian trilobite, based on a data set generously provided by Nigel C. Hughes. Hughes and Chapman (1995) collected the positions of 22 landmarks on the dorsal side of the proetide *Aulacopleura konincki* (Fig. 6.50). We will use data from 70 specimens, covering a wide range from meraspid to holaspid growth stages, investigating the data set using methods from general

Figure 6.50   Landmarks defined on the trilobite *A. konincki*. Adapted from Hughes and Chapman (1995).

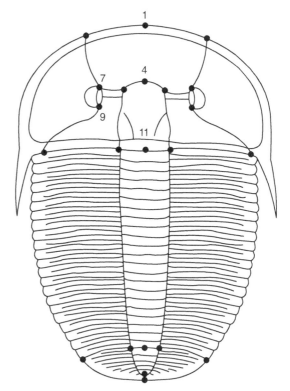

statistics and from morphometrics. Through different visualization methods, we will try to understand and characterize the nature of ontogenetic shape change in this trilobite. Armed with such knowledge for a number of species, it is possible to investigate disparity and evolution through ontogeny. Much more detailed analysis and discussion of this interesting data set are given by Hughes and Chapman (1995), Hughes et al. (1999), and Fusco et al. (2004).

### 6.21.1  Size

As discussed elsewhere in this chapter, size can be defined in many ways. Since this data set consists of landmark positions, we will choose centroid size (section 6.10) as our size measure. Centroid size takes all landmarks into account and is expected to be relatively uncorrelated with shape under the assumption of small, circularly distributed variation in landmark positions (Bookstein 1991).

Figure 6.51 shows a histogram of the centroid sizes. The distribution is clearly non-normal, with a tail to the right, possibly indicating high juvenile mortality.

### 6.21.2  Distance measurements and allometry

In order to search for possible allometries in the cephalon, we use the informal method described in section 6.3. Checking all possible cephalic inter-landmark distances, the analysis indicates that the occipital–glabellar length (landmarks 4–11) is negatively allometric, while the frontal area length (landmarks 1–4) is positively allometric. We therefore plot these two distances against each other in a scatter diagram, as shown in Fig. 6.52. As seen from the statistics, we have a statistically significant allometry between the variates, with allometric coefficient $a = 1.24$. This means that the frontal area is disproportionally long relative to the occipital–glabellar length in large specimens. However, it must be admitted

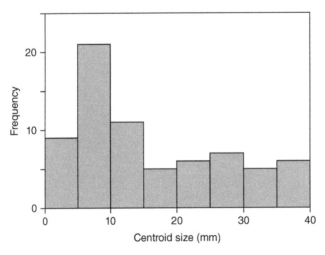

Figure 6.51 Histogram of centroid sizes of 70 specimens of the trilobite *A. konincki.* Data from Hughes and Chapman (1995).

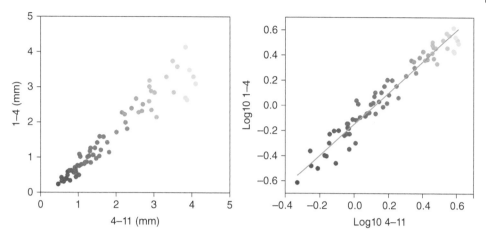

**Figure 6.52** (A) Occipital–glabellar length (4–11) and length of frontal area (1–4). Points are color coded according to centroid size (see below). The point cloud is bending very slightly upwards, indicating a weak positive allometry. The variance of the residual increases with size. (B) Log-transformed measurements with RMA regression line. $r = 0.07$, $a = 1.238$ with standard error 0.035, 95% confidence interval on $a$ is [1.168, 1.309]. The slope (allometric coefficient) is significantly different from 1 ($t$-test, $p < 0.001$).

that the plots in Fig. 6.52 do not show this clearly, and it may be argued that the allometric effect is so small as to be biologically unimportant. We should also remember the multiple testing problem here, as we have deliberately searched for the most allometric distances. In fact, Hughes et al. (1999) refer to the relation between these two variates as isometric.

### 6.21.3 Procrustes fitting of landmarks

Hughes and Chapman (1995) used a robust landmark-fitting method called resistant-fit theta-rho analysis (RFTRA), but we choose normal Procrustes fitting. The color coding in Fig. 6.53 indicates movement of some landmarks as a function of size.

### 6.21.4 Common allometric component analysis

Figure 6.54 shows the scatter plot of specimens resulting from CAC analysis (section 6.17). The CAC has an $R^2$ value of 0.50, which indicates relatively strong covariation between size and shape. The first RSC (RSC 1) explains 24.1% of the remaining variance. The plots of landmark displacements and deformation grids corresponding to CAC and RSC 1 are shown in Fig. 6.55. The CAC is dominated by an elongation of the thorax region, i.e., the addition of thoracic segments during ontogeny. RSC 1 mainly represents a relative widening of the pygidium, producing a more rectangular shape for larger RSC 1 scores.

It is also useful to plot the CAC and RSC 1 scores as a function of centroid size (Fig. 6.56). This shows a somewhat complicated picture. First, the CAC scores increase only for small sizes, before stabilizing at approximately 10 mm. For the smaller sizes, there is a negative correlation between CAC and RSC 1, while for the larger sizes, the RSC 1 scores increase

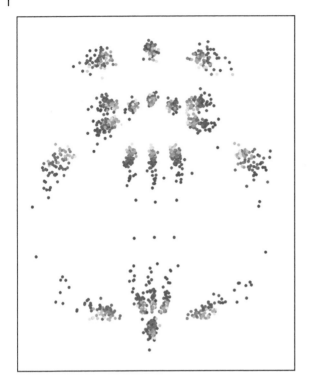

Figure 6.53  Procrustes-fitted landmarks of 70 specimens. Landmarks are colored according to centroid size, from purple (smallest) to yellow (largest).

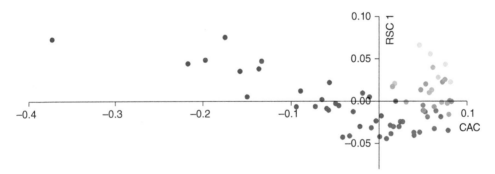

Figure 6.54  Plot of trilobite specimens on the CAC and RSC1 axes. Smaller specimens are purple, and larger are yellow.

again. This indicates two distinct ontogenetic phases – the first dominated by thorax elongation and relative narrowing of the pygidium (captured by the CAC) and then a phase with little continued elongation but instead a reversal of the narrowing of the pygidium. This two-phase ontogeny produces a complicated nonlinear dependency between the CAC and RSC 1. Hughes and Chapman (1995) used this abrupt shift in the ontogenetic trajectory to define the boundary between meraspid and holaspid growth stages.

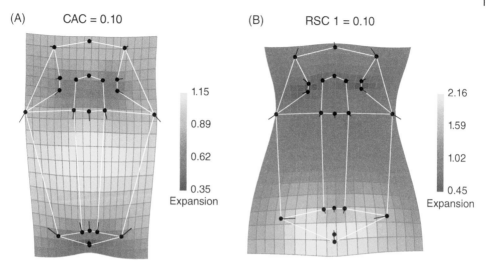

Figure 6.55 Landmark displacements from the mean shape, with TPS deformation grids, corresponding to a common allometric component score of 0.10 (A) and a residual shape component 1 score of 0.10 (B).

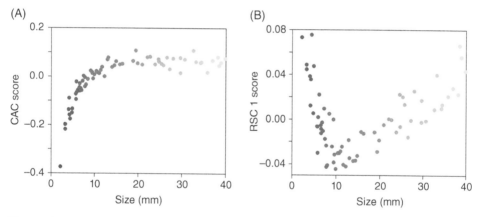

Figure 6.56 (A) CAC scores as a function of centroid size. (B) RSC 1 scores as a function of centroid size. Color coding as above.

## References

Affouard, A., Goëau, H., Bonnet, P., Lombardo, J.-C., Joly, A. 2017. Pl@ntNet app in the era of deep learning. *ICLR: International Conference on Learning Representations, Apr 2017*, Toulon, France.

Aitchison, J. 1986. *The Statistical Analysis of Compositional Data.* Chapman & Hall, New York.

Anderson, T.W. 1984. *An Introduction to Multivariate Statistical Analysis.* 2nd edition. John Wiley, New York.

Benton, M.J., Kirkpatrick, R. 1989. Heterochrony in a fossil reptile: juveniles of the rhynchosaur *Scaphonyx fischeri* from the late Triassic of Brazil. *Palaeontology* 32, 335–353.

Bookstein, F.L. 1984. A statistical method for biological shape comparisons. *Journal of Theoretical Biology* 107, 475–520.

Bookstein, F.L. 1991. *Morphometric Tools for Landmark Data: Geometry and Biology*. Cambridge University Press, New York.

Bookstein, F.L. 2021. Centric allometry: studying growth using landmark data. *Evolutionary Biology* 48, 129–159.

Breiman, L. 2001. Random forests. *Machine Learning* 45, 5–32.

Bruton, D.L., Owen, A.W. 1988. The Norwegian Upper Ordovician illaenid trilobites. *Norsk Geologisk Tidsskrift* 68, 241–258.

Ciampaglio, C. N., Kemp, M., McShea, D.W. 2001. Detecting changes in morphospace occupation patterns in the fossil record: characterization and analysis of measures of disparity. *Paleobiology* 27, 695–715.

Close, R.A., Friedman, M., Lloyd, G.T., Benson, R.B.J. 2015. Evidence for a mid-Jurassic adaptive radiation in mammals. *Current Biology* 25, 2137–2142.

Cristopher, R.A., Waters, J.A. 1974. Fourier analysis as a quantitative descriptor of miosphere shape. *Journal of Paleontology* 48, 697–709.

Crônier, C., Renaud, S., Feist, R., Auffray, J.-C. 1998. Ontogeny of *Trimerocephalus lelievrei* (Trilobita, Phacopida), a representative of the Late Devonian phacopine paedomorphocline: a morphometric approach. *Paleobiology* 24, 359–370.

Davis, J.C. 1986. *Statistics and Data Analysis in Geology*. John Wiley & Sons, Chichester, UK.

Domínguez-Rodrigo, M., Baquedano, E. 2018. Distinguishing butchery cut marks from crocodile bite marks through machine learning methods. *Scientific Reports* 8, 5786.

Dryden, I.L., Mardia, K.V. 2016. *Statistical Shape Analysis*. 2nd edition. John Wiley & Sons, New Jersey.

Dryden, I.L., Walker, G. 1999. Highly resistant regression and object matching. *Biometrics* 55, 820–825.

Dudani, S.A. 1976. The distance-weighted k-nearest-neighbor rule. *IEEE Transactions on Systems, Man, and Cybernetics* 6, 325–327.

Eckart, C., Young, G. 1936. The approximation of one matrix by another of lower rank. *Psychometrika* 1, 211–218.

Ferson, S.F., Rohlf, F.J., Koehn, R.K. 1985. Measuring shape variation of two-dimensional outlines. *Systematic Zoology* 34, 59–68.

Fisher, R.A. 1936. The use of multiple measurements in taxonomic problems. *Annals of Eugenics* 7, 179–188.

Fix, E. Hodges, J.L. 1951. Discriminatory analysis, nonparametric discrimination: consistency properties. *Technical Report No. 4, USAF School of Aviation Medicine, Randolf Field Texas*, 238–247.

Flury, B. 1988. *Common Principal Components and Related Multivariate Models*. John Wiley & Sons, New York.

Foote, M. 1989. Perimeter-based Fourier analysis: a new morphometric method applied to the trilobite cranidium. *Journal of Paleontology* 63, 880–885.

Foote, M. 1992. Rarefaction analysis of morphological and taxonomic diversity. *Paleobiology* 18, 1–16.

Foote, M. 1994. Morphological disparity in Ordovician–Devonian crinoids and the early saturation of morphological space. *Paleobiology* 20, 320–344.

Foote, M. 1997. Sampling, taxonomic description, and our evolving knowledge of morphological diversity. *Paleobiology* 23, 181–206.

Foxon, F. 2021. Ammonoid taxonomy with supervised and unsupervised machine learning algorithms. *PaleorXiv* ewkx9.

Fusco. G., Hughes, N.C., Webster, M., Minelli, A. 2004. Exploring developmental modes in a fossil arthropod: growth and trunk segmentation of the trilobite *Aulacopleura konincki*. *American Naturalist* 163, 167–183.

Golub, G.H., Reinsch, C. 1970. Singular value decomposition and least squares solutions. *Numerical Mathematics* 14, 403–420.

Haines, A.J., Crampton, J.S. 2000. Improvements to the method of Fourier shape analysis as applied in morphometric studies. *Palaeontology* 43, 765–783.

Hammer, Ø. 2004. Allometric field decomposition: an attempt at morphogenetic morphometrics. In Elewa, A.M.T. (ed.), *Morphometrics - Applications in Biology and Paleontology*, pp. 55–65. Springer Verlag.

Hand, D.J., Yu, K. 2001. Idiot's Bayes – not so stupid after all? *International Statistical Review* 69, 385–398.

Hansen, B.B., Bucher, H., Schneebeli-Hermann, E., Hammer, Ø. 2021. The middle Smithian (Early Triassic) ammonoid *Arctoceras blomstrandi*: conch morphology and ornamentation in relation to stratigraphy. *Papers in Palaeontology* 7, 1435–1457.

Hopkins, J.W. 1966. Some considerations in multivariate allometry. *Biometrics* 22, 747–760.

Hopkins, M.J., Pearson, J.K. 2016. Non-linear ontogenetic shape change in *Cryptolithus tesselatus* (Trilobita) using three-dimensional geometric morphometrics. *Palaeontologia Electronica* 19.3.42A.

Hotelling, H. 1933. Analysis of a complex of statistical variables into principal components. *Journal of Educational Psychology* 24, 417–441, 498–520.

Hsiang, A. Y., Brombacher, A., Rillo, M.C., Mleneck-Vautravers, M.J., Conn, S., Lordsmith, S., Jentzen, A., Henehan, M.J., Metcalfe, B., Fenton I.S. 2019. Endless forams: >34,000 modern planktonic foraminiferal images for taxonomic training and automated species recognition using convolutional neural networks. *Paleoceanography and Paleoclimatology* 34, 1157–1177.

Hughes, N.C., Chapman, R.E. 1995. Growth and variation in the Silurian proetide trilobite *Aulacopleura konincki* and its implications for trilobite palaeobiology. *Lethaia* 28, 333–353.

Hughes, C.P., Ingham, J.K., Addison, R. 1975. The morphology, classification and evolution of the Trinucleidae (Trilobita). *Philosophical Transactions of the Royal Society of London B* 272, 537–604.

Hughes, N.C., Chapman, R.E., Adrain, J.M. 1999. The stability of thoracic segmentation in trilobites: a case study in developmental and ecological constraints. *Evolution & Development* 1, 24–35.

Huxley, J.S. 1932. *Problems of Relative Growth*. Mac Veagh, London.

Jackson, J.E. 1991. *A User's Guide to Principal Components*. John Wiley, New York.

Jolicoeur, P. 1963. The multivariate generalization of the allometry equation. *Biometrics* 19, 497–499.

Jolliffe, I.T. 1986. *Principal Component Analysis*. Springer-Verlag.

Jongman, R.H.G., ter Braak, C.J.F., van Tongeren, O.F.R. 1995. *Data analysis in community and landscape ecology*. Cambridge University Press, Cambridge, UK.

Kendall, D.G., Barden, D., Carne, T.K., Le, H. 1999. *Shape and Shape Theory*. John Wiley, New York.

Klingenberg, C.P. 1996. Multivariate allometry. In Marcus, L.F., Corti, M., Loy, A., Naylor, G.J.P., Slice, D.E. (eds.), *Advances in Morphometrics*. *NATO ASI Series A*, 284, 23–50. Plenum Press, New York.

Kowalewski, M., Dyreson, E., Marcot, J.D., Vargas, J.A., Flessa, K.W., Hallmann, D.P. 1997. Phenetic discrimination of biometric simpletons: paleobiological implications of morphospecies in the lingulide brachiopod *Glottidia*. *Paleobiology* 23, 444–469.

Kuhl, F.P., Giardina, C.R. 1982. Elliptic Fourier features of a closed contour. *Computer Graphics and Image Processing* 18, 236–258.

Lallensack, J.N., Romilio, A., Falkingham, P.L. 2022. A machine learning approach for the discrimination of theropod and ornithischian dinosaur tracks. *Interface* 19, 20220588.

Liu, X., Jiang, S., Wu, R., Shu, W., Hou, J., Sun, Y., Sun, J., Chu, D., Wu, Y., Song, H. 2022. Automatic taxonomic identification based on the Fossil Image Dataset (>415,000 images) and deep convolutional neural networks. *Paleobiology* 49, 1–22.

Lohmann, G.P. 1983. Eigenshape analysis of microfossils: a general morphometric method for describing changes in shape. *Mathematical Geology* 15, 659–672.

MacLeod, N. 1999. Generalizing and extending the eigenshape method of shape space visualization and analysis. *Paleobiology* 25, 107–138.

Manabe, M. 1994. Convergence and innovations in aquatic adaptation in Ichthyosauria. PhD thesis, University of Bristol, UK.

Marchant, R., Tetard, M., Pratiwi, A., Adebayo, M., de Garidel-Thoron, T. 2020. Automated analysis of foraminifera fossil records by image classification using a convolutional neural network. *Journal of Micropalaeontology* 39, 183–202.

McGhee, G.R. Jr. 1999. *Theoretical Morphology – The Concept and its Applications*. Columbia University Press, New York.

Mitteroecker, P., Gunz, P., Bernhard, M., Schaefer, K., Bookstein, F.L. 2004. Comparison of cranial ontogenetic trajectories among great apes and humans. *Journal of Human Evolution* 46, 679–698.

O'Higgins, P. 1989. A morphometric study of cranial shape in the Hominoidea. PhD thesis, University of Leeds.

O'Higgins, P., Dryden, I.L. 1993. Sexual dimorphism in hominoids: further studies of craniofacial shape differences in *Pan, Gorilla, Pongo*. *Journal of Human Evolution* 24, 183–205.

Press, W.H., Teukolsky, S.A., Vetterling, W.T., Flannery, B.P. 1992. *Numerical Recipes in C*. Cambridge University Press, Cambridge, UK.

Raup, D.M. 1966. Geometric analysis of shell coiling: general problems. *Journal of Paleontology* 40, 1178–1190.

Raup, D.M. 1967. Geometric analysis of shell coiling: coiling in ammonoids. *Journal of Paleontology* 41, 43–65.

Reeves, J.C., Moon, B.C., Benton, M.J., Stubbs, T.L. 2021. Evolution of ecospace occupancy by Mesozoic marine tetrapods. *Palaeontology* 64, 31–49.

Reyment, R.A. 1991. *Multidimensional Palaeobiology*. Pergamon Press, Oxford, UK.

Reyment, R.A., Jöreskog, K.G. 1993. *Applied Factor Analysis in the Natural Sciences*. Cambridge University Press, Cambridge, UK.

Reyment, R.A., Savazzi, E. 1999. *Aspects of Multivariate Statistical Analysis in Geology*. Elsevier, Amsterdam.

Rohlf, F.J. 1986. Relationships among eigenshape analysis, Fourier analysis and analysis of coordinates. *Mathematical Geology* 18, 845–854.

Rohlf, F.J., Bookstein, F.L. 2003. Computing the uniform component of shape variation. *Systematic Biology* 52, 66–69.

Rohlf, F.J., Marcus, L.F. 1993. A revolution in morphometrics. *Trends in Ecology and Evolution* 8, 129–132.

Rokach, L., Maimon, O. 2014. *Data Mining with Decision Trees*. World Scientific, Singapore.

Seber, G.A.F. 1984. *Multivariate Observations*. John Wiley, New York.

Siegel, A.F., Benson, R.H. 1982. A robust comparison of biological shapes. *Biometrics* 38, 341–350.

Slice, D.E. 1996. Three-dimensional generalized resistant fitting and the comparison of least-squares and resistant-fit residuals. In Marcus, L.F., Corti, M., Loy, A., Naylor, G.J.P., Slice, D.E. (eds.), *Advances in Morphometrics. NATO ASI Series A* 284, 179–199. Plenum Press, New York.

Slice, D.E. 2000. The geometry of landmarks aligned by generalized Procrustes analysis. *American Journal of Physical Anthropology* 114, 283–284.

Swan, A.R.H., Sandilands, M. 1995. *Introduction to Geological Data Analysis*. Blackwell Science.

Thompson, D.W.T. 1917. *On Growth and Form*. Cambridge University Press.

Webster, M., Hughes, N.C. 1999. Compaction-related deformation in well-preserved Cambrian olenelloid trilobites and its implications for fossil morphometry. *Journal of Paleontology* 73, 355–371.

Wills, M.A., Briggs, D.E.G., Fortey, R.A. 1994. Disparity as an evolutionary index: a comparison of Cambrian and recent arthropods. *Paleobiology* 20, 93–130.

Zahn, C.T., Roskies, R.Z. 1972. Fourier descriptors for plane closed curves. *IEEE Transactions, Computers C-21*, 269–281.

# 7

# Directional and spatial data analysis

The orientations and the spatial distribution of fossils on bedding planes (in 2D) or in rock volumes (in 3D) can give valuable information about paleoecology and taphonomy. Shells and bones can be oriented by currents; burrows can be oriented with varying inclinations depending on oxygenation and water energy; dinosaur footprints may indicate the main direction of movement; the distribution of trees in a fossil forest or the distribution of in situ brachiopods on a bedding plane can say something about ecological interactions. In paleontology, we must use whatever data we can get hold of, and directional and spatial data are often relatively easy to collect, especially in the field.

## 7.1 Analysis of directions and orientations in 2D

Directional (or circular) data can take several different forms. Mathematicians prefer *radians*, which are measured counterclockwise from the East, and with a full circle corresponding to $2\pi$ radians. The full circle also corresponds to 360 degrees, or 400 *gradians* or *gons* (rarely used). Degrees and gradians can be measured counterclockwise from the East (the mathematical convention), or clockwise from the North (the geographical, or compass, convention).

Also, we can differentiate between *directions*, which vary from 0° to 360°, and *orientations*, varying from 0° to 180°. Directions are appropriate for vector-like objects with a front and a rear end, such as trilobites or dinosaur footprints. Orientations ("axial" data) can be used for elongated objects with no differentiation between front and back, such as crinoid stems and cylindrical trace fossils.

A peculiar property of circular measurements is that they repeat every full circle. This means that 0° is closer to 350° than to 20°, for example. The usual statistical methods do not work for circular data, and special methods have been devised.

### 7.1.1 Plotting circular data

Faced with a sample of directional or axial data, the first thing to do is to plot. In Fig. 7.1, we show different ways of plotting circular data, in this case directions from umbo to commissure for 33 brachiopods (*Protatrypa*). In a *circular arrow plot*, each measurement is

*Paleontological Data Analysis*, Second Edition. Øyvind Hammer and David A.T. Harper.
© 2024 John Wiley & Sons Ltd. Published 2024 by John Wiley & Sons Ltd.

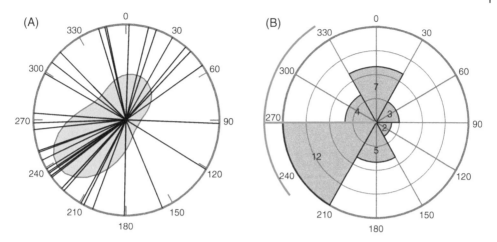

**Figure 7.1** Directions of *Protatrypa* brachiopods, Silurian, Norway. (A) Arrow and kernel density plot. (B) Rose plot with mean direction and its bootstrapped confidence interval (red).

shown as a line from the center of the circle. This is useful for relatively small data sets but becomes cluttered for $n > 50$ or so. A *circular kernel density plot* can be useful for summarizing the distribution. Finally, a *circular histogram* (often called a *rose plot*) can give a good representation of the distribution for larger samples.

In the rose plot (Fig. 7.1B), we have also shown the mean angle with a bootstrapped confidence interval. The circular mean is calculated as

$$\bar{\theta} = \tan^{-1} \frac{\Sigma \sin \theta_i}{\Sigma \cos \theta_i}$$

where the inverse tangent is taken to the correct quadrant.

In a rose plot, the number of points in each histogram bin is usually indicated by the radius of the sector. This means that the area of a sector is proportional to the square of the number of points, which can give an inaccurate visual impression. Some plotting packages, therefore, include an option to instead make sector areas directly proportional to the number of points.

## 7.1.2 Testing for preferred direction

In paleontology and sedimentology, we are often interested in testing whether a sample of fossils or other features on a bedding plane shows some degree of preferred direction. We will discuss two statistical tests for this purpose: Rayleigh's test and Rao's spacing test.

Rayleigh's test (Mardia 1972; Swan and Sandilands 1995) has the following null and alternative hypotheses:

$H_0$: The directions are uniformly distributed (all directions are equally likely).
$H_1$: There is a single preferred direction.

In other words, a small $p$ value signifies a statistically significant preference for one direction. The Rayleigh test assumes a *von Mises* distribution, which can be regarded as a circular version of the normal distribution. The test is therefore not really appropriate for samples with bimodal or multimodal distributions. In the same way as statistical testing for normal distribution (section 4.2), we can test for von Mises distribution with, e.g., Watson's $U^2$ test (Lockhart and Stephens 1985).

The test statistic for Rayleigh's test is $R$, the mean resultant length. This is the length of the sum of all data vectors, divided by sample size:

$$R = \frac{1}{n}\sqrt{\left(\Sigma\cos\theta_i\right)^2 + \left(\Sigma\sin\theta_i\right)^2}$$

Rayleigh's $R$ will vary from around 0 for a uniform distribution to 1 when all directions are equal. Formulas for estimating the significance ($p$ value) are given by Mardia (1972).

Rao's spacing test (Batschelet 1981) for uniform distribution has a test statistic

$$U = \frac{1}{2}\sum_{i=1}^{n}\left|T_i - \lambda\right|$$

where $\lambda = 360/n$ and $T_i = \theta_{i+1} - \theta_i$ for $i < n$ ($T_n = 360 - \theta_n + \theta_1$). This test is non-parametric and does not assume von Mises distribution. The $p$ value can be estimated by interpolation from the probability tables published by Russell and Levitin (1995), or by Monte Carlo simulation.

---

**Example 7.1**

Returning to the *Protatrypa* data set, we find that Rayleigh's test indicates a preferred orientation (Rayleigh's $R = 0.31$, $p = 0.042$). However, Rao's spacing test reports $U = 141$, $p = 0.25$, i.e., not significant. This discrepancy between the two tests is vexing. It could partly be due to the assumptions of Rayleigh's test not being met (the distribution is not unimodal), but also it is known that Rao's spacing test generally has low power (Landler et al. 2018). Although there seems to be a tendency for the brachiopods to have a preferred orientation, the statistical tests are somewhat ambiguous.

---

## 7.2 Analysis of directions and orientations in 3D

Analyzing directions in 3D is not very common in paleontology, but occurs, e.g., in the study of trace fossils (burrows) from CT scans (Fig. 8.6). Such data can be given as "trend and plunge," where the trend is the direction in the horizontal plane (azimuth) and the plunge is the angle down from the horizontal. The trend is taken in the general direction of the plunge.

The convention in geology is to plot such 3D directions as stereographic projections to the lower hemisphere. The principle is shown in Fig. 7.2. The data point is situated on the lower half of a sphere, with a given azimuthal angle in the horizontal (trend) and vertical angle from the horizontal (plunge). This point is plotted in 2D at the intersection of a line from the data point to the zenith (north pole) with the equatorial plane. Such a projection is also known as the equal angle, or Wulff projection.

Figure 7.2    Lower hemisphere, equal-angle projection of a 3D orientation (red point) onto the equatorial plane.

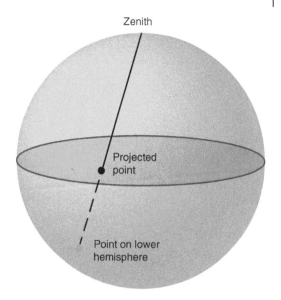

The equal-angle projection tends to spread out points with a large plunge (i.e., close to the south pole of the sphere) and compress points with small plunge. An alternative is the equal-area, or Schmidt, or Lambert projection, which conserves point density.

The Bingham test for uniform distribution of axial data can be used to test for preferred direction (Bingham 1974; Mardia and Jupp 2000, pp. 232–233; Jupp 2001).

---

**Example 7.2**

Gyøry et al. (2018) described peculiar, elongated structures occurring in large numbers in Proterozoic sandstones in Telemark, Norway, ca. 1.2 billion years old (Fig. 7.3). These structures contain carbonate and are up to 25 cm long and were previously described as possibly biogenic.

A wide suite of analysis methods (morphological, geochemical, mineralogical, and magnetic) was applied to these putative fossils, and Gyøry et al. (2018) concluded that an abiotic origin was likely, probably by post-depositional formation of concretions

10 cm

Figure 7.3    Longitudinal section of a Proterozoic carbonate structure from Telemark, Norway. Gyøry et al. (2018).

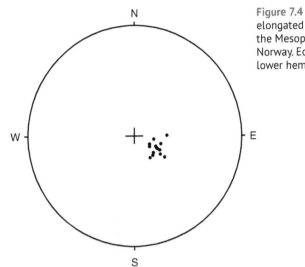

Figure 7.4 Stereo plot of 14 elongated carbonate structures from the Mesoproterozoic of Telemark, Norway. Equal area projection to the lower hemisphere.

during directed groundwater flow. One of the lines of evidence supporting this inter-pretation was the directional analysis of 14 specimens whose trend and plunge were measured in the field. The stereo plot is shown in Fig. 7.4. The 3D directions are strongly concentrated, with a mean trend and plunge of 118° and 71° (the steep plunge is due to the tectonic inclination of the strata). The preferred direction has strong statistical support, with Bingham's $S = 67.6$ and $p < 0.001$ (the $p$ value reported by PAST is $3.7 \cdot 10^{-13}$). Such an extremely small variation in orientation could possibly be due to strong water currents but is more easily explained by directed groundwater flow.

## 7.3   Spatial point pattern analysis

The spatial positions ($x - y$ coordinates) of fossils on a bedding plane can be measured from a digital photograph, or directly in the field using, e.g., a total station or satellite positioning (GNSS, global navigation satellite system). The analysis of such data includes testing for clustering or regular spacing, which can give information about ecological interactions between individuals. Such analysis may also be applied to trace fossils such as shell drill-ings (Rojas et al. 2020). The objects should be small compared with the distances between them, allowing them to be approximated as points.

### 7.3.1   Nearest-neighbor analysis

The null hypothesis of the first test we will describe is that the points are totally randomly and independently positioned – there is no interaction between them. This is sometimes referred to as a *Poisson* pattern, where the counts of points within equally sized boxes should have a so-called Poisson distribution. The human brain easily misinterprets such random patterns, finding structure and clusters of no real significance. Statistical analysis is therefore useful.

One possibility is to use a chi-squared test against the theoretical Poisson distribution, but this involves the somewhat unsatisfactory use of an arbitrary grid over the domain and counting within boxes (quadrats). A more sensitive approach is the so-called *nearest-neighbor* analysis, where the distribution of distances from every point to its nearest neighbor is the basis for the test (Clark and Evans 1954). Given a certain area $A$ and number of points $n$, the expected mean nearest-neighbor distance of a Poisson pattern is

$$\bar{\mu} = \frac{1}{2}\sqrt{\frac{A}{n}}$$

This theoretical mean under the null hypothesis can be compared with the observed mean nearest-neighbor distance $\bar{d}$, to give the so-called *nearest-neighbor statistic*:

$$R = \frac{\bar{d}}{\bar{\mu}}$$

The value of $R$ varies from 0.0 (all points coincide) via 1.0 (Poisson pattern) up to about 2.15 (maximally dispersed points in a regular hexagonal array). Values below 1.0 signify clustering, because the observed mean distance between the points is smaller than expected for a random pattern. Values above 1.0 signify overdispersion, where points seem to stay away from their neighbors (Fig. 7.5).

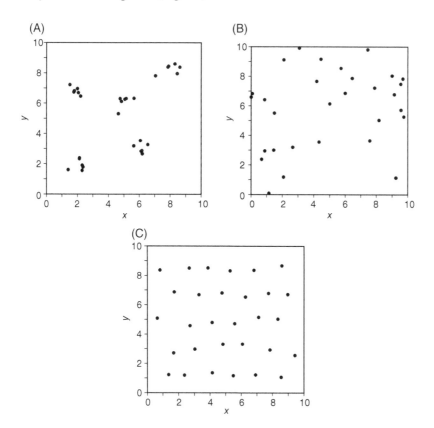

Figure 7.5   Examples of point patterns. (A) Clustered. (B) Poisson pattern. (C) Over dispersed.

---

**Example 7.3**

The cuticula of the Middle Cambrian trilobite *Paradoxides forchhammeri* (Fig. 7.6) is dotted with tiny tubercles. These tubercles are in slightly different positions from specimen to specimen. Figure 7.7 shows the positions of 136 tubercles on a region of the cranidium of a specimen from Krekling, Norway (Hammer 2000).

Figure 7.8 shows a histogram of nearest-neighbor distances, together with the expected Poisson distribution. The observed distances appear to be larger than expected from a random Poisson pattern. The mean nearest-neighbor distance is 0.760 mm, while the expected nearest-neighbor distance for a Poisson pattern is 0.546 mm. The nearest-neighbor statistic is therefore $R = 0.760/0.546 = 1.39$. This indicates an over dispersed point distribution. We used Donnelly's edge correction in this case. The $p$ value (i.e., probability of $R$ obtaining such a large departure from the null hypothesis) is less than 0.001.

This result indicates that although the tubercles are in "arbitrary" positions, they are over dispersed. One possible developmental mechanism for this is so-called lateral inhibition, where the existence of a tubercle hinders the development of new tubercles nearby. A similar mechanism is responsible for the spacing of sensory hairs on the cuticula of modern insects.

**Figure 7.6** The paradoxidid *Paradoxides gracilis* from the Middle Cambrian of Bohemia. Scale bar 1 cm. Courtesy of Lukáš Laibl.

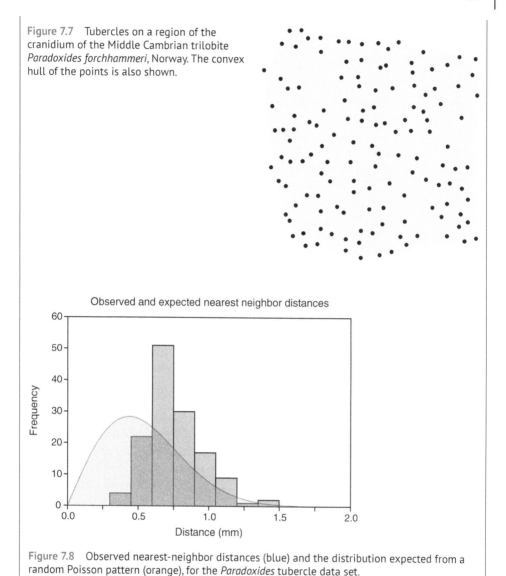

Figure 7.7 Tubercles on a region of the cranidium of the Middle Cambrian trilobite *Paradoxides forchhammeri*, Norway. The convex hull of the points is also shown.

Figure 7.8 Observed nearest-neighbor distances (blue) and the distribution expected from a random Poisson pattern (orange), for the *Paradoxides* tubercle data set.

The formal statistical testing proceeds by taking advantage of the central limit theorem, which lets us assume that the mean values are normally distributed although the underlying distribution of nearest-neighbor distances is exponential. The standard error of the estimate of the mean is then

$$s_e = \frac{\sqrt{(4-\pi)A}}{2\sqrt{\pi}n}$$

The $p$ value can then be estimated by reference to the standardized normal distribution (z test), with

$$z = \frac{\bar{d} - \bar{\mu}}{s_e}$$

One important problem with the nearest-neighbor approach is that points near the edge have fewer neighbors than the others. The distances to their nearest neighbors will therefore on average be a little longer than for the internal points. This *edge effect* makes the statistical test inaccurate. The simplest solution to this problem is to discard the points in a "guard region" around the edge of the domain (they will however be included in the search for nearest neighbors of the interior points). Other corrections, such as Donnelly's correction, are discussed by Davis (1986). Using a Monte Carlo simulation approach for estimating $p$, simulating many random point patterns would also alleviate the edge problem.

Another issue is the estimation of the area within which the points are found. Overestimating this area will give the impression that the points are more clustered than they really are. One objective estimator is the area of the *convex hull*, which is the smallest convex polygon that encloses all the given points (Fig. 7.7). The convex hull may underestimate the area slightly for convex domains, and overestimate it for concave domains.

Finally, the effects of having objects with non-zero size must be considered. Unless the objects can overlap freely, this will necessarily enforce a certain minimal distance to the nearest neighbor, which can influence the result of the analysis.

### 7.3.2 Ripley's K analysis

Nearest-neighbor analysis is simple and transparent, and sufficient for the analysis of many point patterns. However, it only addresses the most local scale. Often, the structure of the pattern varies with scale. Consider, for example, a pattern consisting of separated groups of over dispersed points (Fig. 7.9). In this case, nearest-neighbor analysis reports an over dispersed pattern ($R = 1.18$, $p = 0.004$, Donnelly's edge correction), which misses the clustering at larger scales.

Ripley's $K$ (Ripley 1979) is proportional to point count as a function of distance $d$ from every point, i.e., within circular discs of increasing radius $d$. Define the estimated density (intensity) of the point pattern, with $n$ points in an area $A$, as $\lambda = n/A$. The distance between points $i$ and $j$ is $d_{ij}$. The estimate of Ripley's $K$, as a function of distance, is then computed as

$$K(d) = \frac{1}{\lambda n} \sum_{i=1}^{n} \sum_{j \neq i} I\left(d_{ij} \leq d\right)$$

where the "indicator function" $I$ is one if the argument is true, zero otherwise. Ripley (1979) also suggested an edge correction, giving weights to counts depending on the proportion of the test circle that is inside the rectangular domain.

Figure 7.10A shows Ripley's $K$ for the pattern in Fig. 7.9, together with a 95% confidence interval produced by simulating 1000 random point patterns within a rectangle bounding the points. The $K(d)$ function is defined such that for complete spatial randomness (CSR), $K(d)$ is expected to increase as the area of circles, i.e., $K(d) = \pi d^2$. This is

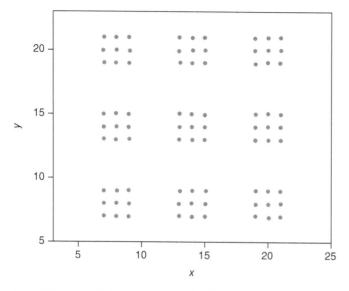

Figure 7.9   A synthetic point pattern that is over dispersed at the local scale, but clustered at larger scales. The clusters are themselves over dispersed at the global scale.

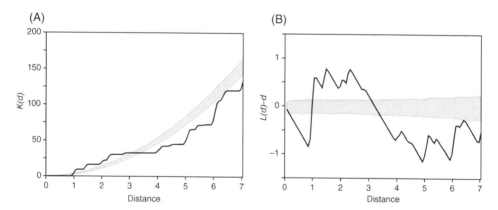

Figure 7.10   Ripley's $K$ analysis of the point pattern in Fig. 7.9, with a 95% confidence interval from Monte Carlo simulation. (A) $K(d)$ is expected to increase as $d^2$ under complete spatial randomness (CSR). (B) $L(d) - d$ has an expectancy of zero under CSR.

shown by the parabolic shape of the confidence interval. The observed $K(d)$ is above the confidence interval for $d$ between ca. 1 and 3, and below it for $d$ larger than ca. 4. This indicates clustering at scales corresponding to circles with a radius around 2, and overdispersion at larger scales.

It is possible to transform $K(d)$ further to make interpretations easier. First, we define a function $L(d)$:

$$L(d) = \sqrt{\frac{K(d)}{\pi}}$$

For CSR, $L(d)$ is expected to increase linearly, as $L(d) = d$. We can therefore plot $L(d) - d$, as shown in Fig. 7.10B, which is expected to have value zero for all $d$. This transformation enhances the features of the curve at the smallest scales, showing fewer points than expected (i.e., overdispersion) below a scale of ca. $d = 1$.

In summary, Ripley's $K$ analysis of the point pattern in Fig. 7.9 indicates overdispersion (points seem to avoid each other) at the smallest scales, $d < 1$, while they are clustered at somewhat larger scales ($d \approx 2$). At the largest scales ($d > 4$), the clusters are themselves over dispersed.

A detail here is that the expected value for $L(d) - d$ under CSR will be zero only if the area is correctly estimated. For more complicated domains than rectangular, the area will usually be over- or underestimated, and this will cause a near-linear trend in $L(d) - d$. The area may be manually adjusted until this trend disappears.

### 7.3.3 Correlation length analysis

While Ripley's $K$ analysis counts all points within discs of increasing radius, giving a cumulative curve, correlation length analysis (e.g., Cartwright and Whitworth 2004; Law et al. 2009; Cartwright et al. 2011) counts points within annular rings of constant width but increasing radius. CLA thus produces a histogram of all pairwise distances between points, i.e., a total of $n(n-1)/2$ distances. Optionally, the curve can be smoothed using a kernel function. Also known as the pair correlation function (PCF), the CLA can be easier to interpret than Ripley's $K$, and in paleontology it has been used, e.g., for the analysis of crinoid attachment discs (Hunter et al. 2020), predatory drilling (Rojas et al. 2020), and trace fossils (Mitchell et al. 2022).

Figure 7.11 shows a correlation length analysis of the point pattern in Figure 7.9. The dominance of pairwise distances $d$ around 1–2 and 6–8 is evident.

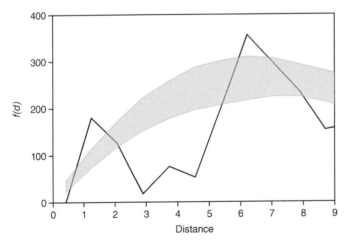

Figure 7.11  Correlation length analysis of the point pattern in Fig. 7.9, with a 95% confidence interval from Monte Carlo simulation. The number of histogram bins was set to 12.

# References

Batschelet, E. 1981. *Circular Statistics in Biology*. Academic Press, London.

Bingham, C. 1974. An antipodally symmetric distribution on the sphere. *Annals of Statistics* 2, 1201–1225.

Cartwright, A., Whitworth, A.P. 2004. The statistical analysis of star clusters. *Monthly Notices of the Royal Astronomical Society* 348, 589–597.

Cartwright, A., Moss, J., Cartwright, J. 2011. New statistical methods for investigating submarine pockmarks. *Computers & Geosciences* 37, 1595–1601.

Clark, P.J., Evans, F.C. 1954. Distance to nearest neighbor as a measure of spatial relationships in populations. *Ecology* 35, 445–453.

Davis, J.C. 1986. *Statistics and Data Analysis in Geology*. John Wiley, Chichester, UK.

Gyøry, E., Hammer, Ø., Jorde, K., Nakrem, H.A., Swajda, M., Domeier, M. 2018. The origin of *Telemarkites* from the Mesoproterozoic of Telemark, Norway. *Norwegian Journal of Geology* 98, 25–39.

Hammer, Ø. 2000. Spatial organisation of tubercles and terrace lines in *Paradoxides forchhammeri* – evidence of lateral inhibition. *Acta Palaeontologica Polonica* 45, 251–270.

Hunter, A.W., Casenove, D., Mayers, C., Mitchell, E.G. 2020. Reconstructing the ecology of a Jurassic pseudoplanktonic raft colony. *Royal Society Open Science* 7, 200142.

Jupp, P.E. 2001. Modifications of the Rayleigh and Bingham tests for uniformity of directions. *Journal of Multivariate Analysis* 77, 1–20.

Landler, L., Ruxton, G.D., Malkemper, E.P. 2018. Circular data in biology: advice for effectively implementing statistical procedures. *Behavioral Ecology and Sociobiology* 72, 128.

Law, R., Illian, J., Burslem, D.F.R.P., Gratzer, G., Gunatilleke, C.V.S., Gunatilleke, I.A.U.N. 2009. Ecological information from spatial patterns of plants: insights from point process theory. *Journal of Ecology* 97, 616–628.

Lockhart, R.A., Stephens, M.A. 1985. Tests of fit for the von Mises distribution. *Biometrika* 72, 647–652.

Mardia, K.V. 1972. *Statistics of Directional Data*. Academic Press, London.

Mardia, K.V., Jupp, P.E. 2000. *Directional Statistics*. John Wiley & Sons.

Mitchell, E.G., Evans, S.D., Chen, Z., Xiao, S. 2022. A new approach for investigating spatial relationships of ichnofossils: a case study of Ediacaran–Cambrian animal traces. *Paleobiology* 48, 557–575.

Ripley, B.D. 1979. Tests of 'randomness' for spatial point patterns. *Journal of the Royal Statistical Society, Ser. B* 41, 368–374.

Rojas, A., Dietl, G.P., Kowalewski, M., Portell, R.W., Hendy, A., Blackburn, J.K. 2020. Spatial point pattern analysis of traces (SPPAT): An approach for visualizing and quantifying site-selectivity patterns of drilling predators. *Paleobiology* 46, 259–271.

Russell, G. S., Levitin, D.J. 1995. An expanded table of probability values for Rao's spacing test. *Communications in Statistics: Simulation and Computation* 24, 879–888.

Swan, A.R.H., Sandilands, M. 1995. *Introduction to Geological Data Analysis*. Blackwell Science, Oxford, UK.

# 8

# Analysis of tomographic and 3D-scan data

Microfocus x-ray tomography (CT) and synchrotron tomography are arguably the most important new technologies advancing paleontology, combining two almost magical abilities: to see *inside* objects and to capture 3D shapes with astonishing accuracy and detail. The resulting images are nothing less than spectacular. However, the processing and analysis of tomographic data can be complex and time-consuming. We have chosen to dedicate a short chapter to this emerging field.

MicroCT and synchrotron scanning have opened up huge possibilities for describing and illustrating fossil morphologies and the function of structures and providing a range of hitherto unobtainable characters for phylogenetic analysis. Applications are wide ranging from understanding the origin of animals in deep time (Cunningham et al. 2015), the evolution of the jaws of vertebrates (Rücklin et al. 2021), and the assembly of enigmatic animals such as the conodonts (Huang et al. 2019) together with the diversification of early flowers (Friis et al. 2019).

## 8.1 The technology of x-ray tomography

A CT machine is basically a glorified x-ray apparatus, producing x-rays that pass through the studied object and then hit a digital detector (x-ray "camera"). In the type of instrument most used in paleontology, the object is rotated very slowly while several thousand x-ray images are collected. From all these images, a complicated "reconstruction" algorithm builds up the 3D shape of the object volume.

In the same way that a 2D digital image consists of pixels, a CT volume file consists of *voxels*. A voxel is a tiny, three-dimensional cube with a given density, shown in grayscale or color. A full CT volume can consist of, say, $2000 \times 2000 \times 2000$, or a total of eight billion voxels. With a 32-bit density value for each voxel, this gives a file size of 32 GB! Obviously, to read, write, display, manipulate, and analyze such an enormous amount of data requires a powerful computer, sophisticated software, and a patient and skilled user.

With all its miraculous powers, CT scanning also has serious limitations that come as a disappointment to many. These problems include limitations on resolution and object size, but the most difficult issue is that the object needs good *x-ray contrast*. If you scan a modern monkey skull, there will be excellent contrast between bone and surrounding air; but, if

*Paleontological Data Analysis*, Second Edition. Øyvind Hammer and David A.T. Harper.
© 2024 John Wiley & Sons Ltd. Published 2024 by John Wiley & Sons Ltd.

you try to scan a calcite fossil inside limestone, there will be no contrast whatsoever, and the fossil will be invisible in x-rays. Typically, you are somewhere in between, and the lack of contrast will need to be counteracted by long scanning times and/or time-consuming post-processing on the computer.

Both contrast and resolution are greatly improved by using a high-flux, monochromatic, and concentrated x-ray source. The *synchrotron*, which is a large particle accelerator, can provide such x-rays, giving sharp, low-noise images even for low-contrast objects. Synchrotron tomography has therefore become popular also in paleontology, despite its high cost. Apart from the often dramatic improvement in quality, synchrotron tomography provides the same type of data as traditional CT.

## 8.2 Processing of volume data

Software for 3D data processing and visualization can be complex and expensive. One of the most popular programs is *Avizo*. Avizo is becoming a standard for scientific 3D visualization, editing, and analysis and it is available at many universities. An important concept in Avizo is that it is based on making a *project* with objects connected in a graph (network). Similar software includes VG Studio MAX, Dragonfly, 3D Slicer, and Simpleware.

### 8.2.1 Volumes and surface meshes

As described above, a CT volume dataset contains the density of every voxel. However, in many cases we are primarily interested in the geometry of the outer surface, not the interior details. We can then convert the voxel data to a *surface mesh* consisting of a large number of connected triangles. Such a polygonal mesh is also produced by 3D scanning methods such as laser scanning, structured light scanning, and photogrammetry.

Surface mesh files can be orders of magnitude smaller than volume files and are, therefore, much more practical for data storage and network transfer. Surface meshes are also required for many purposes such as 3D printing, mechanical simulation (finite element analysis [FEA]), and visualization of shape deformation.

### 8.2.2 Segmentation

Segmentation is the separation of a CT volume into several parts. This is a crucial part of CT post-processing. In Avizo, the parts are called *materials* or *labels*. Segmentation has several purposes. If you have scanned a fossil inside a rock, the contrast can be so poor and the structure so chaotic that a direct 3D visualization does not give an acceptable picture. This is very common. In such cases, a skilled user with great patience and anatomical expertise can mark out the fossil more or less manually. There are many tools available for this in Avizo, including machine learning algorithms that can recognize textures; but, in some cases, it is necessary to draw with a brush tool on every slice. This is tedious work. In the case of fossils, it is quite similar to mechanical preparation and is sometimes referred to as *virtual preparation*.

Another use of segmentation is to make attractive pictures where different parts have different colors. Finally, segmentation is usually necessary in order to make measurements of volumes (Fig. 8.1), directions, grain size distributions, porosity, etc.

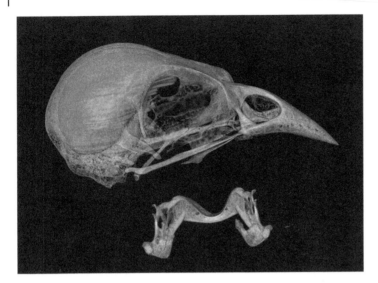

Figure 8.1 Measuring the volume of the braincase in a sparrow requires segmentation of the CT scan. Image by Marina de la Cámara Peña.

### 8.2.3 Landmarks from CT data

CT scans are well suited to morphometric analysis, especially the digitization of 3D landmarks. Software such as Avizo provides efficient tools for placing and editing landmarks on the 3D model and exporting the coordinates in text format (Fig. 8.2). By cutting the model, landmarks can even be placed on internal structures.

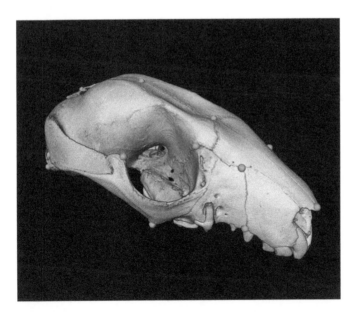

Figure 8.2 CT scan of a skull of the northern brushtail possum (*Trichosurus arnhemensis*), with landmarks.

### 8.2.4 Analysis of volume data

After segmentation, the parts can be subjected to data analysis of different kinds. Avizo and other software can extract parameters such as volume, 3D position, and 3D orientation of the longest axis, for each part, and these numbers can be exported to PAST, R, or Python for statistical analysis.

---

**Example 8.1   Size distribution of Triassic algae**

Our first example of the analysis of volume data concerns a siltstone sample from the Botneheia Formation, Middle Triassic, Spitsbergen, Arctic Norway. The rock contains a large number of unicellular algae (*Tasmanites*), shaped somewhat like smarties, with typical diameters of 0.3–0.4 mm (Vigran et al. 2008). A CT slice is shown in Fig. 8.3. The algae have lower x-ray density than the rock, thus appearing darker.

The algae were segmented out by a thresholding operation, selecting all dark voxels. The exterior region (air) is also dark. It was excluded using a particular function in the Avizo software called "ambient occlusion." In addition, we used automatic functions to filter out very small parts, which were assumed to represent fragmented specimens or other sediment porosity, and very large parts, representing diagenetically fused specimens. Figure 8.4 shows the result of the segmentation, where each alga has been labeled with a separate color. More than 15,000 specimens were identified.

**Figure 8.3**   CT slice through a rock sample with *Tasmanites* algae, Triassic, Spitsbergen. The sample is ca. 5 cm across.

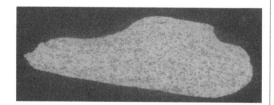

**Figure 8.4**   Results of segmentation and labeling of the siltstone sample, with more than 15,000 specimens of the alga *Tasmanites*.

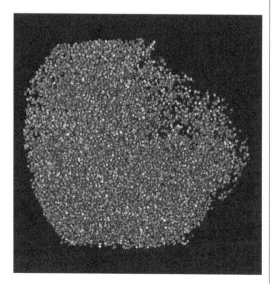

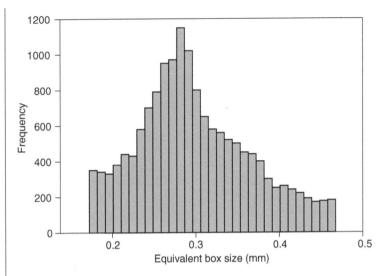

Figure 8.5  Histogram of the sizes of the *Tasmanites* fossils.

Finally, the volumes of all the specimens were measured automatically with Avizo. To get linear rather than volumetric measurements, we calculated the cubic roots of all volumes. This gives the lengths of the edges of cubes of the given volumes (Fig. 8.5).

**Example 8.2    Orientations of Jurassic trace fossils**

Koevoets et al. (2019) studied fossils in cores from the Agardhfjellet Formation, Middle Jurassic to Lower Cretaceous, in Spitsbergen. This formation was deposited in a marine shelf setting. Figure 8.6 shows a CT scan of a 10-cm-long core interval in the Oppdalen Member, of Bathonian to Oxfordian age. Abundant pyritized trace fossils penetrate the core, giving rise to a dense ichnofabric. Koevoets et al. (2019) assigned these traces to the ichnotaxon *Trichichnus*. Kędzierski et al. (2015) suggested that *Trichichnus* represents fossilized filaments from sulfide-oxidizing bacteria within the sediment.

   The orientation of burrows and other elongated trace fossils can say something about the oxygenation in the sediment, as vertical burrowers may be hindered by low oxygen levels. Avizo can extract a number of morphological parameters from segmented CT data, such as volume, elongation, orientation, and position of the objects. Figure 8.7 shows a stereo plot (section 7.2) of the orientations of the traces. There is no preferred orientation in the horizontal plane (Rayleigh's $R = 0.05$, $p = 0.61$; section 7.1), but the traces are predominantly sub-horizontal (mean plunge 25°).

(A)            (B)

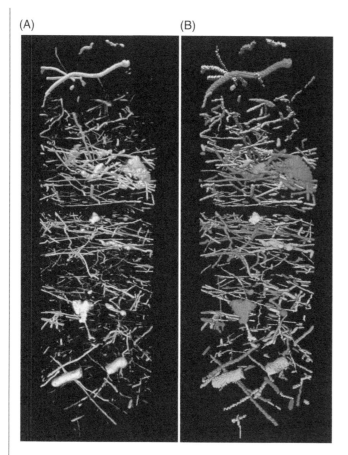

Figure 8.6   (A) CT scan of the DH2 core from Spitsbergen, core depth 724.24–724.34 m. (B) Automatically segmented and labeled with the Avizo software.

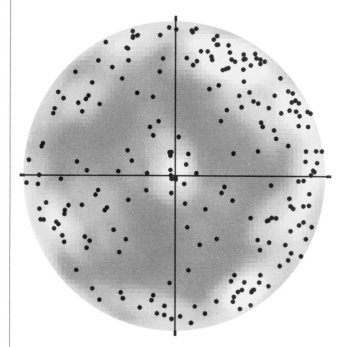

Figure 8.7   Equal-angle stereo plot of the trace fossils in the core sample (Fig. 8.6), with colors for point density.

## 8.3   Functional morphology with 3D data

Tomographic or other 3D scanning methods can provide geometries for physics simulations. This can be very useful for understanding the functional morphology of extinct organisms, their behavior, feeding, and locomotion. One of the most powerful and general of these mathematical modeling methods is the finite element method (FEM). The application of FEM is also known as Finite Element Analysis (FEA). FEM can solve differential equations in complex geometries. In paleontology, it has been used mostly for the structural analysis of skeletal elements, but FEM and other numerical methods are also powerful tools for simulating fluid dynamics around organisms in water and air.

### 8.3.1   Structural analysis – stresses and strains

The forces and deformations in mechanical structures subjected to external forces is a well-developed area of classical mechanics. Fundamentally, the physics in such systems is governed by partial differential equations (PDEs), which can be difficult to solve even in simple geometries. Usually, the problem is tackled by *discretizing*, meaning that the model geometry is divided into a very large number of very small areas or volumes, allowing a finite approximation to the PDEs that can be solved by numerical methods. The approximation will typically be more accurate with smaller elements. One of the most versatile of such methods is the FEM (Zienkiewicz et al. 2005), which allows great flexibility in complex geometries. FEM is mathematically rather complicated and computationally intensive; but, modern software has made its application (FEA) fairly straightforward. However, a good understanding of the basic physics is still required.

FEM is now the method of choice for estimating forces and deformations in physical objects. This is one of the cornerstones of modern engineering and is used for the design and analysis of e.g., buildings, bridges, cars, airplanes, engine parts, and medical implants. The question is usually how we can design a structure that does not break, with minimal use of material, and under functional constraints. Exactly the same challenge applies to biological systems, and FEA is therefore a very relevant tool for analyzing mechanical structure in organisms. In paleontology, FEA has become popular, especially for the study of skeletal elements, particularly in vertebrates but also in invertebrates (Bright 2014, and references therein). Although 2D FEA is possible and can be useful, the analysis is now usually carried out in full 3D.

The first step in the procedure is to acquire the basic 3D shape. This can be done with surface scanning techniques such as photogrammetry or laser scanning. However, this will not provide internal structures that may be crucial for providing an accurate physical model, such as the distribution of cortical and trabecular bone material (which have very different mechanical properties). Tomographic data are therefore usually preferable for the FEA of biological structures. The next step is *meshing*, which is the generation of the 3D elements used for the FEM computations. These are typically tetrahedral elements, connected at their corners (nodes). Meshing can be a complex procedure (Young et al. 2008). The geometry should be captured accurately by the mesh; it must have sufficiently small elements for the FEM to converge to an accurate solution (often, elements are made smaller in regions where higher accuracy is required); and material properties such as Young's modulus, Poisson ratio, and anisotropy may vary over the mesh and must be specified.

Next, the external forces and other physical constraints (boundary conditions) must be imposed. This can also be a complex matter. In particular, the attachment of muscles to bone can be difficult to specify and is often uncertain for extinct animals. For modeling biting, the position of the bitten item along the jaw (proximal or distal) is crucial.

The FEM then estimates forces and displacements at all nodes in the mesh. These are usually reported as *stress* and *strain*, respectively. Both stress and strain are multivalued "tensors" with components along each of the coordinate axes and also components describing shear. The stresses are typically visualized with color coding that summarizes the stress components. A popular method for summarizing the components is *von Mises stress*, which can be related to the yield strength, i.e., how close the structure is to fracture. A related strain measure is sometimes called the von Mises strain.

Bicknell et al. (2023) carried out a comprehensive biomechanical analysis of the frontal raptorial appendages of the large Cambrian predatory arthropod *Anomalocaris*. Their study included a kinematic analysis, i.e., investigating the range of movement; an FEA of the stresses and strains imposed when grasping prey (Fig. 8.8); and a computational fluid dynamics (CFD) analysis (see later) of the drag in outstretched and flexed positions. The analysis indicated that the spines (endites) were adapted for grasping soft prey, not for crushing hard exoskeletons and that the appendages were hydrodynamically efficient when outstretched. These results are compatible with a nektonic, agile, predatory lifestyle for *Anomalocaris*, rather than nektobenthic feeding on trilobites and other shelled benthos.

Clearly, many of the inputs to FEA in paleontology are highly uncertain. This includes incomplete knowledge of the bone geometry, uncertain material properties, lack of information about muscle attachments and forces, and poor knowledge of behavior. Bright (2014) discusses such challenges and how they may be partly resolved, including model validation

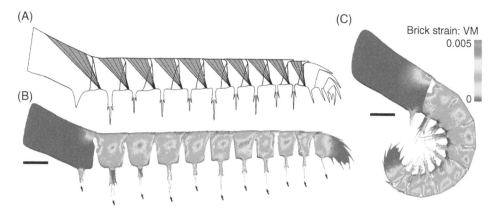

**Figure 8.8** Finite element analysis of the frontal appendage of *Anomalocaris canadensis* when grasping prey (forces applied at the tips of spines). (A) Inferred muscles and muscle attachments. (B, C) "von Mises strains" in outstretched and flexed positions. Bicknell et al. (2023)/The Royal Society/CC BY 4.0.

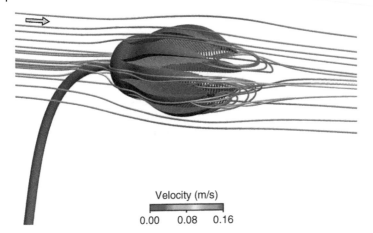

Velocity (m/s)

0.00     0.08     0.16

Figure 8.9   Flow around the Triassic crinoid *Encrinus*, as estimated with CFD. Dynowski et al. (2016)/PLOS/CC BY 4.0.

using biomechanical data from living taxa. Despite the uncertainties, FEA is proving to be a fruitful method in paleontology, perhaps not always producing quantitatively accurate results but at least providing comparative data and possibilities for experimentation and exploration.

### 8.3.2   Computational fluid dynamics

Another exciting use of mathematical modeling in paleontology is CFD, allowing the simulation of water movement past both swimming and sessile organisms, and aerodynamics as applied to flight (Rahman 2017). The computations, based on the Navier-Stokes equations for fluid flow, or simplifications such as Stokes flow, can be very computer intensive, especially for turbulent flow. Although CFD can use FEM, most programs are, instead, based on the so-called finite volume method, which is mathematically distinct from FEM but also works with a polyhedral mesh. Popular programs include the commercial Ansys Fluent package and the free OpenFOAM.

   Interesting applications of CFD in paleontology include the hydrodynamics of swimming ammonoids (Hebdon et al. 2020), the hydrodynamic drag of ichthyosaurs (Gutarra et al. 2019); and filter feeding in Triassic crinoids (Dynowski et al. 2016) (Fig. 8.9).

## References

Bicknell, R.D.C., Schmidt, M., Rahman, I.A., Edgecombe, G.D., Gutarra, S., Daley, A.C., Melzer, R.R., Wroe, S., Paterson, J.R. 2023. Raptorial appendages of the Cambrian apex predator *Anomalocaris canadensis* are built for soft prey and speed. *Proceedings of the Royal Society B* 290, 20230638.

Bright, J.A. 2014. A review of paleontological finite element models and their validity. *Journal of Paleontology* 88, 760–769.

Cunningham, J.A., Vargas, K., Pengju, L., Belivanova, V., Marone, F., Martínez-Pérez, C., Guizar-Sicairos, M., Holler, M., Bengtson, S., Donoghue, P.C.J. 2015. Critical appraisal of tubular putative eumetazoans from the Ediacaran Weng'an Doushantuo biota. *Proceedings of the Royal Society B* 282, 20151169.

Dynowski, J.F., Nebelsick, J.H., Klein, A., Roth-Nebelsick, A. 2016. Computational fluid dynamics analysis of the fossil crinoid *Encrinus liliiformis* (Echinodermata: Crinoidea). *PLoS ONE* 11, e0156408.

Friis, E.M., Crane, P.R., Pedersen, K.R. 2019. *Hedyosmum*-like fossils in the Early Cretaceous diversification of angiosperms. *International Journal of Plant Science* 180, 232–239.

Gutarra, S., Moon, B.C., Rahman, I.A., Palmer, C., Lautenschlager, S., Brimacombe, A.J., Benton, M.J. 2019. Effects of body plan evolution on the hydrodynamic drag and energy requirements of swimming in ichthyosaurs. *Proceedings of the Royal Society B* 286, 20182786.

Hebdon, N., Ritterbush, K.A., Choi, Y.-J. 2020. Computational fluid dynamics modeling of fossil ammonoid shells. *Palaeontologia Electronica* 23, a21.

Huang, J.-Y., Martínez-Pérez, C., Hu, S.-X., Donoghue, P.C.J., Zhang, Q.-Y., Zhou, C.-Y., Wen, W., Benton, M.J., Luo, M., Yao, H.-Z., Zhang, K.-X. 2019. Middle Triassic conodont apparatus architecture revealed by synchrotron X-ray microtomography. *Palaeoworld* 28, 429–440.

Kędzierski, M., Uchman, A., Sawlowicz, Z., Briguglio, A. 2015. Fossilized bioelectric wire – the trace fossil *Trichichnus. Biogeosciences* 12, 2301–2309.

Koevoets, M.J., Hammer, Ø., Little, C.T.S. 2019. Palaeoecology and palaeoenvironments of the Middle Jurassic to lowermost Cretaceous Agardhfjellet Formation (Bathonian–Ryazanian), Spitsbergen, Svalbard. *Norwegian Journal of Geology* 99, 17–40.

Rahman, I.A. 2017. Computational fluid dynamics as a tool for testing functional and ecological hypotheses in fossil taxa. *Palaeontology* 60, 451–459.

Rücklin, M., King, B., Cunningham, J.A., Johanson, Z., Marone, F., Donoghue, P.C.J. 2021. Acanthodian dental development and the origin of gnathostome dentitions. *Nature Ecology & Evolution* 5, 919–926.

Vigran, J.O., Mørk, A., Forsberg, A.W., Weiss, H.M., Weitschat, W. 2008. *Tasmanites* algae – contributors to the Middle Triassic hydrocarbon source rocks of Svalbard and the Barents Shelf. *Polar Research* 27, 360–371.

Young, P.G., Beresford-West, T.B.H., Coward, S.R.L., Notarberardino, B., Walker, B., Abdul-Aziz, A. 2008. An efficient approach to converting three-dimensional image data into highly accurate computational models. *Philosophical Transactions of the Royal Society A* 366, 3155–3173.

Zienkiewicz, O.C., Taylor, R.L., Zhu, J.Z. 2005. *The Finite Element Method: Its Basis and Fundamentals*. 6th edition. Elsevier Butterworth-Heinemann, Oxford, UK.

# 9

# Estimating paleobiodiversity

The concept of biodiversity (a basic measure of which is species richness) is very important both in modern ecology and in paleobiology. Many important discussions in paleontology revolve around biodiversity: How can we recognize mass extinctions in the fossil record as reductions in diversity? What is the pattern of biodiversity recovery after an extinction? Diversity can also possibly be used as an environmental indicator, at least in combination with other information. For example, diversity is typically relatively low in "stressed" environments such as brackish or dysoxic waters. There is also a latitudinal trend in biodiversity, with the highest diversity near the equator (Rosenzweig 1995). Furthermore, diversity can be used to characterize different types of communities. For example, opportunist or pioneer (early succession) communities will typically have relatively low diversity compared with the higher diversities of equilibrium communities.

It must be remembered that biodiversity always applies to a limited area. It has become customary (e.g., Sepkoski 1988; Whittaker et al. 2001) to differentiate between *alpha*, *beta*, and *gamma* diversities. Alpha diversity is the within-habitat diversity, that is, the diversity in a given locality (often represented by a single sample). Beta diversity, or "species turnover," is the variation in taxonomic composition across local (alpha) areas. Gamma diversity is the total diversity over the area of interest, which can be a landscape, a continent, or the Earth as a whole (global diversity). Gamma diversity is a product of alpha and beta diversity. For example, if two continents merge due to continental drift, alpha diversities may stay relatively constant or even increase in some areas due to immigration, but since the distribution of species becomes more homogeneous, beta diversity will decrease and global gamma diversity will also decrease as a result.

Biodiversity may seem an easy parameter to measure – could we not just count the number of species? As it turns out, both the definition and measurement of biodiversity are somewhat ill-defined and rather slippery, and the literature on the subject is vast. Hill (1973) and Magurran (2004) are good starting points, while Pardoe et al. (2021) provide a recent review.

*Paleontological Data Analysis*, Second Edition. Øyvind Hammer and David A.T. Harper.
© 2024 John Wiley & Sons Ltd. Published 2024 by John Wiley & Sons Ltd.

# 9.1 Species richness estimation

Species richness is the fundamental parameter in biodiversity studies and can be defined as the total number of species in an ecosystem. Already at this point, the problems are piling up. First, we must define our species. Even for modern organisms, their taxonomy is in a constant state of flux, and intraspecific variation complicates the delineation of species. Incredibly, the number of species of the modern giraffe varies from one to nine, depending on who you ask! For invertebrates, it can be even worse. On the more practical side, the actual identification of species-defining characters can be uncomfortably subtle. And finally, for paleontological material, all these issues are greatly amplified, as the fossils can be rare, fragmentary, and taphonomically selected. Second, we must define our "ecosystem." What area and what time span are we talking about? It is an almost invariable law in ecology that the number of species increases with the area studied. Similarly, because of migration, the number of species increases with the length of the observed time interval. Once every decade or so, a disoriented, Arctic walrus is spotted near the Oslo harbor, and even, more recently, around the shores of Britain. Should we include the walrus in the marine mammal species richness of the Oslo fjord or indeed the British coastline?

But let us assume that we have our taxonomy worked out, we have delimited our geographic area and our stratigraphic interval, and we have accepted that not all species will be preserved as fossils. We venture out and collect a large number ($n$) of fossils and count the number of observed species ($S_{obs}$) in our sample. Now, the next issue is that our limited sample is unlikely to contain all the fossilized species in the population. If we had collected a larger sample, we would most likely find more species, especially rare ones. This can be a serious problem in paleontology, where the sampling size can be difficult to control – usually, we just collect everything we can find! Therefore, we need methods for estimating the total species richness $S \geq S_{obs}$ from our sample, or at least for comparing species richness in samples of varying sizes.

## 9.1.1 Species richness estimation from single-sample abundance data

The direct count of species in a sample will usually be an underestimate of the species richness even of the preserved part of the biota from which the sample was taken. Species richness will generally increase with sample size, and several indices have been proposed to attempt to compensate for this effect. For example, we could mention *Menhinick's richness index* (Menhinick 1964), which is the number of observed taxa divided by the square root of sample size ($S/\sqrt{n}$), and *Margalef's richness index* (Margalef 1958), defined by $MR = (S - 1)/\ln n$. Whether the number of taxa increases more like the square root or the logarithm of sample size will depend on the relative abundances (see section 9.2), so neither of these attempts at compensating for sample size is universally "correct."

A more stringent approach is to apply statistical sampling theory and try to estimate the number of missing taxa based on the number of rare taxa in the sample. Clearly, if there are many rare taxa, we would expect the number of taxa to increase with further sampling. A classical species richness estimator in this tradition is the Chao-1 estimator (Chao 1984):

$$S_{\text{Chao1}} = S + \frac{n-1}{n} \frac{F_1^2}{2F_2}$$

where $F_1$ is the number of "singleton" species (species observed only once in the sample) and $F_2$ is the number of "doubletons" (observed twice). We note that $S_{\text{Chao1}}$ will be larger than the number of observed species $S$ and that the estimated number of species will increase with the square of $F_1$, which indicates the number of very rare species. A more accurate (unbiased) version of the Chao-1 estimator, which also works when $F_2 = 0$, is

$$S_{\text{Chao1}} = S + \frac{n-1}{n}\frac{F_1(F_1 - 1)}{2(F_2 + 1)}$$

For many paleontological collections, the value $F_1$ is probably underestimated. When counting specimens in a sample, we often observe single occurrences of "problematic" fossils that we are not able to identify, and they are therefore disregarded. Using the Chao-1 estimator in such cases would be problematic.

Confidence intervals for Chao-1 can be estimated analytically; see Colwell (2013) for equations.

An "improved" Chao-1 estimator, also considering the number of taxa observed 3 or 4 times, was suggested by Chiu et al. (2014):

$$S_{\text{iChao1}} = S_{\text{Chao1}} + \frac{n-3}{4n}\frac{F_3}{F_4} \times \max\left[F_1 - \frac{n-3}{2(n-1)}\frac{F_2 F_3}{F_4}, 0\right]$$

Finally, we mention the "squares" estimator (Alroy 2018), which was designed to be more accurate than Chao-1 when abundance distributions are even:

$$S_{\text{sq}} = S + \frac{F_1^2}{n^2 - F_1 S}\sum_{i=1}^{S} n_i^2$$

### 9.1.2 Species richness estimation from multiple-sample presence-absence data

So far, we have looked at species richness estimation based on a single sample with abundance data (counts). If we have taken presence-absence data from a number of randomly selected quadrats (areas of standardized size or samples of standardized weight), a whole new class of species richness estimators is available. This approach is popular in ecology but is perhaps less used in paleontology. Richness estimators based on presence-absence data from several samples include Chao-2 (Chao 1987), first-order jackknife (Heltshe and Forrester 1983), second-order jackknife, and bootstrap (Smith and van Belle 1984). Colwell and Coddington (1994) reviewed these and other species richness estimators and found the Chao-2 and second-order jackknife to perform particularly well.

Let $S$ be the total observed number of species; $L$ the number of species that occur in exactly one sample; $M$ the number of species that occur in exactly two samples; and $n$ the number of samples. We then have

$$\text{Chao2} = S + \frac{L^2}{2M}$$

$$\text{Jackknife1} = S + L\frac{n-1}{n}$$

$$\text{Jackknife2} = S + L\frac{2n-3}{n} - M\frac{(n-2)^2}{n^2 - n}$$

$$\text{Bootstrap} = S + \Sigma(1 - p_i)^n$$

where the sum is taken over all species, and $p_i$ is the proportion of samples containing species $i$.

## 9.2 Rarefaction and related methods

Species richness is unfortunately a function of sample size. Let us consider a hypothetical horizon at one locality, containing one million fossils belonging to 50 species. If you collect only one specimen, you can be sure to find only one species. If you collect 100 specimens, you may find perhaps 20 species, while you may need to collect several thousand fossils in order to find all 50 of them. The actual number of species recovered for a certain sample size will partly be determined by chance and partly by the structure of the fossil population. If all species are equally abundant, then chances are good that many species will be recovered, but any particularly rare species are not likely to be found unless the sample size is very large.

### 9.2.1 Classical rarefaction

Given a collected sample containing $N$ specimens, we can study the effect of sample size on species richness by estimating what the species count would have been if we had collected a smaller sample. For example, given a list of abundances from a sample containing $N = 1000$ fossils, what species count would we expect to see if we had collected only $n = 100$ of these specimens? One way to estimate this is to randomly pick 100 specimens from the original sample and count the number of species obtained. This being a random process, the result is likely to be slightly different each time, so we should repeat the procedure many times in order to get an idea about the range of expected values.

Furthermore, we can carry out this procedure for all possible sample sizes $n < N$. This will give us a *rarefaction curve*, showing the expected number of species $S(n)$ as a function of $n$, with standard deviations or confidence intervals (Fig. 9.1). If the rarefaction curve seems to be flattening out for the larger $n$, we may take this as a rough indication that our original, large sample has recovered most of the species. However, it must be remembered that all this is based on the observed abundances in the particular, original sample. If we had collected another sample of size $N$ from the same horizon, the rarefaction curve would have looked a little different. Ideally, $N$ should therefore be large enough for the sample to provide a reasonable estimate of the total population.

We now have a method for comparing species counts in samples of different sizes. The idea is to standardize on your smallest sample size $n$ and use rarefaction on all the other samples to reduce them to this standard size. The rarefied species counts $S(n)$ for a pair of

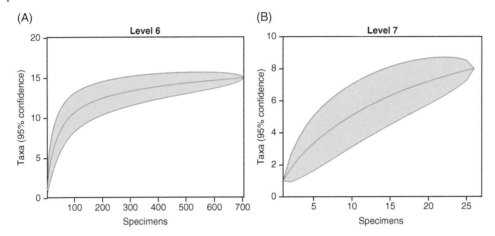

(A)

(B)

**Figure 9.1** Rarefaction curves with 95% confidence intervals for two samples from the Middle Ordovician of Coed Duon, Wales, collected by Williams et al. (1981). (A) Sample 6 ($N = 702$). (B) Sample 7 ($N = 26$).

samples can then be compared statistically using, e.g., a $z$ test with the given standard deviations. This approach is often taken – see, for example, Adrain et al. (2000). Tipper (1979) warns about some possible pitfalls using this technique.

As described above, rarefaction curves can be computed by repeated random selection of specimens from the original sample. This can, however, be very time-consuming, in particular for large $n$ and if we want large numbers of repetitions for accurate estimation of variances. There is fortunately a much quicker way, using a direct, combinatorial formula (Heck et al. 1975; Tipper 1979).

Let $N$ be the total number of individuals in the sample, $s$ the total number of species, and $N_i$ the number of individuals of species number $i$. The expected number of species $E(S_n)$ in a sample of size $n$ and the variance $V(S_n)$ are then given by

$$E(S_n) = \sum_{i=1}^{s} \left[ 1 - \frac{\binom{N-N_i}{n}}{\binom{N}{n}} \right]$$

$$V(S_n) = \sum_{i=1}^{s} \left[ \frac{\binom{N-N_i}{n}}{\binom{N}{n}} \left[ 1 - \frac{\binom{N-N_i}{n}}{\binom{N}{n}} \right] \right] + 2\sum_{j=2}^{s}\sum_{i=1}^{j-1} \left[ \frac{\binom{N-N_i-N_j}{n}}{\binom{N}{n}} - \frac{\binom{N-N_i}{n}\binom{N-N_j}{n}}{\binom{N}{n}\binom{N}{n}} \right]$$

Here, a stack of two numbers inside round brackets denotes the binomial coefficient

$$\binom{N}{n} = \frac{N!}{(N-n)!n!}$$

For small $N$ and $n$, the factorials in the binomial coefficients can be computed by repeated multiplication, but we soon run into problems with arithmetic overflow. The solution to this problem is to use logarithms (Krebs 1989).

---

**Example 9.1**

We return to the Middle Ordovician data set from mid-Wales of Williams et al. (1981) that we used in section 4.10. Horizons 6 and 7 are separated by an unconformity, and we would like to investigate whether the number of taxa has changed substantially across the hiatus. There are 15 taxa in sample 6 and 8 taxa in sample 7. The taxonomic diversity seems to have dropped almost by half across the unconformity, until we notice that sample 6 has 702 fossil specimens while sample 7 has only 26. Clearly, we need to correct for sample size in this case.

The rarefaction curve for sample 6 is shown in Fig. 9.1A, with 95% confidence intervals. The curve flattens out nicely, and it seems that further sampling would not increase the number of taxa substantially. This is as we might hope, since the sample size is much larger than the number of taxa. The curve for sample 7 (Fig. 9.1B) is not as reassuring and indicates that further sampling might have recovered quite a lot more taxa.

To compare the taxon counts using rarefaction, we standardize on the smallest sample size, which is $n = 26$ as dictated by sample 7. From the rarefaction curve of sample 6, we see that the expected mean number of taxa for a sample size of 26 is 6.57 (standard deviation 1.35), compared with the eight taxa in sample 7. This is a small difference, and a permutation test (section 9.6) reports no significant difference in taxon count between the two samples.

---

**Example 9.2**

If we return to the Cambrian Explosion assemblages from Sirius Passet, North Greenland (chapter 3), we can start to look at the biodiversity of the entire fauna (Fig. 9.2). Previous expeditions in the 1980s and 1990s collected material from the scree slopes where the fossils were weathered out of the rock matrix (Fig. 9.3A). These expeditions yielded

(A)

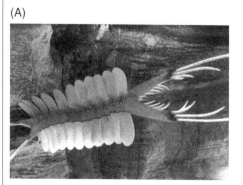

(B)

**Figure 9.2** Two common taxa from the Sirius Passet fauna. (A) *Kerygmachella*. (B) *Pambdelurion*. Models by Esben Horn, 10 tons, Copenhagen.

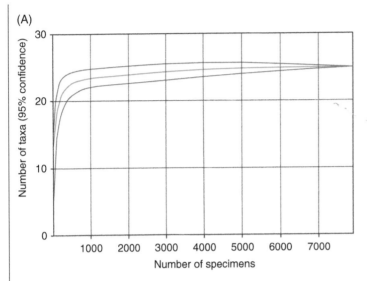

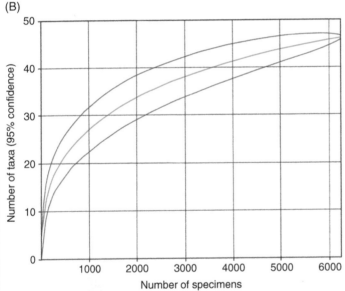

Figure 9.3 Rarefaction curves with 95% confidence intervals for two large collections from the Cambrian Sirius Passet Lagerstätte, North Greenland. (A) Data from scree slope, collected in the 1980s and 1990s, ca. 8000 specimens. (B) In situ, bed-by-bed sampling in 2011, yielding ca. 6000 specimens. Harper et al. (2019)/Geological Society of London/CC BY 4.0.

almost 8000 specimens and some 25 species. The expeditions in 2009 and 2011 collected directly from the rock exposures and yielded successively some 25 species from just over 700 specimens in 2009 and 45 species from nearly 6000 specimens in 2011 (Fig. 9.3B). Clearly, the fauna was underestimated in the screes and reached a limit, for a variety of reasons, but the curves based on material from the exposure are still rising, suggesting more species are yet to be found from this remote Arctic location.

### 9.2.2 Unconditional variance rarefaction

In classic rarefaction, the confidence intervals reduce to zero at the right end of the curve (at $n = N$). This is because when we have recollected all the specimens in the original sample, we will always have recovered the original species count, so the sampling variance is zero. The classical rarefaction variance is called *conditional variance* because it is conditional on the reference sample. While this is entirely logical, it gives the misleading impression of species richness being known exactly at the full sample size.

In contrast, Colwell et al. (2012) described an unconditional rarefaction variance estimate that will not reduce to zero at the end of the rarefaction curve. The computation requires an estimate of total (sampled and unsampled) species richness, such as the Chao1 estimator. This clearly adds some uncertainty to the estimated rarefaction variance.

Figure 9.4 shows unconditional rarefaction curves for the two samples from Coed Duon (compare with Fig. 9.1).

### 9.2.3 Shareholder quorum subsampling

Shareholder quorum subsampling (SQS) was introduced by John Alroy (2010). Chao and Jost (2012) provided analytical solutions for an almost equivalent method, which they called "coverage-based rarefaction." Like classic rarefaction, SQS can be used to standardize species counts across samples of different sizes but standardizing on a fixed "coverage" rather than a fixed sample size. Coverage is defined as the proportion of individuals in the population that are represented by the species recovered in the sample. If all species are recovered in the sample, coverage is 1. If only one species is recovered, but 50% of the individuals in the population belong to this species, coverage is 0.5. SQS gives a fairer sampling of communities with different evenness than classical rarefaction, which suffers from a "compression effect" where differences in richness are artificially dampened. SQS is therefore gradually replacing classic rarefaction in the literature.

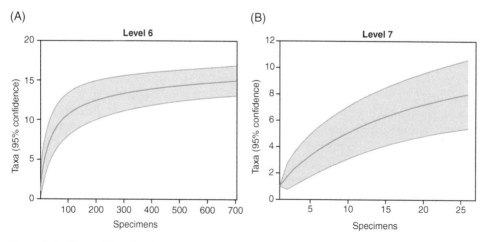

**Figure 9.4** Unconditional variance rarefaction curves with 95% confidence intervals for two samples from the Middle Ordovician of Coed Duon, Wales, collected by Williams et al. (1981). (A) Sample 6 ($N = 702$). (B) Sample 7 ($N = 26$).

For our Middle Ordovician example, sample 6 gives a subsampled species richness (SQS value) of 1.66, while sample 7 gives an SQS value of 1.58, for a coverage of 0.5. These values are clearly much more similar than the raw species richnesses of 15 and 8, respectively.

### 9.2.4  Sample rarefaction

In section 9.1, we discussed the estimation of species richness from multiple samples of presence/absence (incidence) data. It is possible to define a rarefaction procedure where we resample smaller numbers of these quadrats than originally collected, estimating species richness for these smaller data sets. This allows the comparison of species richness across data sets of varying sizes. Colwell et al. (2004) give equations for such sample-based rarefaction curves (called "Mao's tau") with variance estimates.

## 9.3  Diversity curves, origination, and extinction rates

The tabulation of numbers of species, genera, or families through geological time has been an important part of paleontology for the last 50 years. Curves of diversity, origination, and extinction can potentially give information about large-scale processes in the biosphere, including climatic and geochemical change, possible astronomical events (large meteorite impacts, gamma-ray bursts, etc.), and mega-evolutionary patterns and trends.

The important issues of incomplete sampling, preservation bias, chronostratigraphic uncertainty, taxonomical confusion, and choice of taxonomical level (species, genus, family) have been much discussed in the literature (e.g., Smith 2001; Alroy et al. 2008) and will not be treated here. We will instead focus on more technical aspects.

It is common to use the *range-through assumption*, meaning that the taxon is supposed to have been in continuous existence from its first to last appearance. Theoretically, we can then count the number of observed taxa directly as a function of time, such that the diversity may change at any point where one or more taxa originate or disappear (this is known as *running standing diversity*; Cooper and Sadler 2002). However, due to lack of stratigraphic precision, it is more common to count taxa within consecutive, discrete chronostratigraphic intervals, such as stages or biozones. Doing so opens a can of worms when it comes to the definition and calculation of biodiversity. It is obvious that a longer interval will generally contain more taxa in total – there are more taxa in the Mesozoic than in the Cretaceous and more in the Cretaceous than in the Maastrichtian. A possible solution might be to divide the number of taxa by the duration of the interval (if known), but this may not give good results when taxa have long ranges compared with the intervals used. The other side of this coin is that diversity will come out as artificially high when the turnover (origination and extinction) is high in the interval (Fig. 9.6). What we would really like to estimate is the *mean standing diversity S* within each interval, i.e., the average number of taxa at any point in time through the interval.

Alroy (2014) defined the following quantities (Fig. 9.5):

$N_{bt}$: Number of taxa passing through the interval, from below base to above top.
$N_b$: Taxa crossing the base from below but ending within the interval.

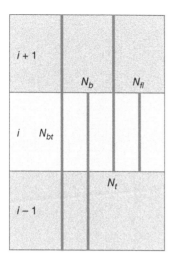

**Figure 9.5** Definitions of terms based on taxon occurrences in an interval *i*, and crossings of base and top. Adapted from Alroy (2014).

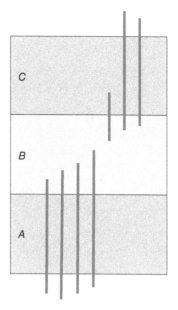

**Figure 9.6** Range chart of seven species. The species count in interval *A* is unquestionably 4. In interval *C* there are altogether three species, but the mean standing diversity is closer to 2.3 due to the one species that disappears in that interval. Interval *B* has a high turnover. The total species count (7) in this interval is much higher than the maximum standing diversity of 4. By letting species that appear or disappear in each interval count as 0.5 units, we get the estimated mean standing diversities of *A* = 4; *B* = 3.5; and *C* = 2.5.

$N_t$: Taxa first found within the interval and crossing the top.

$N_{fl}$: Taxa found only within the interval (first–last).

One early approach (Sepkoski 1975; Foote 2000) was to count taxa that range through the interval as one unit each, while taxa that have their FAD or LAD, but not both, within the interval (single-ended taxa) count as a half unit, i.e., $S = N_{bt} + (N_b + N_t)/2$ (Fig. 9.6). The rationale for this is that if the FADs and LADs are randomly, uniformly distributed within the interval, the average proportion of the interval length occupied by the single-ended taxa is 0.5. This method leaves the taxa confined to the interval ("singletons") behind. It may be reasonable to include also these taxa in the diversity estimate, and it has been suggested (e.g., Hammer 2003) to let them count a third of a unit, based on the expected average proportion of the interval length occupied by such taxa: $S = N_{bt} + (N_b + N_t)/2 + N_{fl}/3$. Doing so will, however, produce some correlation between interval length and estimated diversity, especially when the intervals are long (Foote 2000).

As discussed in section 9.2, diversity increases with the number of specimens. When databases are compiled from the occurrences in many samples, diversity in a given time slice will similarly increase with the number of samples taken in that interval. One should attempt to correct for this, or at least get an idea about the sensitivity of the diversity curve with respect to sampling effort. If abundances are available, direct application of the rarefaction procedure from section 9.2 is one possibility. Another option is to rarefy at the level of *samples*, i.e., to reduce the number of samples in each time slice to a common number (e.g., using Mao's tau, section 9.2). It has been repeatedly shown that such rarefaction can change the conclusions considerably (e.g., Alroy et al. 2001, 2008; Fig. 9.7). Bootstrapping

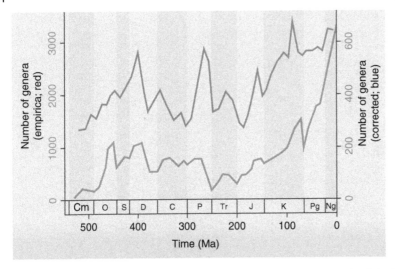

**Figure 9.7** Paleobiodiversity curves based on global data sets such as the Paleobiology Database can appear quite different when corrected for sampling bias. The red curve is the classic unmodified Sepkoski curve; the blue curve was based on drawing collections from a randomly generated set of 65 publications until 16,200 specimens were recovered in each time bin. The Sepkoski curve emphasizes the Great Ordovician Biodiversification Event, the Paleozoic Plateau, the end-Permian Extinction and a subsequent seemingly exponential rise in diversity interrupted by the end-Cretaceous extinction. The corrected curve, on the other hand, shows peaks in the Devonian, Permian, and Cretaceous. Adapted from Benton (2009) and Alroy et al. (2008).

(random resampling with replacement) on the level of samples is another useful procedure for evaluating the robustness of diversity curves (e.g., Hammer 2003).

In summary, the estimation and interpretation of diversity curves are complex, and the results can be controversial. For example, the "Cambrian explosion" and the "great Ordovician biodiversification event" as seen in the Sepkoski curve and many other studies seem to partially break down when comparing several data sets, estimation methods, and taxonomic groups. Instead, a more detailed analysis indicates a continuous diversification trend throughout the early Paleozoic (Servais et al. 2023).

Origination and extinction rates (originations and extinctions per unit of time) are even more complex to estimate. First, it must be understood that they cannot be calculated from the derivative (slope) of diversity: a huge origination rate will not lead to any increase in diversity if balanced by a similar extinction rate. Therefore, frequencies of appearances and disappearances must be calculated separately.

Many measures of origination and extinction rates have been suggested (see Foote 2000 and references therein). In general, it is recommended to normalize for an estimate of standing diversity in order to get a measure of origination/extinction rate per taxon. One reasonably well-behaved estimate (Foote 2000) is the Van Valen metric (Van Valen 1984), which is the number of originations or extinctions within the interval divided by the estimated mean standing diversity as described above. This number could be further divided by interval length to get the origination/extinction rate per unit of time.

According to Foote (2000), a theoretically even more accurate metric is his *estimated per capita rate*. For originations, this is defined as the natural logarithm of the ratio between the number of taxa crossing the upper boundary and the number of taxa ranging through the interval: $R_o = \ln((N_{bt} + N_t)/N_{bt})$. Again, this number could be divided by interval length (discussed by Alroy 2014). For extinctions, the estimated per capita rate is defined as the logarithm of the ratio between the number of taxa crossing the lower boundary and the number of taxa ranging through the interval: $R_e = \ln((N_{bt} + N_b)/N_{bt})$.

The rate equations above are based on counting different types of interval boundary crossings, from the first and last occurrence data. In contrast, Alroy (2008, 2014) proposed equations based on the actual sampled occurrences in three or four consecutive time slices, potentially reducing the bias further. These equations are based on the following definitions (cf. Fig. 9.8):

$t_2^L$: Lower two-timers: Number of taxa sampled in the focal time bin ($i$) and the bin below ($i-1$).
$t_2^U$: Upper two-timers: Taxa sampled in the focal bin and the bin above ($i+1$).
$t_3$: Three-timers: Taxa sampled in three consecutive bins $i-1$, $i$, and $i+1$.
$p$: Part-timers: Taxa sampled below and above a bin but not within it.
$g$: Gap fillers: Taxa found in the bin below ($i-1$) and two bins above ($i+2$), but not in $i+1$. It does not matter if it was found in bin $i$.

The "three-timer" origination and extinction rates (Alroy 2008) are then given as follows:

$$\text{Origination: } \lambda = \ln\frac{t_2^U}{t_3} + \ln s$$

$$\text{Extinction: } \mu = \ln\frac{t_2^L}{t_3} + \ln s$$

Here, $s$ is the per-taxon, per-interval sampling probability, calculated as

$$s = \frac{t_3}{t_3 + p}$$

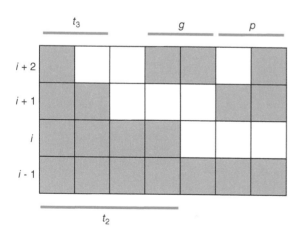

Figure 9.8 Definitions of terms used for Alroy's (2008, 2014) estimators for origination and extinction rates. Here $t_2$ refers to the lower two-timers. Adapted from Alroy (2008, 2014).

where $t_3$ is counted for the bin below $(i-1)$ for origination, and above $(i+1)$ for extinction.

The "gap-filler" extinction rate was suggested by Alroy (2014) to give even more accurate estimates, especially when sampling is poor:

$$\mu = \frac{t_2^L + p}{t_3 + p + g}$$

All this commotion has of course also attracted the Bayesians. A sophisticated framework for estimating evolutionary rates from fossil occurrences has been developed by Silvestro et al. (2014a, b, 2019). This method, implemented in the Python program PyRate, jointly estimates the total life span of each species or lineage (cf. section 13.2), the origination and extinction rates, and the preservation rate through time. As usual, the Bayesian framework produces credible intervals for all these parameters. With simulated data, PyRate produced considerably more accurate evolutionary rates than the traditional methods described above, but this may or may not be the case for real data.

## 9.4 Abundance-based biodiversity indices

Species richness estimators are relatively easy to understand, but this can perhaps not be claimed for most of the other diversity indices, which attempt to incorporate relative abundances of taxa. A central concept is that of *dominance* (or its conceptual inverse, called *evenness*). Consider two communities A and B, both with 10 species and 1000 individuals. In community A, we find exactly 100 individuals belonging to each species, while in community B there are 991 individuals belonging to one species and only one individual in each of the remaining nine species. We would then say that community A has maximal evenness and minimal dominance, while community B has minimal evenness and maximal dominance. We may want to define diversity in such a way that community A gets a higher diversity index than community B. The different indices define evenness in different ways, and the relative contributions from richness and evenness also vary from index to index. Collectively, these indices are referred to as *taxonomic heterogeneity indices*.

The most basic index of dominance is the *Berger-Parker index*, which is simply the number of individuals of the most common taxon divided by the sample size (Magurran 1988). In our example, community A would have a Berger-Parker index of $100/1000 = 0.1$, while community B would have an index of $991/1000 = 0.991$. This index is not totally independent of species richness $S$, because the minimum value of the index (at minimal dominance) is $1/S$. Also, it only considers the abundance of the single most dominant species. Still, the Berger-Parker index is attractive because of its simplicity.

A more commonly used index of dominance is the *Simpson index* (Simpson 1949). The definition of this index has been totally confused in the literature, sometimes being presented as the simple sum of squares of relative abundances, sometimes as its inverse, and sometimes as its complement. This has led to the unfortunate situation that it is no longer sufficient simply to refer to the Simpson index – the mathematical formula must always be given so that the reader knows which version we are talking about. We use the following definitions:

Simpson index of dominance: $D = \sum p_i^2$

Simpson index of diversity: $1 - D = 1 - \sum p_i^2$

where $p_i = n_i/n$ (the proportion of species $i$). Considering the Simpson index of dominance, the most common taxa will clearly contribute disproportionally more to the total sum because the relative frequency of each taxon is squared. The index will be close to 1 if there is a single very dominant taxon. If all taxa are equally common, the Simpson index of dominance will have its minimal value of $\sum (1/S)^2 = 1/S$, like the Berger-Parker index. The Simpson index of dominance indicates the probability that two randomly picked individuals are of the same species.

A popular but more complex diversity index is the Shannon index, or entropy (Shannon and Weaver 1949; Krebs 1989). It is sometimes referred to as the Shannon-Wiener (not Weaver) index:

$$H' = -\sum p_i \ln p_i$$

The "ln" is the logarithm to the base of $e$ (the base of two is sometimes used). Informally, this number indicates the difficulty in predicting the species of the next individual collected.

The lowest possible value of $H'$ is obtained for the case of a single taxon, in which case $H' = 0$. The maximum value is $H_{max} = \ln S$. This means that the Shannon index is dependent not only on the relative abundances, but also on the number of taxa. While this behavior is often desirable, it is also of interest to compute an index of evenness that is normalized for species richness. This can be obtained as $J = H'/H_{max}$, sometimes called (Pielou's) equitability, varying from 0 to 1.

A diversity index called *Fisher's* $\alpha$ (Fisher et al. 1943) is defined by the formula $S = \alpha \ln(1 + n/\alpha)$. The formula does not in itself contain relative abundances, because it is assumed that they are distributed according to a logarithmic abundance model (section 9.7). When such a model does not hold, the Fisher index is not really meaningful, but still, it seems to be an index that performs well in practice.

Most diversity indices rely on the species richness being known, but normally this can only be estimated using the observed number of species (Alatalo 1981; Ludwig and Reynolds 1988). An evenness index that is less sensitive to this problem is the modified Hill's ratio:

$$E5 = \frac{(1/D) - 1}{e^{H'} - 1}$$

Diversity indices must be used with caution. The intricacies of an assemblage composition and structure can hardly be captured by a single number. The biological significance (if any) of a diversity index is not always immediately obvious and must be established in each case. Still, such indices have proven useful for comparison within a stratigraphic sequence or within environmentally or geographically distributed sets of standardized samples.

The calculation of diversity indices is in general straightforward, but a couple of comments should be made.

The Shannon index as calculated from a limited sample according to the formula above represents a *biased estimate* of the diversity of the total population. A precise formula for the expected value is given by Poole (1974), of which the most significant terms are:

$$E(H) = \left(-\sum p_i \ln p_i\right) - \frac{S-1}{2n}$$

$H$ and $E(H)$ will be almost the same for large sample sizes where $n \gg S$.

A similar consideration applies to the Simpson index. An unbiased estimate for the population index based on a limited sample is given by

$$SID = \sum_i \frac{n_i(n_i - 1)}{n(n-1)}$$

Fisher's $\alpha$ was given by an implicit formula above. To solve this simple equation for $\alpha$ symbolically is impossible (the reader is welcome to try!), and so a numerical approach is necessary. In PAST, the value of $\alpha$ is found using the bisection method (Press et al. 1992).

### 9.4.1 Confidence intervals for abundance-based diversity indices

For all the indices above, confidence intervals can be estimated by a bootstrapping procedure: $n$ individuals are picked at random from the sample (or from a pool of several samples), with replacement, and the diversity index is computed. This is repeated say 1000 times, giving a bootstrapped distribution of diversity values. A 95% bootstrap interval can be given as the interval from the lower 2.5% end (percentile) to the upper 2.5% end. This procedure will normally produce confidence intervals biased toward lower diversity values.

Analytical confidence intervals have been derived for some of the abundance-based indices. For the Shannon and Simpson indices ($H$ and $D$), equations for the variances are given in section 9.6. Fisher et al. (1943) gave an equation for the variance of Fisher's $\alpha$. Approximate confidence intervals are then given by, e.g.,

$$H \pm 1.96\sqrt{\text{var}(H)}$$

### 9.4.2 Rarefaction of abundance-based diversity indices

The abundance-based diversity indices described above are generally less sensitive to sample size than simple species counts. Nevertheless, they will normally increase with larger sample sizes, and a procedure similar to classic rarefaction (section 9.2) can sometimes be useful. For most of the diversity indices, such rarefaction must be carried out by "brute force," i.e., by taking large numbers of random subsamples of different sizes and calculating their means and variances. However, Chao et al. (2014) described analytical equations for the rarefaction of the so-called Hill numbers. This is a family of diversity indices, of which the Shannon and Simpson indices are special cases.

**Example 9.3**

Williams et al. (1981) published a database of benthic paleocommunities from the Upper Llanvirn (Middle Ordovician) of Wales. The fauna is a typical representative of the Paleozoic fauna, dominated by suspension-feeding invertebrates such as brachiopods and bryozoans (Fig. 9.9). Table 9.1 shows data from one well-documented section, Coed Duon, with 10 sampled horizons. Samples 1–6 are large (from 156 to 702 specimens), while samples 7–10 are smaller (from 13 to 68 specimens).

The raw taxon count and two adjusted richness indices are given in Fig. 9.10. The raw data (Fig. 9.10A) show a considerable increase in sampled taxa from sample 1 up to sample 6, then returning to low numbers in samples 7–10. However, the sample sizes are small in this upper interval, which is expected to reduce the sampled species richness. The Chao-1 species richness estimator (Fig. 9.10B) shows a similar pattern, but with richness reducing already from sample 5. The Chao-1 estimator suggests much higher numbers of taxa than observed, as high as 27 in sample 4, where the sampled richness was 12.

For the rarefaction analysis (Fig. 9.10C) we disregarded the smallest sample (sample 10), because standardizing on such a small sample size ($N = 13$) would give very large

**Figure 9.9** An Ordovician brachiopod-dominated fauna. The Coed Duon fauna was a variant of this bottom-level type of benthos. McKerrow (1978)/Duckworth.

Table 9.1 The distribution of faunal data through the 10 horizons sampled in Coed Duon Quarry, Wales. Lithologically, there is a transition from ashy siltstones (samples 1–5), through limestone (6), to breccias and conglomerates (7–10) in this Middle Ordovician succession.

| | Hya | Cri | Fen | Pra | Rbr | Tal | Mar | Bas | Ort | Nuc | Gas | Ros | Mac | Sow | Oxo | Tri | Hor | Sal | Dal | Gly | Hes | Sch | Pse |
|---|---|---|---|---|---|---|---|---|---|---|---|---|---|---|---|---|---|---|---|---|---|---|---|
| 10 | 0 | 0 | 0 | 1 | 0 | 0 | 0 | 0 | 0 | 0 | 0 | 0 | 6 | 306 | 0 | 0 | 2 | 2 | 4 | 0 | 0 | 0 | 0 |
| 9 | 0 | 1 | 0 | 0 | 2 | 1 | 0 | 1 | 0 | 0 | 0 | 0 | 6 | 225 | 1 | 0 | 0 | 4 | 2 | 0 | 0 | 0 | 0 |
| 8 | 0 | 0 | 0 | 2 | 4 | 1 | 1 | 1 | 0 | 0 | 0 | 1 | 2 | 288 | 0 | 0 | 0 | 0 | 4 | 0 | 0 | 0 | 0 |
| 7 | 0 | 0 | 1 | 3 | 3 | 10 | 0 | 1 | 0 | 1 | 0 | 1 | 15 | 115 | 0 | 0 | 0 | 1 | 4 | 0 | 1 | 0 | 0 |
| 6 | 3 | 1 | 0 | 4 | 3 | 9 | 1 | 3 | 0 | 0 | 1 | 1 | 13 | 422 | 0 | 0 | 2 | 6 | 4 | 0 | 0 | 0 | 0 |
| 5 | 11 | 37 | 0 | 40 | 18 | 0 | 0 | 330 | 1 | 0 | 0 | 16 | 10 | 190 | 0 | 10 | 4 | 0 | 28 | 0 | 1 | 3 | 3 |
| 4 | 0 | 3 | 0 | 1 | 2 | 0 | 0 | 0 | 0 | 0 | 0 | 0 | 1 | 12 | 0 | 0 | 0 | 0 | 4 | 1 | 2 | 0 | 0 |
| 3 | 0 | 8 | 0 | 1 | 4 | 0 | 0 | 1 | 0 | 0 | 0 | 5 | 4 | 25 | 0 | 1 | 1 | 0 | 15 | 0 | 3 | 0 | 0 |
| 2 | 0 | 3 | 0 | 0 | 5 | 0 | 0 | 1 | 0 | 0 | 0 | 0 | 2 | 36 | 0 | 0 | 0 | 0 | 8 | 0 | 2 | 0 | 0 |
| 1 | 0 | 1 | 0 | 0 | 1 | 0 | 0 | 2 | 0 | 0 | 0 | 0 | 1 | 7 | 0 | 0 | 0 | 0 | 0 | 0 | 1 | 0 | 0 |

Adapted from Williams et al. (1981).

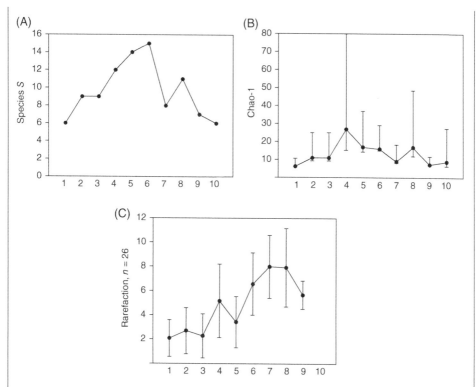

**Figure 9.10** Species richness through the Coed Duon section. (A) Raw taxon counts. (B) Chao-1 richness estimator, with 95% confidence intervals. (C) Unconditional rarefaction to $n = 26$ (see also Fig. 9.2).

confidence intervals. Instead, the rarefaction reduces all samples to $n = 26$, corresponding to the size of sample 7. The rarefaction analysis gives a rather different picture than the raw taxon counts and the Chao-1 estimator, as richness now remains high in the upper interval.

Three different diversity indices are plotted in Fig. 9.11. It is encouraging that the curves support each other quite well – they all indicate that diversity is generally increasing through the section, and there is considerable agreement also in many (but not all) of the details. When the indices correlate this well, it is probably not necessary to present more than one of them in the publication.

In summary, both the species richness, corrected for sample size with rarefaction, and other biodiversity indices show diversity increasing through the section, stabilizing at high levels in the upper part. This is in clear contrast with the raw species richness, which reduces in the upper part.

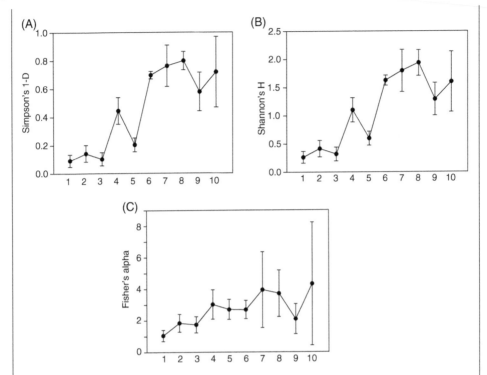

**Figure 9.11** Diversity through the Coed Duon section, with analytical 95% confidence intervals. (A) Simpson index of diversity. (B) Shannon index. (C) Fisher's alpha. All curves correlate well and deviate only in details such as Fisher's alpha not increasing from level 5 to level 6 as the other indices do.

## 9.5 Taxonomic distinctness

Taxonomic diversity and taxonomic distinctness are diversity indices that try to incorporate the taxonomic structure of the community (Warwick and Clarke 1995; Clarke and Warwick 1998). The idea of these indices is that we might like to assign a higher diversity index to a community containing a wide range of higher-level taxa such as families and orders. An interstitial community consisting solely of 1000 nematode species will have high species richness and probably high diversity as calculated with, e.g., the Shannon index (section 9.4), but appears intuitively less diverse than a community with the same number of species and the same relative abundances, but with other groups (copepods, ostracods, etc.) represented as well.

A special advantage of taxonomic distinctness is that it seems to be relatively independent of sample size. This is not the case for most other diversity indices (Clarke and Warwick 1998).

The first step in the procedure is to define a "taxonomic distance" between two taxa (usually species). This can be done as follows. If two species $i$ and $j$ belong to the same genus, they have a distance $w_{ij} = 1$. If they are in different genera but the same family, the distance is $w_{ij} = 2$. If they are in different families but the same order, the distance is $w_{ij} = 3$, etc. Other taxonomic levels can of course be substituted. Contemporary systematicists will undoubtedly shudder at this simplistic definition of taxonomic distance, but we appeal for some pragmatism in this case.

*Taxonomic diversity* is then defined as follows:

$$D_{div} = \frac{\sum\sum_{i<j} w_{ij} n_i n_j}{n(n-1)/2}$$

where $n_i$ is the abundance of species $i$ and $n$ is the total number of individuals as usual.

If all species are of the same genus ($w_{ij} = 1$ for all $i$ and $j$), $D_{div}$ becomes comparable to the Simpson index.

If we want to put less emphasis on abundance distribution, we can divide the taxonomic diversity by a number comparable to Simpson's dominance index. This gives the definition of *taxonomic distinctness*:

$$D_{dis} = \frac{\sum\sum_{i<j} w_{ij} n_i n_j}{\sum\sum_{i<j} n_i n_j}$$

This index is formally appropriate also for transformed abundances (logarithms or square roots). For presence-absence data, taxonomic diversity and taxonomic distinctness reduce to the same number.

---

### Example 9.4

Although the most striking results are likely to be generated by comparisons of taxonomically diverse and morphologically disparate communities, the technique can nevertheless provide useful information when such differences are more subtle. A range of brachiopod-dominated assemblages has been described from the upper slope and outer shelf facies in the Upper Ordovician rocks of the Girvan district, SW Scotland (Harper 2001). Two very distinctive and different cosmopolitan faunas occur in this succession (Fig. 9.12). Near the base of the Upper Ordovician strata, the *Foliomena* brachiopod fauna is a deep-water biofacies dominated by

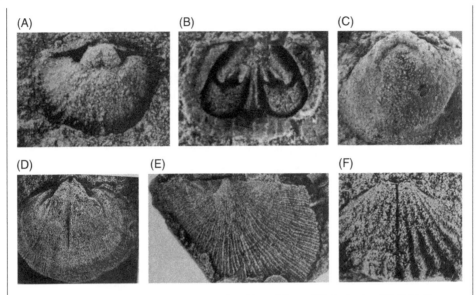

Figure 9.12   Key elements of the *Foliomena* brachiopod fauna ((A) *Dedzetina*; (B) *Nubialba*; (C) *Cyclospira*; most specimens in the order of 5 mm in length) and the *Hirnantia* brachiopod fauna ((D) *Hirnantia*; (E) *Eostropheodonta*; (F) *Rostricelulla*; most specimens in the order of 10–20 mm in length). All specimens from the Upper Ardmillan Group, Girvan.

minute, thin-shelled brachiopods with a widespread distribution. At the top of the succession, the shallower-water *Hirnantia* brachiopod fauna appeared following the first strike of the end Ordovician extinction event and is associated with cooler water habitats. The *Hirnantia* fauna spread to occupy a range of latitudes from the subtropics to the poles. Are both cosmopolitan communities of close-knit groups of similar taxa or are there subtle differences between the composition and structure of the two?

As shown in Table 9.2, data for three localities in the Red Mudstone Member (RMM) with the *Foliomena* fauna are compared with data for two localities in the High Mains Formation with the *Hirnantia* fauna (HMS).

Although the Simpson diversities and taxonomic diversities of the deep-water *Foliomena* faunas (RMM) are somewhat higher than those for the *Hirnantia* faunas (HMS), the taxonomic distinctness is similar or perhaps even slightly lower (Table 9.3), although none of these differences are statistically significant. Life within the cosmopolitan deep-sea communities may have been relatively diverse but inhabited by fewer taxonomic groups than that on the shelf. Moreover, the widespread *Hirnantia* brachiopod faunas may have been drawn from a wider variety of taxa combining to form a disaster fauna in response to the changing environmental conditions associated with the Late Ordovician glaciation.

Table 9.2 Data on the faunas from the Red Mudstone Member, Myoch Formation, and the High Mains Formation, Girvan.

| Species | Genus | Family | Superfamily | Order | RMM-01 | RMM-02 | RMM-03 | HMS-01 | HMS-02 |
|---|---|---|---|---|---|---|---|---|---|
| *albadomus* | Dedzetina | Dalmanellid | Dalmanelloid | Orthide | 7 | 13 | 25 | 0 | 0 |
| *diffidentia* | Neocramatia | Eocramatiid | Davidsonioid | Strophomenide | 1 | 5 | 9 | 0 | 0 |
| *Exigua* | Foliomena | Foliomenid | Strophomenoid | Strophomenide | 3 | 5 | 10 | 0 | 0 |
| *C. sp.* | Christiania | Christianiid | Strophomenoid | Strophomenide | 1 | 5 | 11 | 0 | 0 |
| *inexpecta* | Lingula | Lingulid | Linguloid | Lingulide | 0 | 1 | 1 | 0 | 0 |
| *gracilis* | Sericoidea | Sowerbyellid | Plectambonitoid | Strophomenide | 0 | 2 | 2 | 0 | 0 |
| *C. sp.* | Cyclospira | Dayiid | Dayioid | Atrypide | 0 | 1 | 1 | 0 | 0 |
| *Primadventus* | Sulevorthis | Orthid | Orthoid | Orthide | 0 | 0 | 1 | 0 | 0 |
| *P. sp.* | Petrocrania | Craniid | Cranioid | Craniide | 0 | 0 | 2 | 0 | 0 |
| *actoniae* | Nicolella | Productorthid | Orthoid | Orthide | 0 | 0 | 0 | 0 | 0 |
| *Wattersorum* | Dolerorthis | Hesperorthid | Orthoid | Orthide | 0 | 0 | 0 | 0 | 0 |
| *insularis* | Triplesia | Triplesiid | Triplesioid | Clitambonitide | 0 | 0 | 0 | 0 | 0 |
| *subborealis* | Oxoplecia | Triplesiid | Triplesioid | Clitambonitide | 0 | 0 | 0 | 0 | 0 |
| *magna* | Leptestiina | Leptellinid | Plectambonitoid | Strophomenide | 0 | 0 | 0 | 0 | 0 |
| *E. sp.* | Eoplectodonta | Sowerbyellid | Plectambonitoid | Strophomenide | 0 | 0 | 0 | 0 | 0 |
| *L. sp.* | Leptaena | Leptaenid | Strophomenoid | Strophomenide | 0 | 0 | 0 | 0 | 0 |
| *F. sp.* | Fardenia | Chilidiopsid | Davidsonioid | Strophomenide | 0 | 0 | 0 | 0 | 0 |
| *advena* | Eochonetes | Sowerbyellid | Plectambonitoid | Strophomenide | 0 | 0 | 0 | 3 | 6 |

| | | | | | | | | | |
|---|---|---|---|---|---|---|---|---|---|
| *Comes* | Fardenia | Chilidiopsid | Davidsonioid | Strophomenide | 0 | 0 | 0 | 2 | 1 |
| *Hirnantensis* | Eostropheodonta | Stropheodontid | Strophomenoid | Strophomenide | 0 | 0 | 0 | 17 | 3 |
| *incipiens* | Hindella | Meristellid | Athyridoid | Athyridide | 0 | 0 | 0 | 21 | 113 |
| *G. sp.* | Glyptorthis | Glyptorthid | Orthoid | Orthide | 0 | 0 | 0 | 0 | 1 |
| *Alta* | Plaesiomys | Plaesiomyid | Plectorthoid | Orthide | 0 | 0 | 0 | 0 | 4 |
| *threavensis* | Platystrophia | Platystrophid | Plectorthoid | Orthide | 0 | 0 | 0 | 0 | 2 |
| *sagittifera* | Hirnantia | Schizoporiid | Enteletoid | Orthide | 0 | 0 | 0 | 0 | 11 |
| *matutina* | Eopholidostrophia | Eopholidostrophid | Strophomenoid | Strophomenide | 0 | 0 | 0 | 0 | 1 |
| *R. sp.* | Rostricellula | Trigonorhynchiid | Rhynchonelloid | Rhynchonellide | 0 | 0 | 0 | 0 | 15 |
| *Anticostiensis* | Hypsipthyca | Rhynchotrematid | Rhynchonelloid | Rhynchonellide | 0 | 0 | 0 | 0 | 10 |
| *E. sp.* | Eospirigerina | Atrypid | Atrypoid | Atrypide | 0 | 0 | 0 | 0 | 1 |

Adapted from Harper (2001).

Table 9.3 Simpson diversity (bias corrected), taxonomic diversity, and taxonomic distinctness for the two faunas.

| | RMM-01 | RMM-02 | RMM-03 | HMS-01 | HMS-02 |
|---|---|---|---|---|---|
| Simpson diversity | 0.64 | 0.78 | 0.77 | 0.61 | 0.53 |
| Taxonomic diversity | 3.03 | 3.64 | 3.58 | 2.96 | 2.63 |
| Taxonomic distinctness | 4.76 | 4.66 | 4.66 | 4.84 | 4.94 |

## 9.6 Comparison of diversity indices

Statistical comparison of diversity indices from two or more samples must involve some estimate of how much the index might vary because of incomplete sampling. A mathematical expression for the variance of the Shannon index was given by Hutcheson (1970) and Poole (1974). The Shannon indices of two samples can then be easily compared using a Welch test. A similar test for the Simpson index was given by Brower et al. (1998).

An alternative and more robust approach, which will work for any diversity index, is to perform a permutation test as follows: First, the diversity indices in the two samples $A$ and $B$ are calculated as div($A$) and div($B$). One thousand random pairs of samples $(A_i, B_i)$ are then constructed by random assignment of the individuals to the two samples. For each random pair, the diversity indices div($A_i$) and div($B_i$) are computed. The number of times the difference between div($A_i$) and div($B_i$) exceeds or equals the difference between div($A$) and div($A$) indicates the probability that the observed difference could have occurred by random sampling from one parent population as estimated by the pooled sample.

For comparing more than two samples, the same considerations apply as when comparing the means of several samples. In chapter 4, we noted that it is not entirely appropriate to use pairwise tests, because having many samples implies a large number of pairwise comparisons, producing a danger of spurious detection of inequality. A randomization test for the equality of diversity indices in many samples can be easily constructed.

Hutcheson's (1970) test for the equality of two Shannon indices proceeds as follows. For each of the two samples, calculate the Shannon index $H$, ideally using the unbiased estimate given in section 9.4. Also, compute an approximate variance of each estimate according to the formula

$$\text{var}(H) = \frac{\sum_{i=1}^{S} p_i (\ln p_i)^2 - \left( \sum_{i=1}^{S} p_i \ln p_i \right)^2}{n} + \frac{S-1}{2n^2}$$

Given two samples $A$ and $B$, the $t$ statistic is calculated as

$$t = \frac{H_A - H_B}{\sqrt{\text{var}(H_A) + \text{var}(H_B)}}$$

The degrees of freedom is given by

$$df = \frac{\left( \text{var}(H_A) + \text{var}(H_B) \right)^2}{\left( \text{var}(H_A) \right)^2 / n_A + \left( \text{var}(H_B) \right)^2 / n_B}$$

For the Simpson index, the variance is estimated as (Brower et al. 1998)

$$\text{var}(D) = \frac{4n(n-1)(n-2)\sum p_i^3 + 2n(n-1)\sum p_i^2 - 2n(n-1)(2n-3)\left( \sum p_i^2 \right)^2}{n^2 (n-1)^2}$$

and the test then proceeds as for the Shannon index.

**Example 9.5**

We return to the data set used in section 6.2, from the Middle Ordovician of Coed Duon, Wales. We will compare diversities in samples 1 and 2. Table 9.4 shows some diversity indices and the results of the statistical tests for equal values.

Selecting a significance level of $p < 0.05$, we see that none of the differences in diversity indices are statistically significant. The usefulness of a statistical test is obvious, as, for example, the seemingly large difference in the Shannon index would probably have been overinterpreted without such a test. The large dominance of one taxon makes some of the diversity indices sensitive to random flukes in the sampling because the rarer species can be easily missed.

Hutcheson's (1970) parametric test for equality of the Shannon index reports a $p$ value of 0.12, which is not quite equal to the permutation value of 0.14, but close enough for comfort. At least the two tests agree that the difference is not significant at our chosen significance level. Similarly, the results from the two tests on the Simpson diversity are also similar ($p = 0.18$ and $p = 0.16$).

Table 9.4  Diversity indices for two samples from the Middle Ordovician of Coed Duon, Wales, with statistical comparisons.

| | Sample 1 | Sample 2 | Permutation $p$ | Welch test $p$ |
|---|---|---|---|---|
| Shannon index | 0.256 | 0.400 | 0.14 | 0.12 |
| Simpson diversity | 0.091 | 0.142 | 0.16 | 0.18 |
| Fisher alpha | 1.047 | 1.840 | 0.23 | NA |
| Berger–Parker | 0.953 | 0.926 | 0.21 | NA |

NA = not available.

## 9.7 Abundance models

A mathematical model describing a hypothetical distribution of abundances for different taxa is called an *abundance model*. Several such models and the ecological conditions under which they may hold have been proposed by ecologists. A sample containing many specimens of different taxa will hopefully represent a good estimate of the structure of the population, and we can therefore try to fit the sample to the different abundance models.

The first step in the investigation of the distribution of abundances in a sample is to plot the abundances in descending rank order, with the most abundant species first and the least abundant last. If we also log-transform the abundances, we get a so-called *Whittaker plot* or rank-frequency diagram (Fig. 9.13; Mouillot and Lepretre 2000).

The first abundance model we will consider is the *geometric* model (Motomura 1932). A geometric abundance distribution is characterized by a parameter $k$ in the range 0–1. If the most abundant species has an abundance of $a_1$, then the second most abundant species

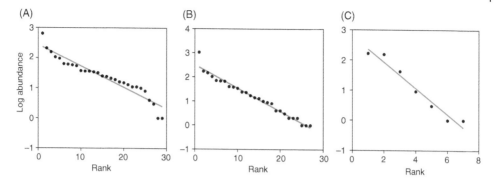

Figure 9.13 Rank-abundance (Whittaker) plots of the three paleocommunities discussed by Peters and Bork (1999). A straight line has been fitted to each of the data sets. (A) Biohermal community. (B) Inter-reef community. (C) Deeper platform community. Adapted from Peters and Bork (1999).

has abundance $a_2 = ka_1$, the third most abundant has abundance $ka_2 = k^2a_1$, and the $i$th most abundant has abundance

$$a_i = ka_{i-1} = k^{i-1}a_1$$

For example, the following set of abundances (sorted by descending rank) follows the geometric model with $k = 0.5$:

$$64, 32, 16, 8, 4, 2, 1$$

Log-transforming the abundances, we get

$$\log a_i = \log a_1 + (i-1)\log k$$

Hence, the log-transformed abundances as a function of the rank $i$ (Whittaker plot) will show a straight line with slope equal to the logarithm of $k$. Since $k < 1$, the slope will be negative. A given sample can therefore be fitted to the geometric model simply by linear regression of the log-transformed abundances as a function of rank number.

The geometric model is sometimes called the *niche preemption model*. Consider a new, empty habitat. Assume that each new species colonizing the habitat is able to grab for itself a certain proportion $k < 1$ of the resources that are left from previous colonists and that the abundance of the species is proportional to the resources it can exploit. For $k = 0.5$, the first species will grab half of the resources, the second will grab half of what is left (1/4), the third will grab half of what is still left (1/8), etc. Each new species will then have an abundance equal to $k$ times the previous one, in accordance with the geometric model. This scenario may seem artificial, but the geometric model does indeed seem to fit communities in early successional stages and in difficult environments.

The *logarithmic series* abundance distribution model was developed by Fisher et al. (1943). The model has two parameters, called $x$ and $\alpha$. The logarithmic series model is normally stated in a somewhat reversed way, with numbers of species as a function of numbers of specimens:

$$f_n = \alpha x^n / n$$

Here, $f_n$ is the number of species counting $n$ specimens. The logarithmic series and the geometric distribution are quite similar in appearance (to the point of being indistinguishable) and are found under similar conditions.

The *log-normal* model (Preston 1962) is perhaps the most popular of the abundance distribution models. It postulates that there are few very common and very rare species, while most species have an intermediary number of individuals. When plotting the number of species as a function of the logarithm of abundance, a normal (Gaussian) distribution will appear. In practice, we usually generate a histogram where the numbers of species are counted within discrete abundance classes. To enforce the log-transformed abundances to be linearly distributed along the horizontal axis in the histogram, the abundance classes need to be exponentially distributed. One possibility is to use classes that successively double in size – in analogy with the frequencies of a musical scale, such abundance classes are referred to as *octaves*:

| Octave | Abundance |
|--------|-----------|
| 1 | 1 |
| 2 | 2–3 |
| 3 | 4–7 |
| 4 | 8–15 |
| 5 | 16–31 |
| 6 | 32–63 |
| 7 | 64–127 |
| ... | |
| $N$ | $2^{n-1}$ to $(2^n - 1)$ |

Many data sets seem to conform relatively well to the log-normal model. It is considered typical for well diversified associations in stable, low-stress, resource-rich environments, with large variation in local habitat (May 1981).

Algorithms for fitting to the logarithmic and log-normal models are described by Krebs (1989).

The *broken stick* model (MacArthur 1957) is also known as the "random niche boundary hypothesis." It does not assume that existing species influence the niche size of new arrivals. If we sort the abundances of the $S$ species in an ascending sequence $N_i$, the broken stick model predicts that

$$N_i = \frac{N}{S}\left(\frac{1}{S} + \frac{1}{S-1} + \frac{1}{S-2} + \cdots + \frac{1}{S-i+1}\right), \; i = 1, 2, \ldots, S$$

*Hubbell's unified neutral theory* (Hubbell 2001) is a simple model of evolution and biogeography that considers the species abundance distribution in one locality and how it changes with time because of deaths, births, and immigration. There are no "niche" effects. Since it makes minimal assumptions, it can be regarded as an ecological null hypothesis. The theory predicts a species abundance distribution in the form of a "zero sum multinomial," which is similar in shape to the log-normal distribution. Hubbell's unified neutral theory has been controversial (e.g., McGill 2003).

A chi-squared metric can be used to evaluate the goodness of fit to most of these models. Hughes (1986) and Magurran (2004) discuss different abundance distribution models in detail.

---

**Example 9.6**

Peters and Bork (1999) studied species abundance distributions in the Silurian (Wenlock) Waldron Shale of Indiana, USA. Three paleocommunities are recognized, occupying different facies: a biohermal, shallow-water community; an inter-reef community; and a low-diversity deeper platform community possibly in oxygen-depleted conditions.

Figure 9.13 shows the rank-abundance plots for the three communities, together with their linear regression lines. The biohermal community is not particularly well described by the straight line and instead clearly shows an S-shaped curve, typical for log-normal distributions. The similarity to a log-normal distribution is confirmed by the octave class plot of Fig. 9.14 and is in accordance with the high heterogeneity of this environment.

The inter-reef community is, however, described rather well by the straight line, in accordance with the geometric or the log-series abundance model. This environment is much more homogeneous than the biohermal one.

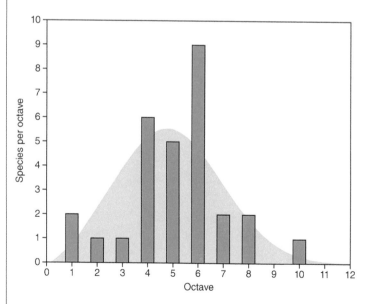

Figure 9.14   Histogram of the numbers of species within power-of-two abundance classes (octaves) for the biohermal community. The distribution has been fitted to a normal distribution (continuous curve).

The deeper platform community is not very well described by the straight line, and it does not look like a log-normal distribution either (the S goes the wrong way!). Peters and Bork (1999) suggested that the broken stick model may be appropriate, but the authors also cautioned against overinterpreting this small data set.

# References

Adrain, J.M., Westrop, S.R., Chatterton, D.E. 2000. Silurian trilobite alpha diversity and the end-Ordovician mass extinction. *Paleobiology* 26, 625–646.

Alatalo, R.V. 1981. Problems in the measurement of evenness in ecology. *Oikos* 37, 199–204.

Alroy, J. 2008. Dynamics of origination and extinction in the marine fossil record. *Proceedings of the National Academy of Sciences USA* 105, 11,536–11,542.

Alroy, J. 2010. Geographical, environmental and intrinsic biotic controls on Phanerozoic marine diversification. *Palaeontology* 53, 1211–1235.

Alroy, J. 2014. Accurate and precise estimates of origination and extinction rates. *Paleobiology* 40, 374–397.

Alroy, J. 2018. Limits to species richness in terrestrial communities. *Ecology Letters* 21, 1781–1789.

Alroy, J., Marshall, C.R., Bambach, R.K., Bezusko, K., Foote, M., Fürsich, F.T., Hansen, T.A., Holland, S.M., Ivany, L.C., Jablonski, D., Jacobs, D.K., Jones, D.C., Kosnik, M.A., Lidgard, S., Low, S., Miller, A.I., Novack-Gottshall, P.M., Olszewski, T.D., Patzkowsky, M.E., Raup, D.M., Roy, K., Sepkoski, J.J. Jr., Sommers, M.G., Wagner, P.J., Webber, A. 2001. Effects of sampling standardization on estimates of Phanerozoic marine diversification. *Proceedings of the National Academy of Science* 98, 6261–6266.

Alroy, J., Aberhan, M., Bottjer, D.J., Foote, M., Fürsich, F.T., Harries, P.J., Hendy, A.J.W., Holland, S.M., Ivany, L.C., Riessling, W., Kosnik, M.A., Marshall, C.R., McGowan, A.J., Miller, A.I., Olszewski, T.D., Patzkowksy, M.E., Peters, S.E., Villier, L., Wagner, P.J., Bonuso, N.E., Borkow, P.S., Brenneis, B., Clapham, M.E., Fall, L.M., Ferguson, C.A., Hanson, V.L., Krug, A.Z., Layou, K.M., Leckey, E.H., Nürnberg, S., Powers, C.M., Sessa, J.A., Simpson, C., Tomašových, A., Visaggy, C.Y. 2008. Phanerozoic trends in the global diversity of marine invertebrates. *Science* 321, 97–100.

Benton, M.J. 2009. The Red Queen and the Court Jester: species diversity and the role of biotic and abiotic factors through time. *Science* 323, 728–732.

Brower, J.E., Zar, J.H., von Ende, C.N. 1998. *Field and Laboratory Methods for General Ecology*. McGraw-Hill, Boston.

Chao, A. 1984. Nonparametric estimation of the number of classes in a population. *Scandinavian Journal of Statistics* 11, 265–270.

Chao, A. 1987. Estimating the population size for capture-recapture data with unequal catchability. *Biometrics* 43, 783–791.

Chao, A., Jost, L. 2012. Coverage-based rarefaction and extrapolation: standardizing samples by completeness rather than size. *Ecology* 93, 2533–2547.

Chao, A., Gotelli, N.J., Hsieh, T.C., Sander, E.L., Ma, K.H., Colwell, R.K., Ellison, A.M. 2014. Rarefaction and extrapolation with Hill numbers: a framework for sampling and estimation in species diversity studies. *Ecological Monographs* 84, 45–67.

Chiu, C.-H., Wang, Y.-T., Walther, B.A., Chao, A. 2014. An improved nonparametric lower bound of species richness via a modified Good–Turing frequency formula. *Biometrics* 70, 671–682.

Clarke, K.R., Warwick, R.M. 1998. A taxonomic distinctness index and its statistical properties. *Journal of Applied Ecology* 35, 523–531.

Colwell, R.K. 2013. EstimateS: Statistical estimation of species richness and shared species from samples. Version 9. User's Guide and application published at: http://purl.oclc.org/estimates.

Colwell, R.K., Coddington, J.A. 1994. Estimating terrestrial biodiversity through extrapolation. *Philosophical Transactions of the Royal Society of London B* 345, 101–118.

Colwell, R.K., Mao, C.X., Chang, J. 2004. Interpolating, extrapolating, and comparing incidence-based species accumulation curves. *Ecology* 85, 2717–2727.

Colwell, R.K., Chao, A., Gotelli, N.J., Lin, S.-L., Mao, C.X., Chazdon, R.L., Longino, J.T. 2012. Models and estimators linking individual-based and sample-based rarefaction, extrapolation and comparison of assemblages. *Journal of Plant Ecology* 5, 3–21.

Cooper, R.A., Sadler, P.M. 2002. Optimised biostratigraphy and its applications in basin analysis, timescale development and macroevolution. *First International Palaeontological Congress (IPC 2002). Geological Society of Australia, Abstracts* 68, 38–39.

Fisher, R.A., Corbet, A.S., Williams, C.B. 1943. The relation between the number of species and the number of individuals in a random sample of an animal population. *Journal of Animal Ecology* 12, 42–58.

Foote, M. 2000. Origination and extinction components of taxonomic diversity: general problems. *Paleobiology* 26, 74–102.

Hammer, Ø. 2003. Biodiversity curves for the Ordovician of Baltoscandia. *Lethaia* 36, 305–313.

Harper, D.A.T. 2001. Late Ordovician brachiopod biofacies of the Girvan district, SW Scotland. *Transactions of the Royal Society of Edinburgh: Earth Sciences* 91, 471–477.

Harper, D.A.T., Hammarlund, E.U., Topper, T.P., Nielsen, A.T., Rasmussen, J.A., Park, T.-Y.S, Smith, M.P. 2019. The Sirius Passet Lagerstätte of North Greenland: a remote window on the Cambrian Explosion. *Journal of the Geological Society* 176, 1023–1027.

Heck, K.L. Jr., van Belle, G., Simberloff, D. 1975. Explicit calculation of the rarefaction diversity measurement and the determination of sufficient sample size. *Ecology* 56, 1459–1461.

Heltshe, J., Forrester, N.E. 1983. Estimating species richness using the jackknife procedure. *Biometrics* 39, 1–11.

Hill, M.O. 1973. Diversity and evenness: a unifying notation and its consequences. *Ecology* 54, 427–432.

Hubbell, S.P. 2001. *The Unified Neutral Theory of Biodiversity and Biogeography*. Princeton University Press, Princeton, NJ.

Hughes, R.G. 1986. Theories and models of species abundance. *American Naturalist* 128, 879–899.

Hutcheson, K. 1970. A test for comparing diversities based on the Shannon formula. *Journal of Theoretical Biology* 29, 151–154.

Krebs, C.J. 1989. *Ecological Methodology*. Harper & Row, New York.

Ludwig, J.A., Reynolds, J.F. 1988. *Statistical Ecology. A Primer on Methods and Computing*. John Wiley and Sons, New York.

MacArthur, R.H. 1957. On the relative abundance of bird species. *Proceedings of the Natural Academy of Science USA* 43, 293–295

Magurran, A.E. 1988. *Ecological Diversity and its Measurement*. Cambridge University Press, Cambridge, UK.

Magurran, A.E. 2004. *Measuring Biological Diversity*. Wiley-Blackwell, Oxford, UK.

Margalef, R. 1958. Information theory in ecology. *General Systematics* 3, 36–71.

May, R.M. 1981. Patterns in multi-species communities. In May, R.M. (ed.), Theoretical Ecology: Principles and Applications, 197–227. Blackwell, Oxford, UK.

McGill, B.J. 2003. A test of the unified neutral theory of biodiversity. *Nature* 422, 881–885.

McKerrow, W.S. (ed.). 1978. *Ecology of Fossils*. Gerald Duckworth and Company Ltd., London.

Menhinick, E.F. 1964. A comparison of some species-individuals diversity indices applied to samples of field insects. *Ecology* 45, 859–861.

Motomura, I. 1932. A statistical treatment of associations. *Zoological Magazine, Tokyo* 44, 379–383.

Mouillot, D., Lepretre, A. 2000. Introduction of relative abundance distribution (RAD) indices, estimated from the rank-frequency diagrams (RFD), to assess changes in community diversity. *Environmental Monitoring and Assessment* 63, 279–295

Pardoe, H.S., Cleal, C.J., Berry, C.M., Cascales-Miñana, B., Davis, B.A.S., Diez, J.B., Filipova-Marinova, M.V., Giesecke, T., Hilton, J., Ivanov, D., Kustatcher, E., Leroy, S.A.G., McElwain, J.C., Oplustil, S., Popa, M.E., Seyfullah, L.J., Stolle, E., Thomas, B.A., Uhl, D. 2021. Palaeobotanical experiences of plant diversity in deep time. 2: how to measure and analyse past plant biodiversity. *Palaeogeography, Palaeoclimatology, Palaeoecology* 580, 110618.

Peters, S.E., Bork, K.B. 1999. Species-abundance models: an ecological approach to inferring paleoenvironment and resolving paleoecological change in the Waldron Shale (Silurian). *Palaios* 14, 234–245.

Poole, R.W. 1974. *An Introduction to Quantitative Ecology*. McGraw-Hill, New York.

Press, W.H., Teukolsky, S.A., Vetterling, W.T., Flannery, B.P. 1992. *Numerical Recipes in C*. Cambridge University Press, Cambridge, UK.

Preston, F.W. 1962. The canonical distribution of commonness and rarity. *Ecology* 43, 185–215, 410–432.

Rosenzweig, M.L. 1995. *Species Diversity in Space and Time*. Cambridge University Press, Cambridge, UK.

Sepkoski, J.J. Jr. 1975. Stratigraphic biases in the analysis of taxonomic survivorship. *Paleobiology* 1, 343–355.

Sepkoski, J.J. Jr. 1988. Alpha, beta or gamma: where does all the biodiversity go? *Paleobiology* 14, 221–234.

Servais, T., Cascales-Miñana, B., Harper, D.A.T., Lefebvre, B., Munnecke, A., Wang, W., Zgang, Y. 2023. No (Cambrian) explosion and no (Ordovician) event: a single long-term radiation in the early Palaeozoic. *Palaeogeography, Palaeoclimatology, Palaeoecology* 623, 111592.

Shannon, C.E., Weaver, W. 1949. *The Mathematical Theory of Communication*. University of Illinois Press, Urbana, IL.

Silvestro, D., Salamin, N., Schnitzler, J. 2014a. PyRate: a new program to estimate speciation and extinction rates from incomplete fossil data. *Methods in Ecology and Evolution* 5, 1126–1131.

Silvestro, D., Schnitzler, J., Li, L. H., Antonelli, A., Salamin, N. 2014b. Bayesian estimation of speciation and extinction from incomplete fossil occurrence data. *Systematic Biology* 63, 349–367.

Silvestro, D., Salamin, N., Antonelli, A., Meyer, X. 2019. Improved estimation of macroevolutionary rates from fossil data using a Bayesian framework. *Paleobiology* 45, 546–570.

Simpson, E.H. 1949. Measurement of diversity. *Nature* 163, 688.

Smith, A.B. 2001. Large-scale heterogeneity of the fossil record: implications for Phanerozoic biodiversity studies. *Philosophical Transactions of the Royal Society of London B* 356, 351–367.

Smith, E.P., van Belle, G. 1984. Nonparametric estimation of species richness. *Biometrics* 40, 119–129.

Tipper, J.C. 1979. Rarefaction and rarefiction – the use and abuse of a method in paleoecology. *Paleobiology* 5, 423–434.

Van Valen, L.M. 1984. A resetting of Phanerozoic community evolution. *Nature* 307, 50–52.

Warwick, R.M., Clarke, K.R. 1995. New 'biodiversity' measures reveal a decrease in taxonomic distinctness with increasing stress. *Marine Ecology Progress Series* 129, 301–305.

Whittaker, R.J., Willis, K.J., Field, R. 2001. Scale and species richness: towards a general, hierarchical theory of species diversity. *Journal of Biogeography* 28, 453–470.

Williams, A., Lockley, M.G., Hurst, J.M. 1981. Benthic palaeocommunities represented in the Ffairfach Group and coeval Ordovician successions of Wales. *Palaeontology* 24, 661–694.

# 10

# Paleoecology and paleobiogeography

The large and exciting fields of paleobiogeography and paleoecology involve the reconstruction and interpretation of ancient communities and environments from the local through the regional to the global level (Brenchley and Harper 1998; Bottjer 2016). Paleobiogeographical and paleoecological units, for example, provinces and communities, are best described, analyzed, and compared using numerical data. Like morphological data, such as the lengths and widths of the shells of brachiopods discussed in chapter 6, spatial data can also be described as distributions of variates. A fossil taxon, such as a genus or species, is variably distributed across a number of localities, regions, or even continents. Many of the techniques we have already used for the investigation of morphological data are thus relevant for the studies of paleocommunities and paleoprovinces. In particular, clustering and ordination techniques are very useful in sorting out the distribution patterns of biotic assemblages. The starting point in most investigations is a matrix of abundances or presence/absence of taxa across a number of localities.

## 10.1   Paleobiogeography

Paleobiogeographical studies have focused on two separate methodologies. The more traditional biogeographic frameworks assume a dispersal model. All organisms have the ability to extend their geographic ranges from a point or points of dispersal. Often, larval stages of animals or seeds of plants can travel considerable distances. Dispersal may be aided by island hopping and rafting. To some extent, these models can be traced back to Alfred Wallace's studies of the animals and plants of the East Indies at a time when both the continents and oceans were supposed to have fixed positions. Multivariate methods using the presence/absence of taxa as variates can thus be developed to group together sites with similar faunas. Such methods have formed the basis of many biogeographical investigations of ancient provinces (see, e.g., Harper and Sandy 2001 for a review of studies of Paleozoic brachiopod provinces and Harper and Servais 2013 for analyses of Lower Paleozoic biotas). On the other hand, vicariance biogeography assumes a more mobilist model based on plate tectonic models. Fragmented and disjunct distributions are created by the breakup of once continuous provinces, divided by, for example, seafloor spreading in oceanic environments or by emergent mountain ranges or rift systems on the continents.

Dispersal (phenetic) models start with a matrix of presence/absence data from which distance and similarity coefficients may be calculated; these indices form the basis for subsequent multivariate cluster and ordination analyses. This matrix, however, can also be analyzed "cladistically" to generate a tree (chapter 14). Assuming a vicariance type model, a fragmented province shared a common origin and common taxa with other provinces. Taxa appearing since the fragmentation event are analogous to autapomorphic characters now defining that particular province. The shared taxa, however, are the synapomorphies within the system that group together similar provinces. Despite the contrast in the respective philosophical assumptions of both methods, both provide a spatial structure where similar biotas are grouped together. For example, Harper et al. (1996) developed a suite of biogeographical analyses of the distribution of early Middle Ordovician brachiopods, based on a range of phenetic and cladistic techniques; their results were largely compatible and matched paleomagnetic data.

A third method, "area cladistics," is specifically designed to describe the proximal positions of continents through time, independently of geological data (Ebach 2003). The method converts a taxonomic cladogram into an area cladogram. Areagrams, however, illustrate geographical isolation marked by evolution after plate divergences (Ebach and Humphries 2002). The conversion of a taxa-based cladogram to an area cladogram is explained in some detail by Ebach et al. (2003) and the technique is not discussed further here.

## 10.2 Paleoecology

Although the science of paleoecology is relatively young, there are many excellent qualitative studies of past environments and their floras and faunas (see Brenchley and Harper 1998 and Benton and Harper 2020 for overviews). Most authors split paleoecology into two parts: autecology, the study of the functional morphology of individual organisms, and synecology, the investigation of associations of organisms. The first is discussed earlier in the book in connection with morphometrics and the second is the focus of this chapter. In many of the older publications, biotas were described in terms of faunal or floral lists of taxa present. These are of course useful (particularly in biogeographical studies) but overestimate the importance of rare elements in the assemblage and underestimate the common taxa. Many papers provide annotated taxa lists with an indication of the relative abundances of fossils, whereas the most useful ones contain numerical data on the absolute abundance of each taxon in each assemblage (Ager 1963). There are, however, many different counting conventions (see Brenchley and Harper 1998 for some discussion). How do we count animals with multipart skeletons, those that molt, fragile colonial organisms, or skeletal fragments? There is no simple solution to these problems, but investigators generally clearly state the counting conventions they have followed when developing a data matrix.

In paleoecological and paleobiographical studies, it is important to differentiate between the so-called Q-mode and R-mode multivariate analysis. In morphological investigations, we generally use Q-mode analysis; the distributions of variables (morphological characters) are used to group together specimens with similar morphologies. Nevertheless, it may be useful to discover which groups of variables vary together; the matrix can be transposed and an R-mode analysis can be implemented to identify patterns of covariation between variates. By analogy, each locality or sample can be characterized by a set of variates (occurring taxa) represented by either presence/absence data (most commonly used in biogeographical analyses) or abundance data (paleoecological analyses). A Q-mode analysis will thus group together localities or samples on the basis of co-occurring taxa, whereas an R-mode analysis will cluster assemblages of taxa that tend to co-occur. Quite often, it is useful to carry out both modes of analysis, and several methods, such as correspondence analysis (CA) and two-way clustering, are specifically designed for simultaneous Q-mode and R-mode analysis.

As noted above, the starting point for most paleoecological studies is a rectangular matrix of localities or samples vs. taxa. Derek Ager in his classic textbook on paleoecology (Ager 1963) provided a page from his field notebook as an example (Fig. 10.1). Here he noted the absolute abundance of complete and fragmented taxa occurring on a bedding plane of the middle Silurian Wenlock limestone (Much Wenlock Limestone Formation of current usage) at Dudley, near Birmingham, UK. This can be converted, simply, into a data matrix for this locality, and results displayed as, e.g., bar charts (Fig. 10.2). Not surprisingly,

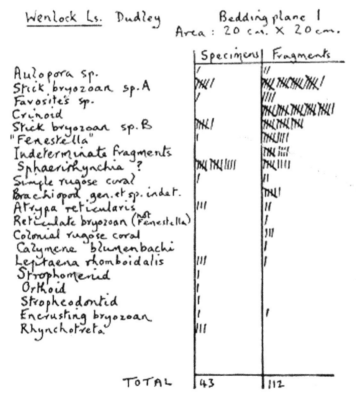

**Figure 10.1** A page from Derek Ager's (1963) field notebook detailing complete and fragmentary fossil specimens from a bedding plane of the Much Wenlock Limestone, Dudley, UK. Ager (1963)/McGraw-Hill.

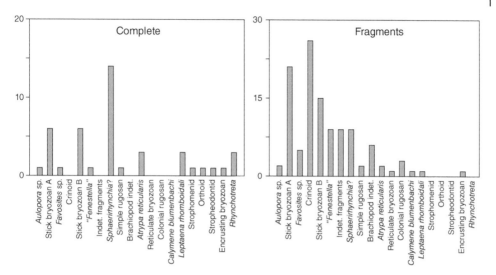

Figure 10.2    Bar charts of the absolute numbers of complete (left) and fragmented (right) fossils from Ager's sample of the Much Wenlock Limestone fossil biota. Data from Ager (1963).

the sets of counts are significantly different (chi-squared test: $\chi^2 = 56.1$; $p < 0.001$). More interestingly, the diversities have similar values and are not significantly different (Shannon's $H = 2.17$ and 2.30, $t$ test $p = 0.49$; Simpson's $D = 0.16$ and 0.13, $t$ test $p = 0.42$), indicating that the complete specimens and fragments capture similar ecological information.

In the study of living communities, the focus is often on the biotic response to environmental factors such as temperature, salinity, pH, oxygenation, substrate type, and nutrient availability. Traditionally, such environmental variables have not been available to paleontologists, but geochemical and sedimentological proxies are becoming increasingly important in studies of paleocommunities and their interactions with changing abiotic environments. The integration of environmental variables with fossil occurrence data requires special methods such as canonical correspondence analysis (CCA) (section 10.12).

Multivariate techniques therefore provide a battery of powerful techniques to investigate the composition and structure of past communities and provinces.

## 10.3    Association similarity indices for presence-absence data

Measuring the similarity (or dissimilarity) between the taxon compositions of two samples is a fundamental operation, being the basis of several multivariate methods such as cluster analysis, multidimensional scaling, and randomization tests for comparing communities. However, a similarity metric, or similarity *index*, can be defined in many ways. The choice of index will depend on the type of data available and what aspects of the taxon compositions we consider important in each case. There are many such indices. Shi (1993) listed and discussed nearly 40 of them!

Any similarity index that ranges from 0 (no similarity) to 1 (equality) can be converted to a dissimilarity (distance) index by subtracting the similarity index from one, and vice versa.

We will first consider binary data, where the presence of a particular taxon is conventionally coded with the number 1 and absence with 0. An astonishing number of similarity indices have been proposed for such data, and we will only go through a few of the more commonly used ones.

Let $M$ be the number of species that are present in both samples and $N$ be the total number of all the remaining species (those that are present in one but not both samples). The *Jaccard* similarity index (Jaccard 1912) is simply defined as $M/(M + N)$, that is, the number of shared taxa divided by the total number of taxa. This means that absences in both samples are ignored. When one of the samples is much larger than the other, the value of this index will always be small, which may or may not be a desirable property.

The *Dice* index, also known as the Sørensen index or the coefficient of community (Dice 1945; Sørensen 1948), is defined as $M/((2M + N)/2) = 2M/(2M + N)$. It is similar to the Jaccard index but normalizes with respect to the average rather than the total number of species in the two samples. It is therefore somewhat less sensitive than Jaccard's index to differences in sample size. Another property to note is that compared with the Jaccard index, the Dice index puts more weight on matches than on mismatches due to the multiplication of $M$ by a factor of two.

*Simpson's coefficient of similarity* (Simpson 1943) is defined as $M/S$, where $S$ is the smaller of the number of taxa in each of the two samples. Unlike the Jaccard and Dice indices, this index is insensitive to the size of the larger sample. Again, this may or may not be a desirable property, depending on the application. Another curious feature of Simpson's index that may or may not be a good thing is that it, in effect, disregards absences in the smaller sample. Only taxa present in the smaller sample can contribute to $M$ and/or to $S$. It could perhaps be argued that this makes Simpson's coefficient of similarity suitable when the sampling is considered incomplete.

The similarity of two samples based on any similarity index can be statistically tested using a randomization method (permutation test). The two samples to be compared are pooled (all presences put in one "bag"), and two random samples are produced by pulling the same numbers of presences from the pool as the numbers of presences in the two original samples. This procedure is repeated say 1000 times, and the number of times the similarity between a random pair of samples equals or exceeds the similarity between the original samples is counted. This number is divided by the number of randomizations (1000), giving an estimate of the probability $p$ that the two original samples could have been taken from the same (pooled) population. If more than two samples are to be compared, they may all be pooled together.

The Raup-Crick similarity index (Raup and Crick 1979; Chase et al. 2011) is based on a somewhat different randomization scheme where the species in the random replicates are selected based on the number of occurrences across all sites participating in the study, not only the two sites that are being compared. The number of such random pairs where the number of shared taxa ($M$) equals or exceeds that of the original samples gives a $p$ value that can be used as a similarity index.

Hurbalek (1982) evaluated 43 different similarity coefficients for presence/absence data and ended up with recommending only four: Jaccard, Dice, Kulczynski, and Ochiai. This result is of course open to discussion. Archer and Maples (1987) and Maples and Archer (1988) discuss how the indices behave depending on the number of variables and whether the data matrix is sparse (contains few presences). We suggest the Dice index as a good starting point.

**Example 10.1**

Smith and Tipper (1986) described the distribution of ammonite genera in the Pliensbachian (Lower Jurassic) of North America. From this study, we will look at presence/absence data for 16 genera in the Upper Pliensbachian, from the Boreal craton, and from four terranes (Wrangellia, Stikinia, Quesnellia, and Sonomia) that were in the process of being accreted to the craton (Fig. 10.3). The data matrix is presented in Table 10.1.

Similarity values can be conveniently shown in a similarity matrix, where the samples enter in both rows and columns. Since the similarity between samples *A* and *B* is

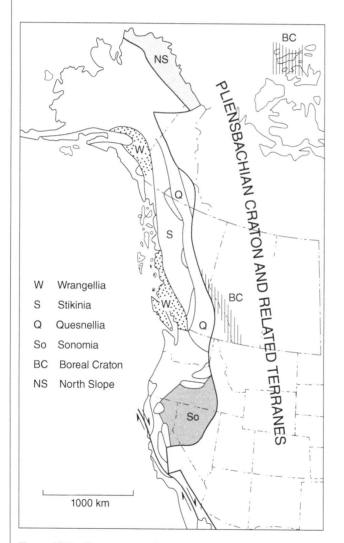

W Wrangellia
S Stikinia
Q Quesnellia
So Sonomia
BC Boreal Craton
NS North Slope

Figure 10.3   The terrane collage of the western Cordillera, USA. Adapted from Smith and Tipper (1986).

Table 10.1   Presence-absence matrix for 16 ammonite genera in the Upper Pliensbachian of five regions in North America.

| | Wrangellia | Stikinia | Quesnellia | Craton | Sonomia |
|---|---|---|---|---|---|
| Amaltheus | 1 | 1 | 1 | 1 | 0 |
| Pleuroceras | 0 | 0 | 0 | 1 | 0 |
| Pseudoamaltheus | 0 | 0 | 0 | 1 | 0 |
| Becheiceras | 1 | 1 | 0 | 0 | 0 |
| Arieticeras | 1 | 1 | 1 | 0 | 1 |
| Fontanelliceras | 1 | 0 | 0 | 0 | 0 |
| Fuciniceras | 1 | 1 | 1 | 0 | 0 |
| Leptaleoceras | 1 | 1 | 0 | 0 | 0 |
| Lioceratoides | 1 | 1 | 0 | 0 | 1 |
| Protogrammoceras | 1 | 1 | 1 | 1 | 1 |
| Aveyroniceras | 1 | 1 | 0 | 0 | 1 |
| Prodactylioceras | 1 | 1 | 0 | 0 | 1 |
| Reynesoceras | 1 | 0 | 1 | 0 | 0 |
| Cymbites | 1 | 0 | 0 | 0 | 0 |
| Fanninoceras | 1 | 1 | 1 | 0 | 1 |
| Phylloceras | 1 | 0 | 0 | 0 | 0 |

Data from Smith and Tipper (1986).

obviously the same as between $B$ and $A$, this matrix will be symmetric. It is therefore only necessary to show half of the matrix cells: either above or below the main diagonal. For compactness, we can then include two different indices in the same matrix, as shown in Table 10.2, where the Dice indices are in the lower left part and the Jaccard indices in the upper right.

Table 10.2   Dice (lower triangle) and Jaccard (upper triangle) similarity indices between the ammonite associations of five regions in the Upper Pliensbachian.

| | Wrangellia | Stikinia | Quesnellia | Craton | Sonomia |
|---|---|---|---|---|---|
| Wrangellia | — | 0.71 | 0.43 | 0.13 | 0.43 |
| Stikinia | 0.83 | — | 0.45 | 0.17 | 0.60 |
| Quesnellia | 0.60 | 0.63 | — | 0.25 | 0.33 |
| Craton | 0.22 | 0.29 | 0.40 | — | 0.11 |
| Sonomia | 0.60 | 0.75 | 0.50 | 0.20 | — |

Data from Smith and Tipper (1986).

Table 10.3  Simpson (lower triangle) and Raup-Crick (upper triangle) similarity indices for the same data set as used in Table 10.2.

|  | Wrangellia | Stikinia | Quesnellia | Craton | Sonomia |
|---|---|---|---|---|---|
| Wrangellia | — | 0.86 | 0.71 | 0.01 | 0.70 |
| Stikinia | 1.00 | — | 0.76 | 0.18 | 0.96 |
| Quesnellia | 1.00 | 0.83 | — | 0.61 | 0.60 |
| Craton | 0.50 | 0.50 | 0.50 | — | 0.19 |
| Sonomia | 1.00 | 1.00 | 0.50 | 0.25 | — |

The two indices give comparable results. Wrangellia and Stikinia are most similar; Sonomia and the craton are least similar. The similarities between the craton and each of the terranes are all relatively small.

The Simpson and Raup-Crick indices are shown in Table 10.3. For the Simpson index, Sonomia and the craton is still the least similar pair, and the craton remains isolated from the terranes. But within the terranes, there are now several maximally similar pairs (value 1.00), showing how the conservativeness of the Simpson index is paid for by a loss in sensitivity.

The Raup-Crick index, which can be loosely interpreted as a $p$ value for equality, shows the same general pattern but with some differences in details. The craton is still relatively isolated, but now Wrangellia and the craton is the least similar pair. Sonomia and Stikinia is the most similar pair using the Raup-Crick index.

## 10.4  Association similarity indices for abundance data

The range of proposed similarity indices for abundance data is at least as bewildering as for presence/absence data. Again, the choice of index should consider the type of data and the particular aspects of association composition that are considered important for the given problem (Legendre and Legendre 1998).

Before going through the different indices, we need to mention that the abundances (counts) are sometimes not used directly but transformed in some way prior to the calculation of the index. First, we may choose to compute relative abundances (percentages), thus normalizing for sample size. In principle, it would probably be a better strategy to ensure that all samples are of the same size to start off with, but in paleontological practice this can often be difficult without throwing away a lot of potentially useful information from rich horizons. In some cases, the absolute abundances may actually be of interest and should contribute to the computation of the similarity index.

Another transformation that is sometimes seen is to take the *logarithms* of the counts. Since zero-valued abundances (taxon absent in sample) are found in most data sets, and the logarithm of zero is not defined, it is customary to add one to all abundances before taking the logarithm. The problem is then neatly swept under the carpet, albeit in a rather arbitrary way. Log-transformation will downweigh taxa with high abundances, which may be a useful operation when the communities are dominated by one or a few taxa. Interesting information from all the rarer taxa might otherwise drown under the variation in abundance of the few common ones.

Even if absolute abundances are used, many similarity indices inherently disregard them, only considering the proportions. This is the most important consideration when choosing a similarity index.

Many of the similarity indices for abundance data are easily visualized by regarding the two sample compositions as two vectors **u** and **v** in $s$-dimensional space, where $s$ is the number of taxa. We will use the somewhat unrealistically simple example of $s = 2$ in our illustrations below.

The *Euclidean distance* index (Krebs 1989) is simply the Euclidean distance (ED) between the two sample compositions, that is, the length of the vector **u** − **v** (Fig. 10.4A).

$$ED = |u - v| = \sqrt{\sum_i \left( u_i - v_i \right)^2}$$

The ED index will generally increase in value as more taxa are added. To compensate for this, we may divide by the square root of the number of taxa ($s$) to get the *mean Euclidean Distance* (MED), although such scaling will not influence the results of most types of analysis.

Now let the vectors $\hat{u}$ and $\hat{v}$ be the abundances normalized to the unit vector length ($\hat{u} = u/|u|$ and $\hat{v} = v/|v|$):

$$\hat{u}_i = \frac{u_i}{\sqrt{\sum_j u_j^2}}$$

$$\hat{v}_i = \frac{v_i}{\sqrt{\sum_j v_j^2}}$$

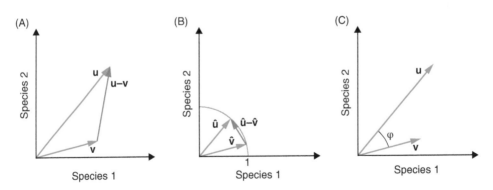

Figure 10.4  Illustration of distance indices calculated between two associations **u** and **v**, with two species. (A) The Euclidean distance. (B) The chord distance is the Euclidean distance between the normalized vectors. (C) The cosine similarity is defined as the cosine of the angle φ between the two vectors.

The ED between these normalized vectors is referred to as the *chord distance* (CRD) (Fig. 10.4B), varying from 0 to $\sqrt{2}$. Absolute abundances are thus normalized away. The CRD was recommended by Ludwig and Reynolds (1988). It can be written in different ways, but the most revealing formulation is probably

$$CRD = |\hat{u} - \hat{v}| = \sqrt{\Sigma(\hat{u}_i - \hat{v}_i)^2}$$

The *cosine similarity* (COS) is the cosine of the angle between the two vectors (Fig. 10.4C). This is easily computed as the inner product of the two normalized vectors:

$$COS = \hat{\mathbf{u}} \cdot \hat{\mathbf{v}} = \Sigma \hat{u}_i \hat{v}_i$$

The cosine similarity (inner product) may be compared with the linear correlation coefficient of section 4.5 and is also related to CRD by the simple equation

$$COS = 1 - CRD^2/2$$

The chi-squared distance, as defined in section 4.10, has interesting statistical properties and allows formal statistical testing. Still, it has not been used much in community analysis. It can be regarded as the square of the ED, after each taxon count difference has been divided by the sum of the corresponding two taxon counts in both samples. This means that the absolute differences between abundant taxa are dampened. The chi-squared distance may also be compared with the so-called *Canberra distance* metric (CM; Clifford and Stephenson 1975; Krebs 1989), ranging from 0 to 1:

$$CM = \frac{\Sigma|u_i - v_i|/(u_i + v_i)}{s}$$

For $u_i = v_i = 0$, we set the corresponding term in the sum to zero (that is $0/0 = 0$ by definition in this case). The Canberra distance reduces to the complement of the Dice index for presence/absence data, and similarly, it disregards absences in both samples.

Continuing this train of thought, the *Bray–Curtis* (BC; Bray and Curtis 1957; Clifford and Stephenson 1975) *distance* measure is defined as

$$BC = \frac{\Sigma|u_i - v_i|}{\Sigma(u_i - v_i)}$$

It reduces to the complement of the Dice index for presence/absence data and disregards absences in both samples. The BC distance does not normalize for abundance as does the Canberra distance and is therefore very sensitive to abundant taxa. Both these indices seem to work best for small samples (Krebs 1989). Incidentally, the *Manhattan distance* (MD; Swan and Sandilands 1995), also known as mean absolute distance (Ludwig and Reynolds 1988), is another scaling of the BC distance and can also be compared with the average ED:

$$MD = \frac{\Sigma|u_i - v_i|}{s}$$

The BC is generally preferable over the MD.

The somewhat complicated *Morisita* similarity index (MO; Morisita 1959) was recommended by Krebs (1989). It is another index of the "inner product" family, effectively normalizing away absolute abundances. It is comparatively insensitive to the sample size.

$$\lambda_u = \frac{\sum u_i (u_i - 1)}{\sum u_i (\sum u_i - 1)}$$

$$\lambda_v = \frac{\sum v_i (v_i - 1)}{\sum v_i (\sum v_i - 1)}$$

$$MO = \frac{2 \sum u_i v_i}{(\lambda_u + \lambda_v) \sum u_i \sum v_i}$$

Finally, we should mention two simple similarity indices that can be regarded as generalizations of the Dice and Jaccard indices of the previous section. The *similarity ratio* (SR; Ball 1966) is defined as

$$SR = \frac{\sum u_i v_i}{\sum u_i^2 + \sum v_i^2 - \sum u_i v_i}$$

For presence/absence data, this reduces to the Jaccard index. Similarly, the *percentage similarity* (PS; Gauch 1982), defined as

$$PS = 200 \frac{\sum \min(u_i, v_i)}{\sum u_i + \sum v_i}$$

reduces to the Dice index for presence/absence data.

Although most of the above were in fact distance measures, they are easily converted to similarity indices by subtracting them from one, taking their reciprocals or their negative logarithms. All these similarity indices can be subjected to the same type of permutation test as described for presence/absence data, allowing statistical testing of whether the similarity is significant.

---

**Example 10.2**

We return yet again to the Middle Ordovician data set of Williams et al. (1981). Figure 10.5 shows a cluster analysis of the ten samples using four different abundance similarity measures. In this data set, most of the samples are numerically dominated by the brachiopod *Sowerbyella*, and the other 22 taxa are rare. The only exception is sample 6, which also has a high abundance of the trilobite *Basilicus*. The dendrogram using the ED measure is totally controlled by these two taxa, which is perhaps not satisfactory. The Bray–Curtis measure, also being very sensitive to the

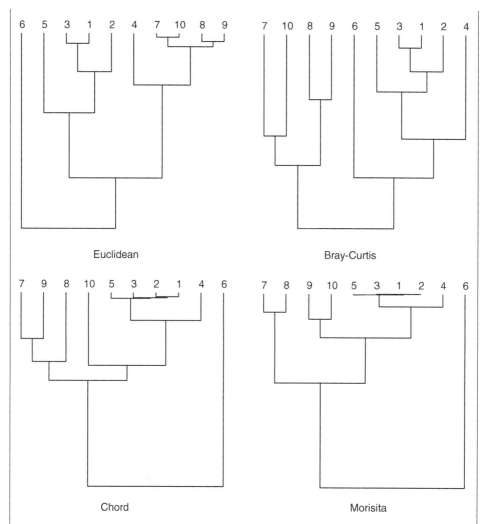

Figure 10.5   Mean linkage clustering (UPGMA) of 10 Middle Ordovician samples in a stratigraphic sequence, using different abundance similarity measures. Data from Williams et al. (1981).

most common taxa, produces a similar branching topology, except that sample 4 has now joined samples 1–3 and 5 in the same cluster, more in line with the stratigraphic sequence.

   The Chord and Morisita distances are included as representatives of the other main class of distance measures, normalizing for absolute abundances. These two dendrograms agree in that samples 1–5 are very similar and that sample 6 is very different from all other samples. Samples 7–10 are assigned to different positions in the two trees.

## 10.5 ANOSIM and PerMANOVA

ANOSIM (Analysis of Similarities) and PerMANOVA (permutational MANOVA) are two non-parametric tests for difference between several groups of multivariate data points. Both are appropriate for taxa-in-samples data, where two or more groups of samples are to be compared. For example, a paleontologist may have collected several samples from each of a number of facies or paleogeographical regions and would like to test for differences in faunal composition between the facies or regions. ANOSIM or PerMANOVA may be used in this situation.

ANOSIM (Clarke 1993; Legendre and Legendre 1998) is based on distances between all pairs of samples, computed from a distance index of choice (sections 10.3 and 10.4). Only the ranks of the distances are used. ANOSIM works by comparing within-group and across-group distances. If two groups were different in their taxonomic compositions, we would expect the distances within each group to be small relative to the distances across groups. A test statistic $R$ is constructed based on this idea, and its significance is estimated by permuting samples across groups.

Let $r_b$ be the mean rank of all distances between groups and $r_w$ be the mean rank of all distances within groups. The test statistic $R$ is then defined as

$$R = \frac{r_b - r_w}{N(N-1)/4}$$

Large positive $R$ (up to 1) signifies dissimilarity between groups. The significance is computed by the permutation of group membership.

The test assumes that the ranked within-group distances have equal median and range. In other words, the distances within the different groups (dispersions) should be equal, in rough analogy with ANOVA which assumes equal within-group variances.

Also, note that ANOSIM in itself cannot be used as a test for a significant *break* within a stratigraphical succession or environmental gradient. Consider a section where the fauna is gradually changing. Splitting the samples into two groups at an arbitrary stratigraphic level is likely to produce a significant difference between the groups, although the transition is gradual.

PerMANOVA (Anderson 2001) is quite similar to ANOSIM and likewise allows a choice of distance measure. It is based directly on the distance matrix rather than on ranked distances. Like ANOSIM, PerMANOVA uses permutation of samples to estimate significance, but using a different test statistic called $F$. In the special case of the ED measure on univariate samples, this $F$ statistic becomes equal to the $F$ statistic of ANOVA.

For PerMANOVA, a total sum of squares is calculated from the distance matrix $d$ as

$$SS_T = \frac{1}{N} \sum_{i=1}^{N-1} \sum_{j=i+1}^{N} d_{ij}^2$$

where $N$ is the total number of samples. With $n$ samples within each group, a within-group sum of squares is given by

$$SS_W = \frac{1}{n} \sum_{i=1}^{N-1} \sum_{j=i+1}^{N} d_{ij}^2 \varepsilon_{ij}$$

where the indicator function $\varepsilon$ is 1 if samples $i$ and $j$ are in the same group, 0 otherwise. Let the among-group sum of squares be defined as $SS_A = SS_T - SS_W$. With $a$ groups, the PerMANOVA test statistic is then given by

$$F = \frac{SS_A/(a-1)}{SS_W/(N-a)}$$

As in ANOSIM, the significance is estimated by permutation across groups.

---

**Example 10.3**

Consider the data set from the Middle Ordovician of Coed Dunn, Wales. Is there a significant difference between the faunas in the lower (samples 1–6) and upper (samples 7–10) parts of the section?

Using the Bray–Curtis distance measure, which is common practice for ANOSIM, we get a high value for the test statistic: $R = 0.96$. The estimated probability of equality from 10,000 permutations is $p = 0.005$. In fact, most distance measures based on abundances (including Euclidean, chord, and Morisita) yield significant differences at $p < 0.05$.

To conclude, the fauna is significantly different in the lower and upper parts of the section. The nature of this change, whether gradual or abrupt, is not addressed by the test.

---

## 10.6 Principal coordinates analysis

Like PCA (section 6.2), principal coordinates analysis (PCoA) is an ordination method used to reduce a multivariate data set down to two or three dimensions for visualization and explorative data analysis. The technique was described by Gower (1966). Reyment et al. (1984) discussed applications in morphometrics, while Pielou (1977) and Legendre and Legendre (1998) presented applications in the analysis of ecological data. The criterion for positioning the points in PCoA is that the Euclidean distancess in the low-dimensional space should reflect the original distances as measured in the multidimensional space. In other words, if two data points are similar, they should end up close together in the PCoA plot.

An interesting aspect of PCoA is that the user can freely choose the multidimensional distance measure. All the association similarity indices described earlier in this chapter can be used with PCoA, for example. This makes the method quite powerful for the analysis of ecological data sets (taxa in samples), but it is not the most used ordination method in ecology despite being regularly described in the literature. Within the ecological community, PCoA is presently somewhat overshadowed by correspondence analysis (CA) and non-metric multidimensional scaling (NMDS), described later in this chapter.

As in PCA, each principal coordinate axis has associated with it an eigenvalue, indicating the amount of variation in the data explained by that axis. One annoying feature of PCoA is that some distance measures (see below) can produce negative eigenvalues (Gower and Legendre 1986). Usually, these negative eigenvalues relate to the least important PCoA axes, and we can disregard them. In some cases, however, large negative eigenvalues will occur, and the PCoA analysis should then be regarded as suspect.

Another interesting connection between PCA and PCoA is that when PCoA is used with the negative ED measure, it will produce an ordination identical to that of PCA. However, since PCoA only operates on the distance matrix, it does not have access to the original variables, and a biplot showing how the variables relate to the ordination aces cannot be produced directly.

For ecological data sets where the taxa have unimodal responses to an environmental gradient, PCoA will tend to place the points in a "horseshoe" with incurving ends. Podani and Miklós (2002) studied this effect and also compared the performance of a number of distance measures when used with PCoA.

PCoA is also known as metric multidimensional scaling (MMDS) (see also section 10.7).

One algorithm for PCoA was briefly described by Davis (1986). Let **A** be the symmetric matrix of similarities between all objects, but with each element squared. It may sometimes be useful to raise the similarities to a higher, even exponent such as 4 to reduce horseshoe effects (Podani and Miklós 2002).

For compositional data, Reyment and Savazzi (1999) suggest transforming the similarity matrix to centered log-ratios (see section 4.3).

Let $\bar{a}$ be the mean value of all elements in **A**. The mean value of row $j$ in **A** is called $s_j$, and the mean value of column $k$ in **A** is $c_k$. Now, set up a matrix **Q** with the following elements:

$$q_{jk} = a_{jk} + \bar{a} - s_j - c_k$$

The eigenvectors of **Q** make up the principal coordinates, in a similar way as for PCA. If there are $N$ objects, there will be $N-1$ principal coordinates (the last eigenvalue will be zero). If there were fewer than $N-1$ variables in the original data set, the number of non-zero eigenvalues will reduce further (i.e., the dimensionality of the ordination space will be lower).

### 10.6.1 Metric distance measures and the triangle inequality

Denoting a given distance measure by |.|, the triangle inequality states that

$$|x+y| \le |x| + |y|.$$

If the triangle inequality holds, the distance measure is called *metric*. With PCoA, the use of non-metric distance measures can produce negative eigenvalues and potentially make the analysis invalid. However, many non-metric distance measures are useful and can work well with PCoA if care is taken to spot large negative eigenvalues.

---

**Example 10.4**

We return to the ammonite data set of Smith and Tipper (1986) from section 10.3. Figure 10.6 shows an average linkage (UPGMA) cluster analysis of the data set, using the Dice similarity coefficient. The craton splits out quite clearly, making a

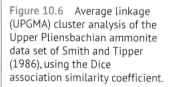

Figure 10.6   Average linkage (UPGMA) cluster analysis of the Upper Pliensbachian ammonite data set of Smith and Tipper (1986), using the Dice association similarity coefficient.

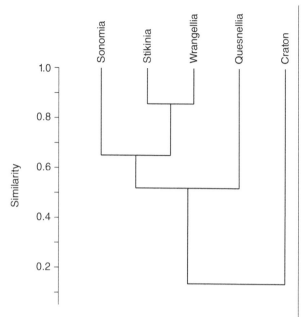

dichotomy between the craton on the one hand and the terranes on the other. Smith and Tipper (1986) referred to these as the Boreal and the Sonomian faunas, respectively. Within the Sonomian group, the Stikinian and Wrangellian terranes are most similar to each other.

Table 10.4 shows the eigenvalues and percentages of variation associated with the PCoA axes using the Dice coefficient. First, we note that $N$ data points will in principle occupy a region ("manifold") with a maximum of $N - 1$ dimensions. For example, two data points will lie on some straight line (one dimension), three data points on some flat surface (two dimensions), while four data points will in general occupy a three-dimensional region. In this case, we have five data points, so at most only four PCoA axes will be informative. In fact, with the Dice coefficient, the PCoA reports only three non-zero, positive eigenvalues. The first two axes, which we will use for plotting, represent 92.6% of the variation in the data, meaning that very little information is lost in the two-dimensional ordination plot.

Table 10.4   Eigenvalues associated with the PCoA axes for the ammonite data set.

| Axis | Eigenvalue | Percent of total |
|------|------------|------------------|
| 1 | 0.6036 | 75.50 |
| 2 | 0.1369 | 17.12 |
| 3 | 0.0586 | 7.32 |
| 4 | 0 | 0 |

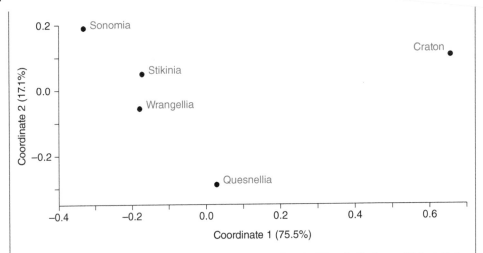

**Figure 10.7** PCoA analysis of the ammonite data set, using the Dice similarity coefficient. Note that we have taken care to use the same distance per unit on both axes, in order to retain the correct relative distances between the points.

Figure 10.7 shows the scatter plot using the first two principal coordinates. This plot is in good accordance with the dendrogram, with the craton in relative isolation from the Sonomian group. Unlike the dendrogram, the PCoA plot does not enforce any (artificial?) hierarchy on the data set. Also, the relative distances between terranes are much more evident. In the dendrogram, Quesnellia appears to be relatively close to Sonomia, but this is an artifact caused by the way the distance from Quesnellia to the (Sonomia, Stikinia, Wrangellia) cluster is computed by the average linkage method.

## 10.7 Non-metric multidimensional scaling

Non-metric (or non-parametric) multidimensional scaling (NMDS) is very similar to PCoA (section 10.6) in that it attempts to ordinate the data in a low-dimensional space such that observed distances are preserved (Kruskal 1964). Just like PCoA, NMDS starts with a distance matrix computed using any distance measure and then positions the data in a two-dimensional scatter plot or a three-dimensional plot in such a way that the Euclidean distancess between all pairs of points in this plot reflect the observed distances as faithfully as possible. Again, like PCoA, NMDS is attractive because it allows any distance measure to be used, including association distance measures based on presence/absence.

The difference between PCoA and NMDS is that the latter transforms the distances into their *ranks* and compares these ranked distances with the ranks of the EDs in the ordination plot. In this way, the absolute distances are discarded. If the distance between data points *A* and *B* is the fourth largest of the distances between any pair using the original distance measure, NMDS attempts to place *A* and *B* in the plot such that their ED is still the fourth largest in this new low-dimensional space. This can be a good idea if the absolute values of the distances do not carry any well-understood meaning. For example, consider an association similarity index such as the Dice index, and three samples *A*, *B*, and *C*. The Dice similarity between *A* and *B* is 0.4, between *A* and *C* it is 0.3, and between *C* and *D* it is 0.2. These numbers do not mean so much in isolation, but at least it is reasonable to say that *A* and *B* are more like each other than *A* and *C* are, while *C* and *D* is the least similar pair. The NMDS method assumes this but nothing more. Because NMDS has a more relaxed optimization criterion than PCoA, it has more freedom to find a good solution.

Unfortunately, ordination based on ranked distances is a riskier operation than using the absolute distances as in PCoA. There is no algorithm to compute the ordination directly, and instead we must use an iterative technique where we try to move the points around the plot until it seems we cannot get a better solution. The quality of the result can be assessed by measuring how much the ranked distances in the ordination deviate from the original ranked distances (this deviation is called *stress*). The initial positions for the points can be randomly chosen, or we can try an educated guess based on PCoA, for example. The bad news is that in most cases of any complexity, the ordination result will depend on the starting positions, and we are therefore never guaranteed to have found the best possible solution. The program should be run a few times to see if a better ordination can be obtained (PAST automatically executes 10 runs with random starting positions and one run using PCoA).

One way of visualizing the quality of the result is to plot the ordinated distances against the original distances in a so-called *Shepard plot*. Alternatively, the ordinated and original *ranked* distances can be used. Ideally, these ranks should be the same, and all the distances should therefore plot on a straight line ($y = x$). Note that there will be a lot of points in the Shepard plot, because of all possible distances being included. For *N* original data points, there will be a total of ($N^2 - N$)/2 points in the Shepard plot.

In NMDS, the number of ordination dimensions is chosen by the user and is normally set to two or three. If the original data set was two dimensional and we ordinate it into two dimensions, we should in theory be able to find a perfect NMDS solution where all ranks are preserved, assuming that the distance measure is well behaved. If we are reducing from a much higher number of dimensions, we cannot in general expect such a perfect solution to exist.

We should mention that the naming of NMDS and PCoA is somewhat confused in the literature. Some authors refer to PCoA as metric multidimensional scaling or MMDS, and include NMDS and MMDS (PCoA) in the general term multidimensional scaling, abbreviated MDS. Others use "MDS" in the sense of NMDS only, to the exclusion of MMDS.

Both metric and non-metric multidimensional scaling were thoroughly treated by Cox and Cox (2000). Minchin (1987) recommended NMDS for use in ecology, while Digby and Kempton (1987) recommended against it.

Most implementations of NMDS use different versions of a somewhat complicated algorithm developed by Kruskal (1964). A simpler and more direct approach was taken by Taguchi and Oono (2005), who simply move the points according to the direction and distance to their individual optimal position in ordination space. The individual preferences of the points will generally conflict with the configuration corresponding to an overall optimum, and therefore these movements are iterated until their positions have stabilized.

---

**Example 10.5**

We will turn to a biogeographic problem and try NMDS on the Hirnantian brachiopod data set of Rong and Harper (1988), using the Dice similarity index. Rong and Harper (1988) and others have developed the biogeographic distribution of Hirnantian brachiopod faunas, assigning them to three separate provinces: the low-latitude Edgewood; the mid-latitude Kosov; and the high-latitude Bani provinces (Fig. 10.8). Figure 10.9 shows the result of one particular run of the program – the points will have different positions each time. The Shepard plot (ranks) is given in Fig. 10.10, showing how the original (target) and obtained ranks correlate. The vertical stacks of points form because many of the distances in the original data set are equal (tied ranks).

  The stress value of 0.20 is relatively high (values below 0.10 are considered "good"), indicating that a considerable amount of important information has been lost in the reduction of dimensionality. Still, the results of the ordination are not unreasonable. Note how the high-latitude samples from the Bani province (1–5) plot together at the bottom left, and the samples from the low-latitude Edgewood province (31–35) cluster in the upper center. The mid-latitude Kosov province occupies a large area in the middle and at the right of the plot.

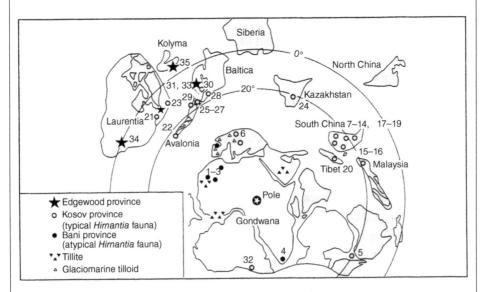

Figure 10.8   The distribution of the late Ordovician Hirnantian brachiopod fauna. Adapted from Ryan et al. (1995).

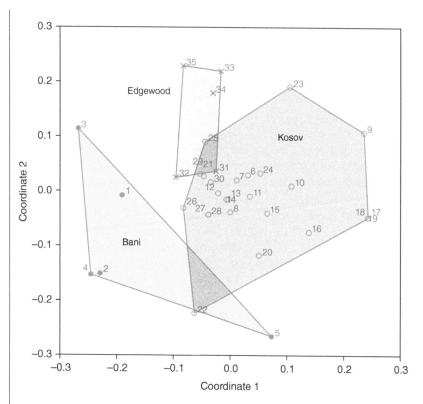

**Figure 10.9** Non-metric multidimensional scaling of Hirnantian brachiopod samples using the Dice similarity index. Data from Rong and Harper (1988).

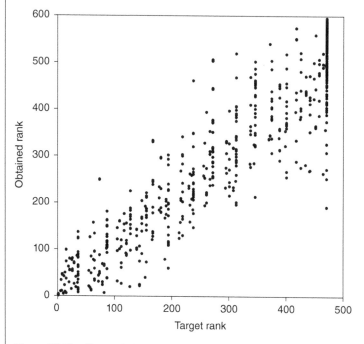

**Figure 10.10** Shepard plot from the NMDS analysis of Fig. 10.9.

## 10.8   Correspondence analysis

Correspondence analysis (CA) is presently the most popular ordination method for discovering groupings, geographic trends, and underlying environmental gradients from taxonomic counts in a number of samples (Greenacre 1984; Jongman et al. 1995; Legendre and Legendre 1998). As in principal components analysis, the basic idea is to try to plot samples and/or taxa in a low-dimensional space such as the two-dimensional plane, in some meaningful way. However, the comparison with PCA must not be taken too literally, because CA has a very different criterion for the placement of the samples and taxa than the maximization of variance used by PCA.

CA tries to position both samples and taxa in the same ordination space, in a way that maximizes correspondence between the two. This means that taxa should be placed close to the samples in which they are found, and samples should be placed close to the taxa that they contain. This will only be possible if there is a structure in the data set, such that there is a degree of localization of taxa to samples.

Let us consider an example. Samples of terrestrial macrofauna have been collected along a latitudinal gradient, from Central America to North America. From the abundances of taxa in the samples alone, we want to see if the samples can be ordered in a nice overlapping sequence where taxa come and go as we move along. This sequence can then be compared with the known geographical distribution to see if it makes sense. Consider three samples: one from Costa Rica, one from California; and one from Alaska. Some taxa are found in both the first two samples, such as coral snakes and opossums, and some taxa are found in both the last two samples, such as bald eagles, but no taxa are found in both the first and the third samples. On the basis of these data alone, the computer can order the samples in the correct sequence: Costa Rica, California, Alaska (or vice versa – this is arbitrary).

This whole procedure hinges on there being some degree of overlap between the samples. For example, consider a data set with one sample from Central Africa containing elephants and zebras, one sample from Germany with badgers and red foxes, and one sample from Spitsbergen with polar bears and reindeer. From the taxonomic compositions alone, the ordering of the three samples is quite arbitrary.

Of course, for fossil samples, we are often uncertain about the position of the samples along the gradient, or we may not even know if there is a gradient at all. CA is an example of so-called *indirect ordination*, where the samples are ordered only according to their taxonomic content, without *a priori* knowledge of their environmental or geographic positions. We then hope that the gradient (if any) and the positions of the samples along it will come out of the analysis.

In contrast with PCA, CA is ideal for the ordination of samples and taxa where the taxa have a *unimodal* response to the environmental gradient. This means that each taxon is most abundant at a certain temperature, for example, and gets rarer both for colder and warmer temperatures. Going along the gradient, taxa then come and go in an overlapping sequence (Fig. 10.11). Specialists may have very narrow response curves (niches), while generalists can cope with a wider temperature range. PCA, on the other hand, assumes a linear response, meaning that a certain taxon is supposed to get more and more abundant the colder it is (down to −273 °C, at least!), or the warmer it becomes. While this may be a reasonable approximation for short segments of the gradient, it is obviously not biologically reasonable for long gradients that accommodate several unimodal peaks for different taxa.

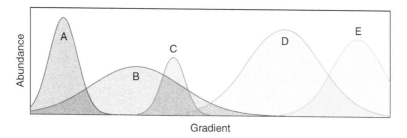

Figure 10.11   Five species A–E displaying unimodal responses to an environmental gradient such as temperature, occurring in an overlapping sequence. Generalists with high tolerance to the environmental parameters have wide response curves (species B and D), while specialists with low tolerance have narrow responses (species A and C).

Once the primary gradient has been set up, it is possible to mathematically remove all information in the data that is explained by this axis and try to ordinate the samples based on the remaining (residual) information. This will produce a second ordination axis, and the samples and taxa can be shown in a two-dimensional scatter plot. If we are lucky, it may be possible to interpret the two axes in terms of two independent environmental gradients, such as temperature and salinity. Each axis will have associated with it an *eigenvalue*, indicating the success of ordination along it. This procedure can be repeated to produce a third and higher order CA axes, but these will explain less and less of the variation in the data and are rarely informative.

The results of a CA may be presented in different ways. A *biplot* is a scatter plot of both samples and taxa in the same ordination space, spanned out by two of the CA axes (usually the first and the second). In the *relay plot* (Fig. 10.13), we attempt to show the sequence of unimodal response curves along the gradient. For each species, we plot the abundances in the different samples sorted according to the ordinated sequence. All these species plots are stacked in the order given by the ordination of the species.

CA can be carried out using several different algorithms, which should all give similar results.

*Reciprocal averaging* is probably the easiest to understand (Jongman et al. 1995; Legendre and Legendre 1998). It is an iterative procedure, consisting of the following steps:

1) Assign initial random positions along the axis to samples and taxa.
2) Calculate the position of each sample as the abundance-weighted mean of the positions of the taxa it contains.
3) Order the samples according to these positions.
4) Calculate the positions of each taxon as the abundance-weighted mean of the positions of the samples in which it is found.
5) Order the taxa according to these positions.

Points 2–5 are iterated until the samples and taxa no longer move.

Alternatively, CA can be done using an analytical (direct) method based on eigenanalysis (Davis 1986). A third, more modern method (Jongman et al. 1995; Legendre and Legendre 1998) is based on the singular value decomposition (SVD).

**Example 10.6**

To demonstrate the potential power of CA when data are relatively complete along an extended environmental gradient, we will use an occurrence table given by Culver (1988). Sixty-nine genera of Recent benthic foraminiferans were collected from the Gulf of Mexico in 14 different depth zones ranging from "Marsh" and "Bay" down to deeper than 3000 m. Abundances are given as "constancy scores," which are numbers from 0 to 10 indicating the proportion of samples within a depth zone where the genus was found.

The CA of this data set results in a first ordination axis explaining 33.1% of the variation and a second axis explaining 24.5%. The scatter plot of CA scores for the depth zones is given in Figure 10.12. The depth classes seem to be ordered in a good sequence, from shallow (left) to deep (right). However, there seems to be a strong arch effect (next section), and the points are compressed together at the right end of the first axis. This makes the diagram somewhat difficult to read.

The relay plot is shown in Figure 10.13. In order to make the diagram clearer, the positions of samples (the relay values) have been log-transformed, thus reducing the compression effect. It is evident from this figure how the different genera occupy different positions on the environmental gradient, in an overlapping sequence. Some genera such as *Cibicides* and *Bolivina* have relatively broad responses (wide tolerances). They are therefore not good indicator species, and their positions along the gradient are not well defined. On the other hand, genera such as *Parrella*, *Marginulina*, and *Millamina* have much narrower optima.

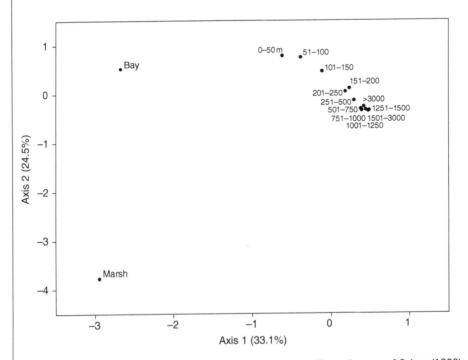

**Figure 10.12** Correspondence analysis of the benthic foraminiferan data set of Culver (1988). Only the scores for depth classes are shown. Note the arch effect and the compression of the axis scores at the right (deep) end of the gradient.

**Figure 10.13** Relay plot from correspondence analysis of the Culver (1988) data set. Both genera (vertically) and samples (horizontally) are ordered according to their positions on the first CA axis. Abundances in the samples are shown as vertical bars. Data from Culver (1988).

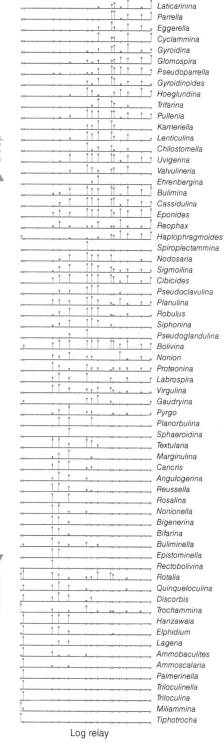

Deep

Shallow

Log relay

It is worth repeating that the CA has not used any information about the water depths of sampling. The ordination is based purely on the taxon occurrence data. Clearly, if these had been fossil samples where depth assignments were not known a *priori*, it would still be possible to place them along a (hypothetical) environmental gradient by CA. If axis 1 could be linked with bathymetry from other evidence, a depth sequence for the assemblages could be erected.

## 10.9   Detrended correspondence analysis

The commonly used algorithms for CA have two related problems. The first is a tendency to compress the ends of the ordination axis, squeezing together the samples and taxa at each end for no obviously useful reason. The second is the *arch effect*, similar to the horseshoe effect for PCoA. This happens when the primary underlying environmental gradient "leaks" into the second ordination axis, instead of being linked to only the first axis as it should. Typically, the samples and taxa end up on a parabolic arch, with the "real" primary gradient along it. Although the mathematical reasons for these phenomena are now well understood (Jongman et al. 1995), they can be annoying and make the plots difficult to read.

One "classic" way of fixing these problems involves rather brutish geometrical operations on the arch, stretching and bending it until it looks better. The two steps are *rescaling*, stretching out the ends of the axis to remove the compression effect, and *detrending*, hammering the arch into a straighter shape. This fudging has raised some eyebrows (Wartenberg et al. 1987), but it must be admitted that such *detrended correspondence analysis* (DCA) has shown its worth in practice, now being a popular method for indirect ordination of ecological data sets (Hill and Gauch 1980).

The rescaling procedure involves standardizing the sample and taxon scores to have variances equal to one. In other words, the widths of the species response curves ("tolerances") are forced to become equal. The detrending step of DCA consists of dividing the first CA axis into segments of equal length. Within each segment, the ordination scores on the second axis are adjusted by subtracting the average score within that segment. This is repeated for the third and fourth axes.

---

**Example 10.7**

Figure 10.14 shows the result of DCA on the Culver (1988) data set, making the picture much clearer than in Fig. 10.12. The first axis now explains about 55% of the total variation. The depth sequence along axis 1 is in accordance with the known answer, except at the deeper end where some depth zones are somewhat out of order. Note that the first axis has been reversed compared with the results from the standard CA (Fig. 10.12) – the polarity of a CA axis is arbitrary. The interpretation of axis 2 is not as obvious.

The scatter plot of genera is shown in Fig. 10.15. Depth zones and genera could have been plotted in the same diagram (biplot), but we choose to split them to avoid clutter. Shallow- and deep-water genera are immediately recognized, in the same areas of the plot where the corresponding depth zones were placed.

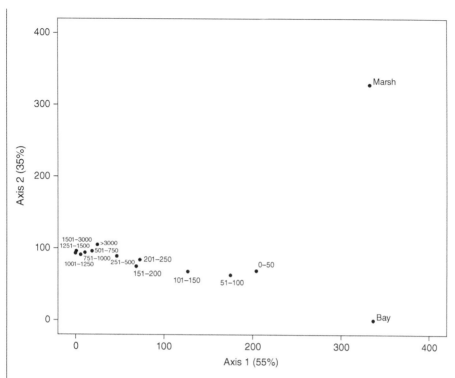

Figure 10.14   Detrended correspondence analysis of the data set of Culver (1988). Compare with Figure 10.12. The depth sequence is reproduced along axis 1.

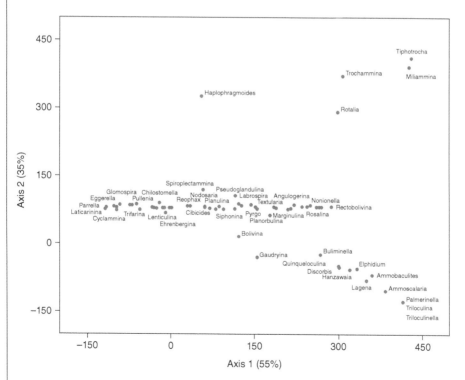

Figure 10.15   Detrended correspondence analysis of the data set of Culver (1988), showing the ordination of genera. Some genera have been removed or slightly displaced for readability. Compare with Fig. 10.14: Marsh genera to the upper right; Bay genera lower right; Deeper genera towards the left.

## 10.10   Seriation

Seriation of a presence-absence occurrence matrix (Burroughs and Brower 1982; Brower in Gradstein et al. 1985; Brower and Kyle 1988) is a simple method for ordering samples in a sequence according to their taxonomic content and/or ordering taxa in a sequence according to their distribution in the samples. Seriation is therefore yet another ordination method.

Seriation comes in two versions: unconstrained and constrained. In unconstrained seriation, the computer is allowed to freely reorder both rows (samples) and columns (taxa) with the aim of concentrating presences optimally along the diagonal in the matrix. When you go down the rows in such a "diagonalized" matrix, the taxa contained in the samples will come and go in an orderly sequence. In other words, the algorithm has found a corresponding, simultaneous ordering of both samples and taxa. This ordering may be found to make sense in terms of biogeography, an environmental gradient, or a stratigraphic sequence. Our example from section 10.8 may be repeated here: consider three samples of Recent terrestrial macrobiota, one from Central America, one from California, and one from the state of Washington. Some taxa are found in both the first two samples, such as opossums and coral snakes, and some taxa are found in both the last two samples, such as gray squirrels, but no taxa are found in both the first and the third samples. On the basis of these data alone, the computer can order the samples in the correct sequence. This is, however, dependent on the overlapping taxonomic compositions of the samples – without such overlap, the sequence is arbitrary.

Unconstrained seriation proceeds iteratively, in a loop consisting of the following steps:

1) For each row, calculate the mean position of presences along the row (this position will generally be a non-integer "column number").
2) Order the rows according to these means.
3) For each column, calculate the mean position of presences along the column (this position will generally be a non-integer "row number").
4) Order the columns according to these means.

The process is continued until the rows and columns no longer move.

Both conceptually and algorithmically, unconstrained seriation is almost equivalent to CA (section 10.8) of binary data. The main practical differences are that seriation can only be used for binary data and that CA can ordinate along more than one axis. CA is a more "modern" and sophisticated method, but on the other hand the simplicity of seriation is quite appealing.

In constrained seriation, samples that have a known order, for example, because of an observed stratigraphic superposition within a section, are not allowed to move. The algorithm can reorder only taxa (columns). Constrained seriation can be used as a simple method for quantitative biostratigraphy, producing a stratigraphic sequence of overlapping taxon ranges. However, the more sophisticated methods of chapter 13 are probably to be preferred for this purpose.

According to Ryan et al. (1999), the results of seriation can depend on the initial order of rows and columns in the matrix, and these authors describe another, more complicated algorithm that is likely to get closer to the globally optimal solution.

The success of a seriation can be quantified using a seriation index, which takes its maximum value of 1.0 when all presences are compactly placed along the diagonal. A high seriation index means that there is some structure in the data that allows the taxa and samples to be ordered in a corresponding sequence. If the presences were randomly distributed in the original data matrix, the seriation algorithm would not be able to concentrate the presences along the diagonal, and the seriation index would be low.

Let $R_i$ be the total range of presences in column $i$, from the highest row to the lowest row, inclusive. $A_i$ is the number of embedded absences in column $i$, that is, the number of zeros in between the ones. The seriation index SI is then defined as

$$\text{SI} = 1 - \frac{\sum A_i}{\sum R_i}$$

(note that the formula given by Brower in Gradstein et al. (1985) is incorrect). For a perfectly seriated matrix with no embedded absences, we have SI = 1. In less well seriated matrices, the number of embedded absences will increase, and the value of SI will drop.

Whether there is any structure in the data that indicates an inherent ordering can be tested statistically by a randomization method. For unconstrained seriation, the seriation index of the original data matrix is first computed. The rows and columns are then randomly reordered (permutated), and the seriation index of the shuffled matrix is noted. This procedure is repeated say $N = 1000$ times, and the permutations for which the seriation index equals or exceeds the seriation index of the original data are counted ($M$). The number $M/N$ now represents a probability that a seriation index as high as the one observed for the original data could have occurred from a random data set. A low probability can be taken as evidence for a significant structure in the data.

---

**Example 10.8**

A consistent faunal change with increasing water depth has been reported from the lower Silurian at many localities around the world (Fig. 10.16). Using data from Cocks and McKerrow (1984), we have made a taxon occurrence matrix for 14 taxa in five generic communities. Will the seriation procedure be able to place the communities (and the taxa) in the expected sequence along the depth gradient?

Table 10.5 shows the seriated data matrix, with rows and columns arranged to concentrate presences along the main diagonal. The seriation has placed the five communities in the correct order, from shallow (*Lingula* community) to deep (*Clorinda* community).

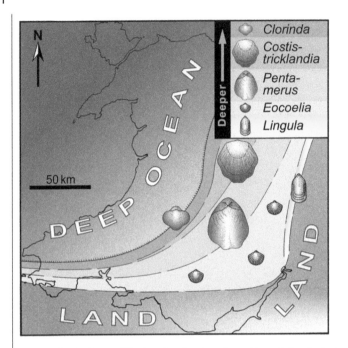

Figure 10.16   Lower Silurian, depth-related, brachiopod-dominated paleocommunities distributed across Wales and the Welsh Borderland. Image courtesy of John Murray.

Table 10.5   Unconstrained seriation of lower Silurian benthic marine communities.

|  | Lingula | Eocoelia | Pentamerus | Stricklandia | Clorinda |
| --- | --- | --- | --- | --- | --- |
| *Lingula* | 1 | 1 | 0 | 0 | 0 |
| Bivalves | 1 | 1 | 1 | 1 | 0 |
| *Pentamerus* | 0 | 1 | 1 | 0 | 0 |
| *Stegerhynchus* | 1 | 1 | 1 | 1 | 0 |
| *Eocoelia* | 1 | 1 | 1 | 1 | 1 |
| *Tentaculites* | 1 | 1 | 1 | 1 | 1 |
| *Eoplectodonta* | 0 | 1 | 1 | 0 | 1 |
| Streptelasmids | 0 | 1 | 1 | 1 | 1 |
| *Stricklandia* | 0 | 0 | 1 | 1 | 0 |
| *Leptostrophia* | 0 | 1 | 0 | 1 | 1 |
| *Atrypa* | 0 | 0 | 1 | 1 | 1 |
| *Clorinda* | 0 | 0 | 0 | 0 | 1 |
| *Glassia* | 0 | 0 | 0 | 0 | 1 |
| *Aegiria* | 0 | 0 | 0 | 0 | 1 |

# 10.11   Nonlinear dimensionality reduction

All the ordination methods we have looked at so far, including PCA, PCoA, NMDS, and CA, are basically dimensionality reduction techniques that find a flat plane (for a 2D ordination) or a Euclidean 3D space (for a 3D ordination) embedded in the original high-dimensional space. This subspace is constructed so that when we project the point cloud onto it, the structure is preserved as far as possible. Such a structure could be defined, e.g., in terms of variance maximization, as in PCA, or in terms of pairwise distances, as in PCoA or NMDS.

However, it is possible (and indeed very common) that the data points do not reside on a linear, flat subspace, but on some curved but smooth (locally flat) surface. Such curved spaces residing in a higher-dimensional space are called *manifolds*. Ordination to a curved manifold is called *nonlinear dimensionality reduction* (NLDR) or manifold learning. NLDR is a central area of research in "machine learning," and a very large number of algorithms have been proposed. We will present only a few of them here as an introduction to the field.

Consider the synthetic point cloud in Fig. 10.17, where 800 points are arranged within a rectangular strip folded into a spiral manifold. This so-called "Swiss roll" is a standard test case for NLDR (e.g., Tenenbaum et al. 2000). We want to reduce the dimensionality from 3D to 2D (typically, the original data set will have a much larger dimensionality). Figure 10.18 shows the result from PCoA with ED. This linear ordination is basically a projection onto the plane giving maximal variance, and it shows the spiral structure of the manifold. This result is easy to interpret and may be exactly what we want. But in many applications, we might like to instead display the positions of the points on the curved

**Figure 10.17**   A synthetic data set with 800 points in 3D. The points are arranged in a spiral strip (manifold) with random scatter.

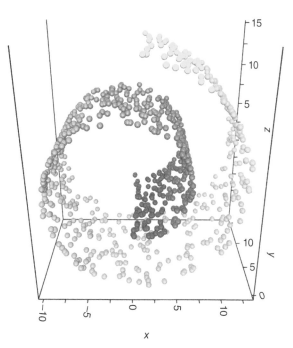

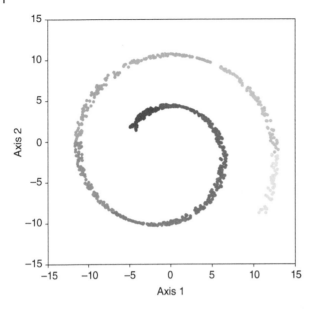

Figure 10.18  PCoA ordination of the Swiss roll data set. In this plot, which is a linear projection of the original point cloud, the yellow (lower right) and blue (center) points are closer than the yellow and green (left) points. This does not reflect the distances along the manifold.

manifold. This is the purpose of NLDR. In our example, the yellow points in the outermost part of the spiral should then be placed closer to the green points at the lower left than to the blue points in the inner whorl, whereas the PCoA gives the opposite result.

### 10.11.1  ISOMAP

The first NLDR method we will discuss is ISOMAP (Tenenbaum et al. 2000). The procedure is relatively straightforward:

1) Produce a network (graph) where we connect each point to its $k$ nearest neighbors, as given by the selected distance measure. The parameter $k$ must be specified by the user and must be large enough to ensure that the graph contains a single connected component (no isolated islands). An alternative method connects points closer than a specified radius.
2) Find the shortest paths between all pairs of points in the network. These path lengths represent "geodesic" distances along the manifold, in analogy with the distances of the shortest paths around the curved globe (great circles).
3) Now that we have an estimate of all pairwise distances along the manifold, we use traditional, linear PCoA to produce the ordination.

The idea here is that when $k$ is small compared with the total number of points, the larger distances between points will be disregarded in Step 1. This forces the paths generated in Step 2 to step along close points, tracing out the manifold. If $k$ is set too high, these paths may "short circuit" the manifold by bridging across large gaps, and the algorithm will not produce the desired result. For an even larger $k$, the graph will be almost fully connected, and the result will resemble that of PCoA.

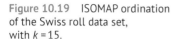

Figure 10.19 ISOMAP ordination
of the Swiss roll data set,
with $k = 15$.

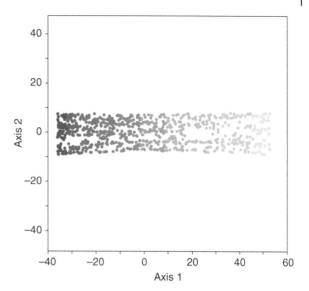

The ISOMAP ordination of the Swiss roll data set, with $k = 15$, is shown in Fig. 10.19. This is an impressive result: the manifold has been almost perfectly unfolded. ISOMAP does not always work this well. Moreover, this result depended on the selection of $k$-values smaller than 10 or larger than 15 do not give a good ordination. For real-world data, we do not know *a priori* what the manifold looks like, and therefore it will not be obvious whether a particular value of $k$ has been successful. This somewhat uncomfortable dependence on parameters is typical for most NLDR methods.

## 10.11.2 Spectral embedding

Belkin and Niyogi (2003) suggested an NLDR method called spectral embedding, or Laplacian eigenmaps. The algorithm starts with producing an adjacency matrix $\mathbf{W}$ for a neighborhood graph. The neighborhood is defined by the $k$ nearest neighbors or a specified radius, just like in ISOMAP. The entries in $\mathbf{W}$ are 0 or 1 depending on whether two points are connected. Alternatively, the entries for the connected points can be weighted by a "kernel function" inspired by heat diffusion:

$$W_{ij} = e^{\frac{\left\| x_i - x_j^2 \right\|}{t}}$$

where the norm is according to the selected distance measure. The parameter $t$ is selected by the user, e.g., 1/50 of the largest distance between any pairs of points. Then compute the "graph Laplacian" $\mathbf{L} = \mathbf{D} - \mathbf{W}$, where $\mathbf{D}$ is a diagonal matrix containing the row (or column, as $\mathbf{W}$ is symmetric) sums of $\mathbf{W}$. The scores on the two ordination axes are given by the two eigenvectors corresponding to the second smallest and third smallest eigenvalue of $\mathbf{L}'$.

While this algorithm is simple, it is not so easy to understand intuitively. Belkin and Niyogi (2003) give a thorough theoretical discussion. In practice, spectral embedding tends

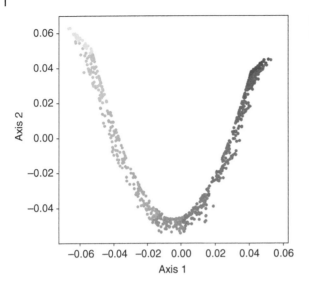

Figure 10.20   Spectral embedding of the Swiss roll data set.

to emphasize local rather than global structure and may therefore work better for discovering clusters. This stems from the fact that the matrix **W** only contains information about the local neighborhood, in contrast with ISOMAP which is based on the total shortest (geodesic) distances along the manifold between any two points.

Figure 10.20 shows the spectral embedding of the Swiss roll data set, with $k = 15$ and without kernel weighting. The manifold has been partly unfolded, but there is a residual arch shape.

### 10.11.3   UMAP

UMAP is an advanced method for NLDR, quickly becoming very popular after its first description by McInnes et al. (2018). It is often applied to large genomic data sets but can also be used for ecological taxa-in-samples data. The method is claimed to be good at preserving both local and global structures, so that it can both separate clusters and show their relative global distances. For the (very complex) mathematical details, see McInnes et al. (2018).

Figure 10.21 shows a UMAP ordination of the Swiss roll data. The unfolding of the spiral manifold is partly successful, but with a residual horseshoe shape that perhaps brings the innermost (blue) and outermost (yellow) points a little too close together. Other analysis parameters would produce somewhat different results.

## 10.12   Canonical correspondence analysis

Ecologists working on modern ecosystems often study the relationships between community compositions and abiotic factors. This requires environmental data such as temperature, depth, salinity, pH, and grain size in addition to the species counts. In paleontology,

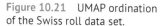

Figure 10.21 UMAP ordination of the Swiss roll data set.

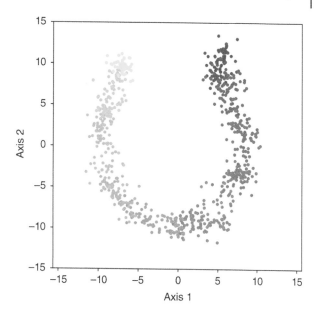

the collection of such data is much more challenging, but modern methods in organic and inorganic geochemistry and sedimentology can at least provide environmental proxy variables that can be included in paleoecological analyses.

One way to include environmental data in an ecological ordination is simply to carry out a simple ordination such as PCoA or NMDS on the samples with species counts and then to compute the correlations between the environmental variables and the ordination scores on each axis. These correlation coefficients are used to draw vectors from the origin in the ordination plot, one for each environmental variable, resulting in a biplot showing relationships between species occurrence and environmental data. An alternative, related method is a regression of each environmental variable onto the ordination axis. This is the method used in the *envfit* procedure of the *vegan* package in R (Oksanen et al. 2013). Because the supplementary environmental data are not contributing to the ordination, this is called *passive projection*.

In contrast, canonical correspondence analysis (CCA; Legendre and Legendre 1998) is a CA where the ordination axes are constrained to be linear combinations of the environmental variables. CCA is an example of direct gradient analysis, where the gradient in environmental variables is known from direct measurements and the species abundances (or presences/absences) are considered to be a response to this gradient. This powerful method is now perhaps the industry standard for analysis of combined species occurrence and environmental data sets. As in indirect (unconstrained) CA, CCA produces an ordination biplot of both samples and species, and, in addition, the environmental variables can be included in the plot ("triplot"). Each ordination axis has two associated eigenvalue percentages, one with respect to the variance of the constrained data set, and one with respect to the total variance. In addition, the associations between species occurrence data and environmental variables can be assessed with permutation tests.

**Example 10.9**

Bjørklund et al. (2019) studied the radiolarians in a core from Andfjord in northern Norway, spanning the Pleistocene-Holocene transition (see also sections 11.1 and 13.1). The radiolarians were assigned to 34 species and counted in 1 cm thick core slices. In addition, eight different sedimentological and geochemical variables were collected for each slice. They include ice-rafted debris (IRD), $CaCO_3$ and total organic carbon (TOC) content, oxygen and carbon isotopes (the latter both from shelly and organic carbon), and the biomarkers brassicasterol (from open marine phytoplankton) and IP25 (from diatoms associated with seasonal sea ice; Cabedo-Sanz et al. 2013).

The result from the CCA is shown in Fig. 10.22. Axis 1 explains 44% of the total variance, while axis 2 explains 11%. The main differences in the species composition (constrained by the environmental variables) are therefore distributed along the first CCA axis. Some groups of samples, identified by cluster analysis (section 13.1), are shown with polygons. The Holocene samples are all on the negative side of axis 1, while the cold Younger Dryas assemblage (YDA) samples are on the positive side, with a large

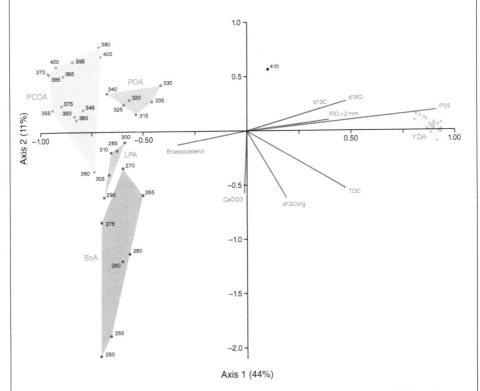

**Figure 10.22** Canonical correspondence analysis (CCA) of the Bjørklund et al. (2019) data set. Species scores are not included in the plot. See text for details. Bjørklund et al. (2019)/ Norwegian Polar Institute.

separation between these two main groups. The vectors for the environmental variables are quite informative. The IRD, IP25 (sea ice proxy), and oxygen isotope ($\delta^{18}O$)
vectors all point to the right, toward the YDA samples, all strongly supporting cold
conditions. On the other hand, the open-marine (i.e., less ice cover) proxy brassicasterol
points to the left, toward the warmer Holocene samples.

The interpretation of the second CCA axis is somewhat less obvious. The Boreal
assemblage (BoA), which corresponds to the youngest part of the core, has strongly
negative scores on this axis. The carbonate and partly the TOC vector point in the negative axis 1 direction and could be interpreted as indicating high organic productivity
(alternatively, they reflect reduced clastic sediment input). The environmental interpretation of the carbon isotope signals is less clear-cut.

## 10.13   Indicator species

After finding differences in the species compositions between groups of sites, it is often of
interest to identify particular species that are indicative of the groups. The indicator species
approach (Dufrene and Legendre 1997) is one way to quantify the degree of association
between species and groups.

For each species $i$ in group $j$, define the *specificity* as

$$A_{ij} = \frac{N_{ij}}{N_i}$$

where $N_{ij}$ is the mean number of individuals of species $i$ across sites in group $j$, and $N_i$ is the
sum of the mean numbers of individuals of species $i$ over all groups. If a species is found in
one group only, its specificity will be 1 (the maximum possible value) in that group and 0 in
all other groups.

Similarly, define the *fidelity* as

$$B_{ij} = \frac{Nsites_{ij}}{Nsites_j}$$

where $Nsites_{ij}$ is the number of sites in group $j$ where species $i$ is present, and $Nsites_j$ is the
total number of sites in group $j$. If a species is found in all the sites in a group, its fidelity
will be 1 in that group.

The indicator value of species $i$ in group $j$ is then a value from 0 to 100 (percentage), combining specificity and fidelity:

$$INDVAL_{ij} = 100 A_{ij} B_{ij}$$

The statistical significances ($p$ values) of the indicator values are estimated by random
reassignments (permutations) of sites across groups. These $p$ values can optionally be
corrected for multiple testing using, e.g., a Bonferroni correction.

## 10.14  Network analysis

Network plots and network analysis are used increasingly in paleontology, mainly for the study of fossil communities at different scales (e.g., Kiel 2017; Rojas et al. 2017; Muscente et al. 2018; Xu et al. 2022). A network, also known as a graph, is a collection of nodes connected by lines (also known as links or edges). For example, the nodes may represent species or higher-ranking taxa, and two nodes will be connected by an edge if the two species occur together in at least one sample, in the same stratigraphic unit, or in the same biogeographical area. The edges may be unweighted, or have a weight determined by, e.g., the number of localities where the two species co-occur. Conversely, the nodes may represent samples (communities), and the edges link samples that share species. In this case, the edges may be weighted by the number of shared species, or by one of the community similarity measures described in sections 10.3–10.4.

When plotting such a network, we normally want connected nodes to be placed close together. This reduces clutter and makes it easier to see groups (clusters) of nodes. Many methods have been developed for this purpose – a simple but robust solution is the Fruchterman-Reingold algorithm, which simulates physical forces that pull together connected nodes and push apart disconnected ones (Fruchterman and Reingold 1991). This may be compared with ordination methods such as PCoA, NMDS, or UMAP (sections 10.6, 10.7, and 10.11), which also attempt to plot similar items close together. High-quality network plotting is available in several programs and libraries, including Gephi and iGraph. Basic graph plotting is also available in PAST (Fig. 10.23).

Many extensions and further analyses are possible. The thicknesses or colors of edges can be made proportional to the edge weights. The plotted sizes of nodes can be made to vary with their connectedness (number of connected edges). Special algorithms such as the

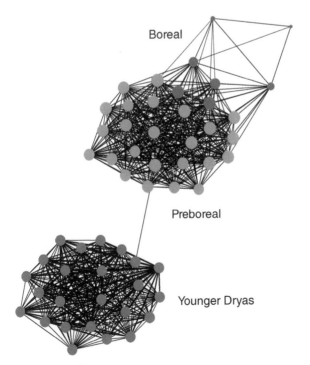

Boreal

Preboreal

Younger Dryas

Figure 10.23  Network plot of radiolarian assemblages in a core from a fiord in northern Norway, spanning the Pleistocene–Holocene transition. Each node represents a sample. Edges are drawn between any two nodes with a Bray–Curtis similarity higher than a 50% cut-off point. Edge thickness is scaled by similarity value. Node sizes are scaled by number of connected edges. Data from Bjørklund et al. (2019).

Louvain algorithm have been developed for partitioning the network into subgraphs of tightly connected nodes – this is in effect a type of cluster analysis.

A *bipartite network* (bigraph) consists of two types of nodes, and all edges connect across the two types (there are no edges connecting nodes of the same type). For example, we could have samples as one node type and species as the other, connected by edges if a given sample contains a given species. Such a bipartite network can give a good picture of the distribution of species on samples (e.g., Rojas et al. 2017; Xu et al. 2022).

Several metrics have been suggested for describing different aspects of network structure, such as the degree of connectedness. For biogeographical data, such metrics can indicate cosmopolitanism (high connectedness) versus endemism (Huang et al. 2018). For example, the average degree (AD) is calculated as the total number of edges divided by the number of nodes in a network. Graph density (GD) is the number of edges divided by the number of possible edges (for a graph with $n$ nodes, there are $n(n-1)/2$ possible edges).

---

**Example 10.10**

We return to the example of the Hirnantian brachiopod faunas discussed in sections 9.5 and 10.7. In the intervening 30 or so years since the paleobiogeography of these faunas was first recognized (Rong and Harper 1988), much new data have been collected from this interval and a more precise chronostratigraphy has been developed for the Hirnantian Stage. Network analysis has identified many more Edgewood faunas and a new Cathay assemblage of roughly the same age (Rong et al. 2020). Both are apparently marginally younger than the classic *Hirnantia* fauna and represent a recovery fauna following the main phase of the extinction (Fig. 10.24).

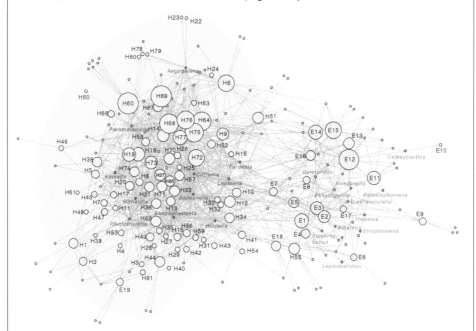

Figure 10.24   Bipartite network analysis of the distribution of brachiopod genera during the Hirnantian and early Silurian. The cool-water *Hirnantia* fauna is indicated in pale blue and the marginally younger, warmer-water Edgewood-Cathay assemblage is in pink. Key genera (dots) are indicated, and the sites (circles) are numbered. Rong et al. (2020)/with permission of Elsevier.

## 10.15   Size-frequency and survivorship curves

Analyses of size distributions are commonly the starting point for many paleoecological investigations. Although size is not linearly related to age, the shape of size histograms can give useful information about the population dynamics of an assemblage or whether the sample is *in situ* or transported (Brenchley and Harper 1998). There is a range of possibilities from (1) right (positively) skewed distributions (Fig. 10.25A) indicating high infant mortality through (2) normal (Gaussian) distribution (Fig. 10.25B) indicating either preservation of a certain age group or transported (sorted) assemblages, to (3) left (negatively) skewed distributions (Fig. 10.25C) more typical of high senile mortality. These patterns may be modified by, e.g., seasonal spawning giving rise to age cohorts and multimodal distributions (Fig. 10.25D).

In a classic study on adult woolly mammoths, Kurtén (1954) estimated the true ages at death for 67 specimens, based on tooth shape and wear (Fig. 10.26A). The data indicate

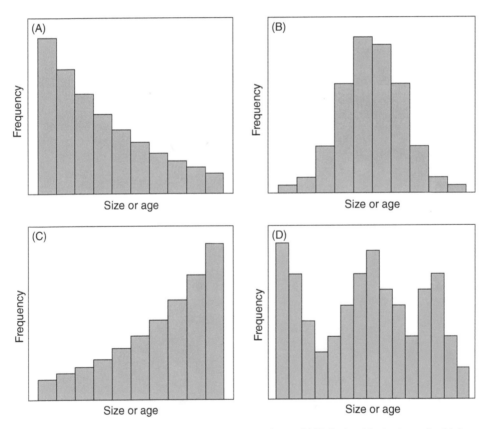

**Figure 10.25**   Schematic selection of size–frequency charts. (A) Right (positively skewed) – high juvenile mortality. (B) Normal (Gaussian) – one age group or transported. (C) Left (negatively) skewed – high senile mortality. (D) Multimodal distribution (age cohorts). Adapted from Brenchley and Harper (1998).

(A)

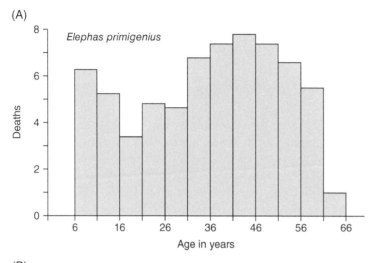

(B)

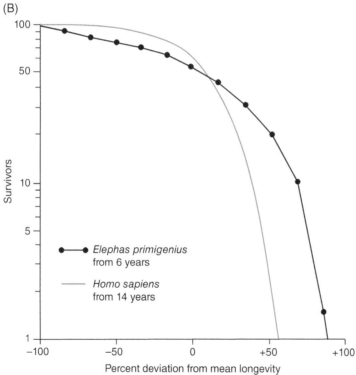

Figure 10.26   Histogram of age at death (A) and survivorship curve (B) for a sample of 67 mammoths (age based on wear of dentition). Survivorship curve of adult humans for comparison. Adapted from Kurtén (1954).

fairly constant mortality, possibly with a slight increase at the end of the mammoth's life span. Mortality patterns can also be displayed by survivorship curves, plotting the number of survivors in the sample as a function of age or size (Fig. 10.26B).

## 10.16   Case study: Devonian paleobiogeography

In conclusion we can expand the use of network analysis (section 10.14) to more global investigations and compare its results with those of clustering and ordination techniques. With the exception of Savage et al. (1979), who investigated the brachiopods, there have been limited numerical analyses of Devonian faunal distributional data from high latitudes; most of the attention is focused on the tropics. The Devonian faunas of South Africa were situated near the south pole and are generally referred to the Malvinoxhosan Bioregion (Malvinokaffric Realm of previous authors), associated with high-latitude Gondwana. Penn-Clarke and Harper (2021) interrogated a large data set consisting of 205 brachiopod genera from within the Malvinoxhosan Bioregion, as well as depocenters across greater West Gondwana; a total of 17 localities were included. The proximities and relationships between these faunas, some close to the South Pole, were never investigated in detail before this study. Cluster analysis and NMDS ordination together with network analysis all identify three groups: the high-latitude Malvinoxhosan Bioregion, the temperate latitude Columbian–West Africa Bioregion, and the intermediate Amazonian Bioregion (Figs. 10.27 and 10.28).

The cluster analysis of the localities (Fig. 10.27A) used the Jaccard similarity coefficient (section 10.3). The dendrogram is relatively poorly resolved, as shown by the long within-group branches compared with short across-group branches, and also some of the bootstrap support values are low. This is common for biogeographic studies. However, other clusters are clearly supported, and the overall groupings fit with the paleogeography.

The NMDS ordination (Fig. 10.27B) with a minimal spanning tree (lines connecting the closest points) gives a similar picture as the cluster analysis, but perhaps more clearly. The Shepard plot and the stress value (0.139) indicate that the distances in the ordination plot reflect the original distances well – i.e., little information is lost in the ordination.

The bipartite network, showing both localities and genera, was plotted with Gephi (Fig. 10.28). A locality node is connected to a genus node if the genus is found in the locality. In this plot, nodes are repelled from each other but attracted by edges using the ForceAtlas2 algorithm, similar to the Fruchterman-Reingold algorithm mentioned in section 10.14. In effect, this can be viewed as a simple ordination method where localities sharing many taxa are placed closer together. The number of shared taxa, as a similarity index, has somewhat different properties than the Jaccard index used for clustering and NMDS above. Also, we may note that localities that plot far away from the others in the NMDS ordination (Accraian, Zorritas, and Southern Peru) plot much closer to the main groups in the network graph, due to the physical model of the ForceAtlas2 algorithm.

This large, high-quality data set thus generates consistent results when investigated by clustering and ordination methods together with those of network analysis. The bioregions can be related to latitude and to continental regions and form a framework to test further hypotheses on the distribution and evolution of Early and Middle Devonian faunas. In this case, the main influence may be due to latitude linked to regional climate variation (Penn-Clarke and Harper 2021).

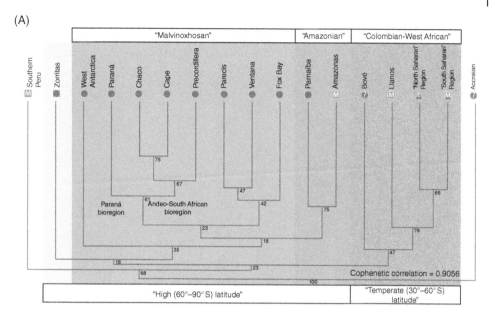

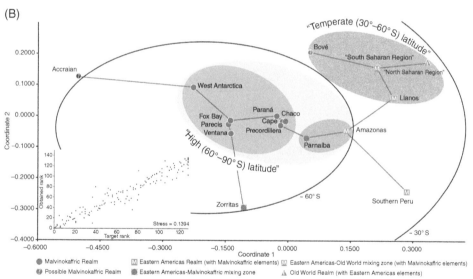

**Figure 10.27** Cluster analysis and NMDS of the Lower-Middle Devonian data set. (A) Results of the cluster analysis using the Jaccard coefficient and the UPGMA method, with bootstrap supports. (B) Non-metric multidimensional scaling (NMDS). The localities are linked by a minimum spanning tree. The Shepard plot indicates the stress value for the analysis (here 0.139) and the reliability of the ordination. Both analyses identify the three bioregions and relate these to latitude. Penn-Clarke and Harper (2021)/Geological Society of America.

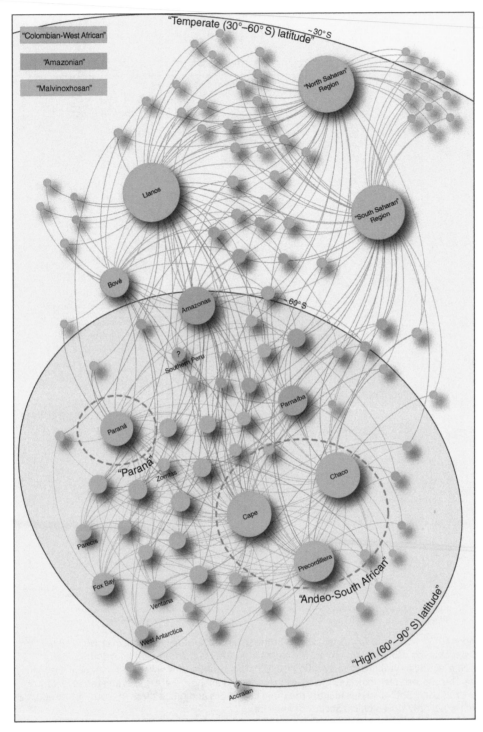

**Figure 10.28** Bipartite network of the key localities associated with southern and western Gondwana during the Early–Middle Devonian. Localities are shown as labeled nodes, while taxa are unlabeled. Sizes of nodes are proportional to their numbers of connected edges. Two main bioregions are recognized, Colombia and West Africa (purple), and Malvinoxhosan (orange), with the intermediate Amazonian bioregion shown in green. Penn-Clarke and Harper (2021)/Geological Society of America.

# References

Ager, D.V. 1963. *Principles of Paleoecology*. McGraw-Hill, New York.

Anderson, M.J. 2001. A new method for non-parametric multivariate analysis of variance. *Austral Ecology* 26, 32–46.

Archer, A.W., Maples, C.G. 1987. Monte Carlo simulation of selected similarity coefficients (I): effect of number of variables. *Palaios* 2, 609–617.

Ball, G. H. 1966. *A Comparison of Some Cluster Seeking Techniques*. Stanford Research Institute, California.

Benton, M.J., Harper, D.A.T. 2020. *Introduction to Paleobiology and the Fossil Record*. 2nd edition. Wiley-Blackwell, Chichester, UK.

Belkin, M., Niyogi, P. 2003. Laplacian eigenmaps for dimensionality reduction and data representation. *Neural Computation* 15, 1373–1396.

Bjørklund, K.R., Kruglikova, S.B., Hammer, Ø. 2019. The radiolarian fauna during the Younger Dryas–Holocene transition in Andfjorden, northern Norway. *Polar Research* 38. https://doi.org/10.33265/polar.v38.3444.

Bottjer, D.J. 2016. *Paleoecology. Past, Present and Future*. Wiley Blackwell.

Bray, J.R., Curtis, J.T. 1957. An ordination of the upland forest communities of southern Wisconsin. *Ecological Monographs* 27, 325–349.

Brenchley, P.J., Harper, D.A.T. 1998. *Palaeoecology: Ecosystems, Environments and Evolution*. Routledge, Oxford, UK.

Brower, J.C., Kyle, K.M. 1988. Seriation of an original data matrix as applied to palaeoecology. *Lethaia* 21, 79–93.

Burroughs, W.A., Brower, J.C. 1982. SER, a FORTRAN program for the seriation of biostratigraphic data. *Computers and Geosciences* 8, 137–148.

Cabedo-Sanz, P., Belt, S.T., Knies, J., Husum, K. 2013. Identification of contrasting seasonal ice conditions during the Younger Dryas. *Quaternary Science Reviews* 79, 74–86.

Chase, J.M., Kraft, N.J.B., Smith, K.G., Vellend, M., Inouye, B.D. 2011. Using null models to disentangle variation in community dissimilarity from variation in α-diversity. *Ecosphere* 2, 1–11.

Clarke, K.R. 1993. Non-parametric multivariate analysis of changes in community structure. *Australian Journal of Ecology* 18, 117–143.

Clifford, H.T., Stephenson, W.S. 1975. *An Introduction to Numerical Classification*. Academic Press, London.

Cox, T.F., Cox, M.A.A. 2000. *Multidimensional Scaling*. 2nd edition. Chapman & Hall, New York.

Cocks, L.R.M., McKerrow, W.S. 1984. Review of the distribution of the commoner animals in Lower Silurian marine benthic communities. *Palaeontology* 27, 663–670.

Culver, S.J. 1988. New foraminiferal depth zonation of the northwestern Gulf of Mexico. *Palaios* 3, 69–85.

Davis, J.C. 1986. *Statistics and Data Analysis in Geology*. John Wiley, Chichester, UK.

Dice, L.R. 1945. Measures of the amount of ecological association between species. *Ecology* 26, 297–302.

Digby, P.G.N., Kempton, R.A. 1987. *Multivariate Analysis of Ecological Communities*. Chapman & Hall, London.

Dufrene, M., Legendre, P. 1997. Species assemblages and indicator species: the need for a flexible asymmetrical approach. *Ecological Monographs* 67, 345–366.

Ebach, M.C. 2003. Area cladistics. *Biologist* 50, 1–4.

Ebach, M.C., Humphries, C.J. 2002. Cladistic biogeography and the art of discovery. *Journal of Biogeography* 29, 427–444.

Ebach, M.C., Humphries, C.J., Williams, D. 2003. Phylogenetic biogeography deconstructed. *Journal of Biogeography* 30, 1285–1296.

Fruchterman, T. M. J., Reingold, E. M. 1991. Graph drawing by force-directed placement. *Software: Practice and Experience* 21, 1129–1164.

Gauch, H.G. 1982. *Multivariate Analysis in Community Ecology*. Cambridge University Press, Cambridge, UK.

Gower, J.C. 1966. Some distance properties of latent root and vector methods used in multivariate analysis. *Biometrika* 53, 325–338.

Gower, J.C., Legendre, P. 1986. Metric and Euclidean properties of dissimilarity coefficients. *Journal of Classification* 3, 5–48.

Gradstein, F.M, Agterberg, F.P., Brower, J.C., Schwarzacher, W.S. 1985. *Quantitative Stratigraphy*. D. Reidel, Dordrecht, Holland.

Greenacre, M.J. 1984. *Theory and Applications of Correspondence Analysis*. Academic Press, London.

Harper, D.A.T., Mac Niocaill, C., Williams, S.H. 1996. The palaeogeography of Early Ordovician Iapetus terranes: an integration of faunal and palaeomagnetic constraints. *Palaeogeography, Palaeoclimatology, Palaeoecology* 121, 297–312.

Harper, D.A.T., Sandy, M.R. 2001. Paleozoic brachiopod biogeography. In Carlson, S.J., Sandy, M.R. (eds.), *Brachiopods Ancient and Modern. Paleontological Society Papers* 7, 207–222. Paleontological Society, USA.

Harper, D.A.T., Servais, T. (eds.). 2013. Early Palaeozoic biogeography and palaeogeography. *Geological Society, London, Memoirs* 38, 1–496.

Hill, M.O., Gauch Jr., H.G. 1980. Detrended correspondence analysis: an improved ordination technique. *Vegetatio* 42, 47–58.

Huang, B., Jin, J., Rong, J.-Y. 2018. Post-extinction diversification patterns of brachiopods in the early–middle Llandovery, Silurian. *Palaeogeography, Palaeoclimatology, Palaeoecology* 493, 11–19.

Hurbalek, Z. 1982. Coefficients of association and similarity based on binary (presence-absence) data: an evaluation. *Biological Reviews* 57, 669–689.

Jaccard, P. 1912. The distribution of the flora of the alpine zone. *New Phytologist* 11, 37–50.

Jongman, R.H.G., ter Braak, C.J.F., van Tongeren, O.F.R. 1995. *Data Analysis in Community and Landscape Ecology*. Cambridge University Press, Cambridge, UK.

Kiel, S. 2017. Using network analysis to trace the evolution of biogeography through geologic time: a case study. *Geology* 45, 711–714.

Krebs, C.J. 1989. *Ecological Methodology*. Harper & Row, New York.

Kruskal, J.B. 1964. Nonmetric multidimensional scaling: a numerical method. *Psychometrika* 29, 115–131.

Kurtén, B. 1954. Population dynamics – a new method in paleontology. *Journal of Paleontology* 28, 286–292.

Legendre, P., Legendre, L. 1998. *Numerical Ecology*. 2nd edition. Elsevier, Amsterdam.

Ludwig, J.A., Reynolds, J.F. 1988. *Statistical Ecology. A Primer on Methods and Computing*. John Wiley and Sons, New York.

Maples, C.G., Archer, A.W. 1988. Monte Carlo simulation of selected binomial similarity coefficients (II): effect of sparse data. *Palaios* 3, 95–103.

McInnes, L., Healy, J., Melville, J. 2018. UMAP: Uniform manifold approximation and projection for dimension reduction. https://arxiv.org/abs/1802.03426.

Minchin, P.R. 1987. An evaluation of relative robustness of techniques for ecological ordinations. *Vegetatio* 69, 89–107.

Morisita, M. 1959. Measuring of interspecific association and similarity between communities. *Memoirs Faculty Kyushu University, Series E* 3, 65–80.

Muscente, A.D., Prabhu, A., Zhong, H., Eleish, A., Meyer, M.B., Fox, P., Hazen, R.M., Knoll, A.H. 2018. Quantifying ecological impacts of mass extinctions with network analysis of fossil communities. *PNAS* 115, 5217–5222.

Oksanen, J., Blanchet, F.G., Kindt, R., Legendre, P., Minchin, P.R., O'hara, R.B., Simpson, G.L., Solymos, P., Stevens, M.H.H., Wagner, H. 2013. Vegan, community ecology package, version 2.

Penn-Clarke, C.R., Harper, D.A.T. 2021. Early–Middle Devonian brachiopod provincialism and bioregionalization at high latitudes: a case study from southwestern Gondwana. *GSA Bulletin* 133, 819–836.

Pielou, E.C. 1977. *Mathematical Ecology*. 2nd edition. Wiley, New York.

Podani, J., Miklós, I. 2002. Resemblance coefficients and the horseshoe effect in principal coordinates analysis. *Ecology* 83, 3331–3343.

Raup, D., Crick, R.E. 1979. Measurement of faunal similarity in paleontology. *Journal of Paleontology* 53, 1213–1227.

Reyment, R.A., Blackith, R.E., Campbell, N.A. 1984. *Multivariate Morphometrics*. 2nd edition. Academic Press, London.

Reyment, R.A., Savazzi, E. 1999. *Aspects of Multivariate Statistical Analysis in Geology*. Elsevier, Amsterdam.

Rojas, A., Patarroyo, P., Mao, L., Bengtson, P., Kowalewski, M. 2017. Global biogeography of Albian ammonoids: a network-based approach. *Geology* 45, 659–662.

Rong, Jia-yu, Harper, D.A.T. 1988. A global synthesis of the latest Ordovician Hirnantian brachiopod faunas. *Transactions of the Royal Society of Edinburgh: Earth Sciences* 79, 383–402.

Rong, J., Harper, D.A.T., Huang Bing, Li Rongyu, Zhang Xiaole, Chen Di. 2020. The latest Ordovician Hirnantian brachiopod faunas: new global insights. *Earth-Science Reviews* 208, 103280.

Ryan, P.D., Harper, D.A.T., Whalley, J.S. 1995. *PALSTAT, Statistics for Palaeontologists*. Chapman & Hall, New York.

Ryan, P.D., Ryan, M.D.C., Harper, D.A.T. 1999. A new approach to seriation. In Harper, D.A.T. (ed.) *Numerical Palaeobiology*, 433–449. John Wiley, Chichester, UK.

Savage, N.M., Perry, D.G., Boucot, A.J. 1979. A quantitative analysis of Lower Devonian brachiopod distribution. In Gray, J., Boucot, A.J. (eds.). *Historical Biogeography, Plate Tectonics and the Changing Environment*, 169–200. Oregon State University Press, Corvallis, Oregon.

Shi, G.R. 1993. Multivariate data analysis in paleoecology and paleobiogeography – a review. *Palaeogeography, Palaeoclimatology, Palaeoecology* 105, 199–234.

Simpson, G.G. 1943. Mammals and the nature of continents. *American Journal of Science* 241, 1–31.

Smith, P.L., Tipper, H.W. 1986. Plate tectonics and paleobiogeography: early Jurassic (Pliensbachian) endemism and diversity. *Palaios* 1, 399–412.

Sørensen, T. 1948. A method of establishing groups of equal amplitude in plant sociology based on similarity of species content. *Det Kongelige Danske Videnskab. Selskab, Biologiske Skrifter* 5(4), 1–34.

Swan, A.R.H., Sandilands, M. 1995. *Introduction to Geological Data Analysis*. Blackwell Science, Oxford, UK.

Taguchi, Y.-H., Oono, Y. 2005. Relational patterns of gene expression via non-metric multidimensional scaling analysis. *Bioinformatics* 21, 730–740.

Tenenbaum, J.B., de Silva, V., Langford, J.C. 2000. A global geometric framework for nonlinear dimensionality reduction. *Science* 290, 2319–2323.

Wartenberg, D., Ferson, S., Rohlf, F.J. 1987. Putting things in order: a critique of detrended correspondence analysis. *American Naturalist* 129, 434–448.

Williams, A., Lockley, M.G., Hurst, J.M. 1981. Benthic palaeocommunities represented in the Ffairfach Group and coeval Ordovician successions of Wales. *Palaeontology* 24, 661–694.

Xu, H.P., Zhang, Y-C., Yuan, D.-X., Shen, S.Z. 2022. Quantitative palaeobiogeography of the Kungurian–Roadian brachiopod faunas in the Tethys: implications of allometric drifting of Cimmerian blocks and opening of the Meso-Tethys Ocean. *Palaeogeography, Palaeoclimatology, Palaeoecology* 601, 111078.

# 11

# Calibration – estimating paleoenvironments

Understanding climate in the geological past (paleoclimatology) is of fundamental importance in climate research. In these times of dramatic anthropogenic climate change (atmospheric $CO_2$ levels exceeding 420 ppm, compared with preindustrial levels around 280 ppm), paleoclimatology can provide key information on both the mechanisms and possible consequences of rapid global warming.

One of the fundamental methods in paleoclimatology is to sample modern communities of organisms together with associated temperature data. Assuming that the modern communities are to some degree controlled by temperature, these data can be used to estimate paleotemperature based on the composition of fossil communities. Often, the modern samples are called "core-top" samples, because they can be taken from the most recent layer in a core sample, while the fossil samples are "downcore." This approach, known as *calibration*, can also be used for the reconstruction of other environmental parameters such as pH, salinity, and water depth.

In the terminology of machine learning, the core-top samples and the associated modern temperatures constitute the training data, and the job of the calibration algorithm is to produce a robust model relationship between the community data and the temperature data based on the training set. Finally, the fossil data can be plugged into the model to estimate paleotemperatures.

Clearly, almost any machine learning method could be applied to the calibration problem, but we will focus on three classical procedures: the modern analog technique (MAT), weighted averaging (WA), and weighted averaging partial least squares (WA-PLS).

## 11.1 Modern analog technique

The idea behind the MAT (Overpeck et al. 1985; Flower et al. 1997) is to search for modern (core-top) samples that have similar taxonomic compositions as a downcore sample, and then to use the corresponding environmental measurements to estimate the paleoenvironment of the downcore sample. In the machine learning community, this technique is known as K-nearest-neighbors regression.

For each downcore sample, the procedure consists of two steps. First, one or more modern analogs are selected. The taxonomic composition of the downcore sample is compared to all the modern samples using a distance measure. Many of the distance measures described in section 10.4 will work, but the squared chord distance is often used for MAT. In PAST, only samples closer to the downcore sample than a user-selected value are selected ($d < d_{max}$), and a maximum of N such samples are included as modern analogs. Additionally, samples can be removed from the set of modern analogs by the "jump method": for each downcore sample, modern samples are sorted by ascending distance, and when the distance increases by more than a user-selected percentage, the subsequent modern analogs are discarded.

Second, the reconstructed environmental variable (e.g., temperature $T$) is computed as a weighted average of the values of the modern analogs. The values can be weighted equally, inversely proportional to faunal distance, or inversely proportional to ranked faunal distance.

The averaging of several modern samples for reconstruction increases the robustness and decreases the variance of the estimate, but it can also potentially reduce the resolution. This is the usual trade-off between accuracy and over-fitting in data modeling and forecasting. A good way of setting the parameters for MAT is cross-validation, where the measured temperature for each modern sample is compared with the estimated temperature for the same sample. Carrying out this procedure for all modern samples, we can compute the "root mean square error of prediction" (RMSEP) as the square root of the average squared prediction error.

---

**Example 11.1**

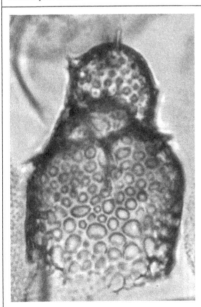

Figure 11.1 The nassellarian radiolarian *Lithomelissa setosa*, a warm-water indicator in the North Atlantic. Length ca. 60 μm. Bjørklund et al. (2019)/Norwegian Polar Institute.

Bjørklund et al. (2019) studied the radiolarian assemblages in a core from a fiord in northern Norway, spanning the Pleistocene–Holocene transition (Fig. 11.1). The radiolarians were assigned to 34 species and counted in 1-cm thick core slices. The same species had previously been counted at 160 sites in the modern North Atlantic (Kruglikova et al. 2010). For each of these core-top sites, temperatures were recorded at 200 m depth in spring (April–June), when radiolarian productivity peaks. These measured temperatures vary from −0.8 °C to 7.6 °C. We reconstructed the downcore temperatures using the MAT technique, with the squared chord distance, inverse distance weighting, a maximum of six nearest samples, and no distance thresholding. These parameters were partly chosen in order to minimize the cross-validated mean prediction error (RMSEP), which is 0.74 °C in this case (Fig. 11.2A).

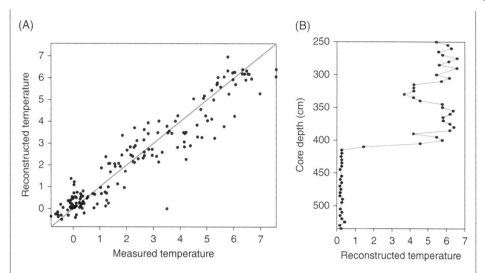

**Figure 11.2** Reconstruction of temperatures in the JM99-1200 core from Andfjorden, Norway, based on radiolarians and implementing the modern analog technique. (A) Measured versus reconstructed temperatures in core-top samples from the North Atlantic. The *x* = *y* line is shown in red. (B) Reconstructed downcore temperatures in the JM99-1200 core.

The downcore temperature curve is shown in Fig. 11.2B. There is a dramatic temperature increase from just over 0 °C in the lower part of the core to around 5 °C in the upper part. This jump in temperature is observed around 410-cm core depth, which is close to the base of the Holocene (termination of the Younger Dryas) according to radiocarbon dates and other data from the core. An interesting dip in temperature is observed around 330-cm core depth. The radiocarbon ages from the core indicate that this may represent the so-called Preboreal Oscillation, a cold phase in the earliest Holocene. It certainly seems that the radiolarian fauna can work as a sensitive temperature proxy in this case.

## 11.2 Weighted averaging

Weighted averaging (WA) is a robust and relatively simple method for reconstructing past environmental parameters (e.g., temperature, pH) from a fossil assemblage, based on a training set of modern samples (ter Braak and Barendregt 1986; Birks et al. 1990). The algorithm can be described as follows, based on Birks et al. (1990).

For the modern (training) set, we have $x_i$, the value of the measured environmental parameter at site $i$, and the $n \times m$ matrix **Y** with $y_{ik}$ the abundance of taxon $k$ at site $i$. There are $n$ sites and $m$ taxa. Moreover, a "+" replacing a subscript means summation over that subscript.

Calculate the optimum environmental value for each taxon as an average of the environmental values across sites in the training set, weighted by the abundance of the taxon in each site:

$$\hat{u}_k = \sum_{i=1}^{n} \frac{y_{ik} x_k}{y_{+k}}$$

Optionally, calculate a tolerance for each taxon:

$$\hat{t}_k = \sqrt{\sum_{i=1}^{n} y_{ik} \left( x_i - \hat{u}_k \right)^2 \Big/ y_{+k}}$$

For each taxon, we then calculate a reconstructed environmental value for each site in the training set, as a weighted mean of the optimum values:

$$\hat{x}_i = \sum_{k=1}^{m} y_{ik} \hat{u}_k \Big/ y_{i+}$$

Alternatively, we can use a tolerance-weighted mean, with lower weights for taxa with wide tolerances:

$$\hat{x}_i = \frac{\sum_{k=1}^{m} y_{ik} \hat{u}_k \Big/ \hat{t}_k^2}{\sum_{k=1}^{m} y_{ik} \Big/ \hat{t}_k^2}$$

We have now taken averages twice, leading to a reduction in the range of reconstructed environmental values. To compensate for this reduction (deshrinking), we carry out a linear regression of the original and reconstructed values, i.e., we fit the model (section 4.6)

$$\hat{x}_i = a + b x_i + e_i$$

Birks et al. (1990) discuss the use of this form of regression versus inverse regression, with the inferred values as the independent variable. A model II regression is a possible third alternative. The deshrinking produces the final reconstructed values:

$$x'_i = \left( \hat{x}_i - a \right) / b$$

For a new (downcore) sample $y_{0k}$, the reconstructed environmental value $x_0$ is estimated as for $x'_i$ in the above equation, with or without tolerance weighting.

---

**Example 11.2**

The WA reconstruction results for the Bjørklund et al. (2019) example are shown in Fig. 11.3, without tolerance weighting and with RMA deshrinking. The cross-validated root mean square error of prediction (RMSEP) is 1.04 °C. The downcore temperature curve is similar to the MAT curve (section 11.1), but with a somewhat larger total temperature range and a less marked return to high temperatures after the negative excursion around 330-cm core depth.

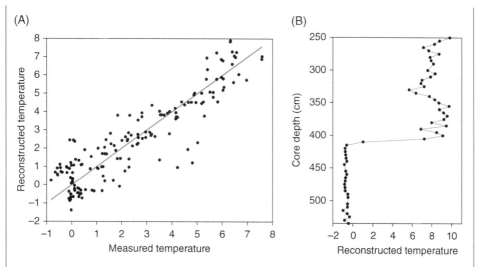

**Figure 11.3** Reconstruction of temperatures in the JM99-1200 core from Andfjorden, Norway, based on radiolaria and with the WA algorithm. (A) Measured versus reconstructed temperatures in core-top samples. (B) Reconstructed downcore temperatures.

## 11.3 Weighted averaging partial least squares

WA-PLS is another calibration method, like MAT and WA. It was first described by ter Braak and Juggins (1993) and ter Braak et al. (1993). The algorithm is described below, following ter Braak and Juggins (1993). For notation, see the description of WA (section 11.2).

**Step 0**: Subtract the weighted mean from the environmental variable:

$$x_i = x_i - \sum_i y_{i+} x_i / y_{++}$$

**Step 1**: Take the centered environmental variable $x_i$ as initial site scores $r_i$. Follow steps 2–7 for each PLS component $p$:

**Step 2**: Calculate new species scores by weighted averaging of the site scores:

$$u_k^* = \sum_i \frac{y_{ik} r_i}{y_{+k}}$$

**Step 3**: Calculate new site scores $r_i$ by weighted averaging of the species scores:

$$r_i = \sum_k \frac{y_{ik} u_k^*}{y_{i+}}$$

**Step 4**: For the first PLS component, go to step 5. For second and higher components, make the new site scores $r_i$ uncorrelated with previous components by orthogonalization, according to ter Braak (1987), Table 5.2b.

**Step 5**: Take the site scores $r_i$ and the species scores $u_k^*$ as the new PLS component consisting of two vectors $\mathbf{r}^p$ and $\mathbf{u}^p$. In the original algorithm, site scores are standardized in step 5. In the PAST implementation, this standardization is not carried out, in order to facilitate reconstruction for new samples (ter Braak, pers. comm. 2019).

**Step 6**: Do a weighted multiple regression of $x_i$ on the components $\mathbf{r}$ obtained so far using weights $y_{i+}/y_{++}$. The regression coefficients are $a_0 \dots a_p$. Take the fitted values as current estimates $\hat{x}_i$ (as used for calculating RMSE). Go to step 2 with the residuals of the regression as the new site scores $r_i$.

**Reconstruction**: After steps 2–6 have been iterated the specified number of times, a full PLS model has been constructed. Reconstruction of the environmental value $x_0$ from a new sample $y_{0k}$ is then computed as follows (in addition it must be remembered to add back the mean value subtracted in step 0). First, calculate updated species optima:

$$\hat{u}_k = a_0 + \sum_p a_p u_k^p$$

Then, the reconstructed $x_0$ is calculated as the weighted sum

$$x_0 = \sum_k y_{0k} u_k / y_{0+}$$

The method is cross-validated with the leave-one-out procedure (jackknifing), which is the basis for the RMSEP value. The number of PLS components should be set in order to minimize the RMSEP. Sometimes the minimal value is obtained already with only one component when the method is equivalent to two-way WA.

---

### Example 11.3

We return to the radiolarian data set from Bjørklund et al. (2019). The WA-PLS algorithm was run with the number of components varying from 1 to 6. The resulting RMSEP values are shown in Table 11.1.

The smallest RMSEP value is reported using four components (RMSEP = 0.79 °C), so we select this as our best value (Fig. 11.4A). The resulting temperature curve (Fig. 11.4B)

Table 11.1 Errors of prediction as a function of a number of components included in the WA-PLS analysis of the Bjørklund et al. (2019) data set.

| Number of components | RMSEP (°C) |
| --- | --- |
| 1 | 1.03 |
| 2 | 0.85 |
| 3 | 0.81 |
| 4 | 0.79 |
| 5 | 0.82 |
| 6 | 0.88 |

Bjørklund et al. (2019).

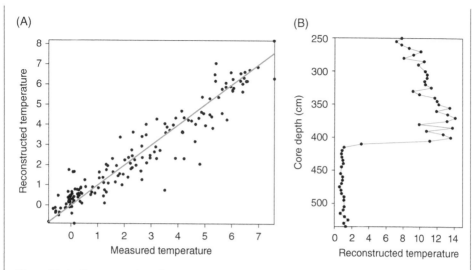

Figure 11.4 Reconstruction of temperatures in the JM99-1200 core from Andfjorden, Norway, based on radiolaria and with the WA-PLS algorithm. (A) Measured versus reconstructed temperatures in core-top samples. (B) Reconstructed downcore temperatures.

supports the rapid temperature increase at 410-cm depth shown by the MAT and WA results above, but there are differences in the details. Most importantly, the warmest temperatures just after the Younger Dryas reach 14 °C. This seems an unrealistically high temperature at 200-m water depth north of the Arctic Circle, and much higher than the MAT temperatures, which never exceed 7 °C. This is partly due to extremely high abundances in the core (up to 88%) of the warm-water radiolarian *Lithomelissa setosa*. The samples in the core-top data set never contain such high abundances of warm-water species, and the measured temperatures are never above 8 °C. Because MAT temperatures are calculated as the averages of modern sample temperatures, the MAT method can never go beyond the modern temperatures. WA-PLS, in contrast, can extrapolate from the core-top data. We believe that in this case, this extrapolation has gone astray because the extreme blooms of warm-water species in the core were partly caused by other factors than temperature.

## 11.4 Which calibration method?

In the examples above, MAT seemed to give the "best" results among the three methods, with a low RMSEP value (0.74 °C) for the training set and a reasonable temperature range for the downcore reconstruction. WA also gave a reasonable downcore result, but with a somewhat higher RMSEP value (1.04 °C). WA-PLS gave a low RMSEP value (0.79 °C) but seemed to extrapolate to excess temperatures for some downcore samples. It is not possible to give a general recommendation about any method being superior for all data sets. Rather, we recommend trying several methods and assessing the results critically.

Newer machine learning methods such as decision trees/random forests and neural networks (section 6.20) can also be applied to the calibration problem. For example, Salonen et al. (2019) tested several methods for temperature reconstruction from pollen samples. They obtained the best cross-validation results for ensemble regression tree methods, followed by MAT, WA, and WA-PLS. Neural networks performed slightly less well.

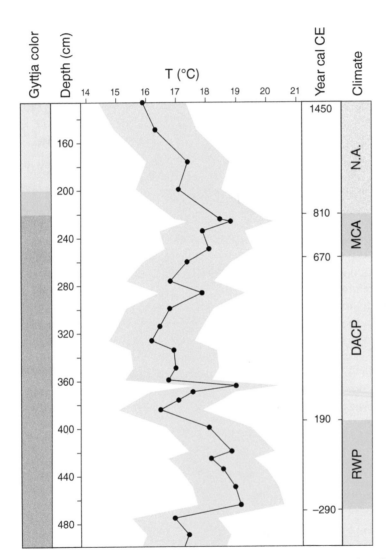

**Figure 11.5**  Temperature reconstruction through the Upper Holocene based on WA-PLS calibration of chironomid assemblages, Lake Uddelermeer, the Netherlands. 95% confidence interval. RWP = Roman Warm Period. DACP = Dark Age Cold Period. MCA = Medieval Climate Anomaly. N.A. = Not available. Data from Gouw-Bouman et al. (2019).

## 11.5  Case study: Late Holocene temperature inferred from chironomids

Gouw-Bouman et al. (2019) used WA-PLS calibration to reconstruct Late Holocene pale-otemperatures through a core from Lake Uddelermeer, the Netherlands. The study was based on chironomid assemblages (nonbiting midges), with a large training set consisting of modern chironomid samples and associated July air temperatures from 274 lakes in Norway and the Alpine region (Heiri et al. 2011). Confidence intervals were constructed by bootstrapping samples in the training set. Radiocarbon ages were obtained from 19 samples through the core, in addition to $^{210}$Pb ages in the uppermost part, giving a detailed age model.

The reconstructed temperature curve is shown in Fig. 11.5. Gouw-Bouman et al. (2014) divided the core into intervals based on temperature.

The slightly higher temperatures from 2240 to 1760 cal. yr. BP (i.e., 290 BCE to 190 CE) can be interpreted as the so-called Roman Warm Period (RWP), the end of which was roughly contemporaneous with the collapse of the Roman Empire and subsequent decrease in population in NW Europe. The following cool climate interval from ca. 190 to 670 CE (390–255-cm core depth) can be correlated with the Dark Ages Cold Period (DACP), although the DACP is usually placed at ca. 400–765 CE. Finally, the warmer interval from 670 to 810 CE corresponds to the Medieval Climate Anomaly, a phase of strong population growth and the emergence of Medieval kingdoms in Europe. The upper part of the core consists of sediments deposited under eutrophication. The reconstructed temperatures above 220-cm core depth are therefore considered unreliable.

In conclusion, the chironomid assemblages at Lake Uddelermeer seem to have responded to climate events that were of profound cultural importance in Europe from Roman and well into Medieval times.

## References

Birks, H.J.B., Line, J.M., Juggins, S., Stevenson, A.C., ter Braak, C.J.F. 1990. Diatoms and pH reconstruction. *Philosophical Transactions of the Royal Society of London B* 327, 263–278.

Bjørklund, K.R., Kruglikova, S.B., Hammer, Ø. 2019. The radiolarian fauna during the Younger Dryas–Holocene transition in Andfjorden, northern Norway. *Polar Research* 38, 3444.

ter Braak, C.J.F. 1987. Ordination. In: Jongman, R.H.G., ter Braak, C.J.F., van Tongeren, O.F.R. (eds), *Data Analysis in Community and Landscape Ecology*, pp. 91–173. Pudoc.

ter Braak, C.J.F., Barendregt, L.G. 1986. Weighted averaging of species indicator values: its efficiency in environmental calibration. *Mathematical Biosciences* 78, 57–72.

ter Braak, C.J.F., Juggins, S. 1993. Weighted averaging partial least squares (WA-PLS): an improved method for reconstructing environmental variables from species assemblages. *Hydrobiology* 269/270, 485–502.

ter Braak, C.J.F., Juggins, S., Birks, H.J.B., van der Voet, H. 1993. Weighted Averaging Partial Least Squares regression (WA-PLS): definition and comparison with other methods for species-environment calibration. In Patil, G.P., Rao, C.R. (eds.), *Multivariate Environmental Statistics*, 525–560. Elsevier, Amsterdam.

Flower, R.J., Juggins, S., Battarbee, R.W. 1997. Matching diatom assemblages in lake sediment cores and modern surface sediments samples: the implications for lake restoration, management objectives and biodiversity. *Hydrobiologia* 344, 27–40.

Gouw-Bouman, M.T.I.J., van Asch, N., Engels, S., Hoek, W.Z. 2019. Late Holocene ecological shifts and chironomid-inferred summer temperature changes reconstructed from Lake Uddelermeer, the Netherlands. *Palaeogeography, Palaeoclimatology, Palaeoecology* 535, 109366.

Kruglikova S.B., Bjørklund K.R., Dolven J.K., Hammer Ø., Cortese G. 2010. High-rank polycystine radiolarian taxa as temperature proxies in the Nordic seas. *Stratigraphy* 7, 265–281

Overpeck, J.T., Webb, I.I.I.T., Prentice, I.C. 1985. Quantitative interpretation of fossil pollen spectra: dissimilarity coefficients and the method of modern analogs. *Quaternary Research* 23, 87–108.

Salonen, J.S., Korpela, M., Williams, J.W., Luoto, M. 2019. Machine-learning based reconstructions of primary and secondary climate variables from North American and European fossil pollen data. *Scientific Reports* 9, 15805.

# 12

# Time series analysis

A time series is a data set consisting of discrete data points collected (sampled) through time. In this chapter, we will only consider *univariate* time series, meaning a series of single numbers collected at successive points in time. Examples of such time series are stock-market curves, temperature records through time, and curves of biodiversity through Earth's history. The data points can be collected at even intervals (each second or each year) or at uneven intervals. Some methods of analysis can only handle evenly sampled time series. Time series analysis depends on reasonably accurate dates for the samples. This can of course be a problem in paleontology, where stratigraphic correlation and radiometric dating usually have considerable error bars.

From a mathematical viewpoint, there is nothing special about univariate time series. They are simple functions $y_i = f(t_i)$ where the $t_i$ represent time values and could be studied using methods outlined in chapter 4, such as regression (curve fitting). However, some aspects of time series analysis make it natural to consider it a subject of its own. Particular questions involving time series analysis include the identification of non-random structures such as periodicities or trends and the comparison of two time series to find an optimal alignment. A special challenge is that time series are usually autocorrelated (smooth), meaning that successive values are more similar than values further apart. This violates the assumptions of some statistical methods.

Perhaps the most famous example of time series analysis in paleontology is the paper by Raup and Sepkoski (1984) claiming a 26 million year periodicity in extinction rates during the last 250 million years. Although such an extinction cycle now seems less obvious (e.g., Erlykin et al. 2017, 2018; but see also e.g., Rampino et al. 2021), the identification of periodicities remains an exciting part of paleontology. Phenomena such as orbital (Milankovitch) cycles, annual isotopic cycles, and cyclicity in solar and tectonic activity may be reflected in the stratigraphic and paleontological records (Fig. 12.1), but the signal is often so subtle that it can only be detected using special analytical tools. A good overview of time series analysis in stratigraphy is provided by Weedon (2003).

*Paleontological Data Analysis*, Second Edition. Øyvind Hammer and David A.T. Harper.
© 2024 John Wiley & Sons Ltd. Published 2024 by John Wiley & Sons Ltd.

**Figure 12.1** Limestone-marl alternations in the type section for the Hauterivian Stage (Lower Cretaceous, La Charce, Drôme, France). The thick limestone units represent 100 kyr cycles, corresponding to one of the Milankovitch periodicities (orbital eccentricity).

## 12.1 Spectral analysis

A time series such as a diversity or paleotemperature curve can always be viewed as a sum of superimposed parts, and we can imagine many ways of decomposing the time series into such components in order to bring out structure in the data. For example, we have already seen how we can use linear regression to split bivariate data into a linear part and "the rest" (the residual).

A particularly useful way of detecting structure in a time series is to decompose it into a "sum of sines" (Fourier series). The calculation of such decomposition is usually called *spectral analysis*. It can be shown mathematically that for any function $F(t)$ of total duration $T$, we can find values $a_i$ (amplitudes) and $\omega_i$ (phases) such that

$$F(t) = \sum_{i=0}^{\infty} a_i \cos\left(2\pi\, it/T + \omega_i\right)$$

In other words, the time series can be viewed as a sum of periodic, sinusoid functions, each with its own amplitude and phase, and with a period (duration of one cycle) determined by the integer-valued index $i$. Let us look at these components individually (Fig. 12.2). For $i = 0$, the component reduces to a constant

$$a_0 \cos\left(0 + \omega_0\right) = a_0 \cos\omega_0$$

This constant will simply be identical to the mean value of the time series. We usually force this "nuisance parameter" to zero by subtracting the mean from the time series prior

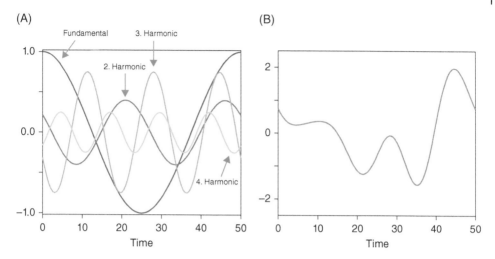

Figure 12.2   (A) The first four harmonics within a time window of length 50 units, each with a different amplitude and phase. (B) The sum of these four sinusoids. Conversely, the time series to the right can be decomposed into the sinusoids to the left.

to spectral analysis. For $i = 1$, the component consists of exactly one period of a sinusoid within the time window $T$:

$$a_1 \cos\left(2\pi t/T + \omega_1\right)$$

This component is known as the *fundamental*. It has a period of $T$ and consequently a frequency (periods per time unit) of $f_1 = 1/T$. For $i = 2$, the component consists of exactly two periods within the time window. This component is called the *second harmonic*, with a period of $T/2$ and a frequency of $f_2 = 2f_1$. Higher order harmonics will all have frequencies at multiples of the fundamental frequency, that is $f_i = if_1$.

A plot of the amplitudes $a_i$ as a function of frequencies $f_i$ is called a *spectrogram*. Often, we plot the squares of $a_i$, giving the so-called *power spectrum* of $F$ (Fig. 12.4). The phases $\omega_i$ are usually of less interest and are not plotted. From the discussion above, it should be clear that the lowest directly observable frequency is $f_1$, corresponding to only one full cycle through the observed time series, and the resolution in the spectrogram (the distance in frequency between consecutive values) is also $f_1$. If your time series is long, you have a large $T$ and therefore a small $f_1 = 1/T$, giving you good spectral resolution.

The amplitudes usually diminish for the higher harmonics, so we can cut off the spectrogram at some high frequency without losing too much information. The discrete sampling of the original continuous time series also sets a limit on the highest frequency that can exist in the decomposition (the Nyquist frequency). For an evenly sampled time series, the highest observable frequency has a period corresponding to two times the interval between two consecutive samples.

The main attraction of this form of decomposition is that a sinusoid can be regarded as the simplest type of periodic function. The analysis consists of calculating how strong a presence (energy) we have of different sinusoidal components at different frequencies. If you have one or several sinusoidal periodicities present in your time series, they will turn up as narrow peaks in the spectrogram and can be easily detected. Such near-sinusoidal periodicities are typical for Milankovitch cycles in climatic records. Cycles with other shapes, such as a sawtooth-like curve, can themselves be decomposed into a harmonic series of sinusoids with a certain fundamental frequency, and such a harmonic series can also be identified in the spectrogram as a series of evenly spaced peaks.

There is one basic problem with the spectral analysis approach as outlined earlier. While it is correct that any time series of finite duration can be decomposed into a sum of sinusoids in a harmonic series, the frequencies of these harmonics do not necessarily correspond to the "real" frequencies in the time series from which our observation window was taken. Suppose that there is a real Milankovitch cycle in nature with a period of 41,000 years and that we have observed a time window of 410,000 years. There will then be precisely 10 cycles within the window, and our spectral analysis will pick this out as the tenth harmonic. But we will normally not be that lucky. If our observation window was 400,000 years long, there would be 9.76 cycles within our window, and this does not correspond precisely to any whole-numbered harmonic in the analysis. The result is that the 41,000 years cycle will "leak" out to harmonic numbers 9 and 10 and, to a smaller extent, to harmonics even farther away. Instead of a narrow, well-defined peak, we get a broader lobe and lateral noise bands (side lobes). This phenomenon reduces the resolution of the spectrogram.

Most real-time series will generate a spectrogram full of peaks. Even a sequence of random, uncorrelated numbers (white noise) will produce a large number of spectral peaks, and a single sinusoid can also give rise to multiple peaks because of spectral leakage. Only the strongest peaks will correspond to interesting periodicity in the time series, and we need to be able to quantify their statistical significance. One way of proceeding is to formulate the null hypothesis that the time series is *white noise*. Like white light, white noise contains all frequencies in equal proportion, but with random phases. With a given window length and sampling density, we can calculate the probability that the observed peak could have occurred in such a random sequence. If this probability is low, we have a statistically significant peak. Also, a *white noise line* can be calculated for any significance level. Any peak that rises above the $p = 0.05$ white noise line can be considered significant at that level. While the white noise line method uses a null hypothesis of an uncorrelated, random signal, the alternative *red noise* method operates with respect to a somewhat autocorrelated sequence, which is more realistic for most time series (Schulz and Mudelsee 2002). But that ominous sharpshooter from Texas (section 1.2) is now lurking in the distance – while it may be improbable to observe a strong spectral peak at a particular frequency, it is not so improbable to see a strong peak at any frequency (Smith and Bailey 2018; Vachula and Cheung 2023). This can to some extent be alleviated by correction for multiple testing, e.g., using the "critical false alarm" significance level in the REDFIT algorithm (Schulz and Mudelsee 2002).

Apparently significant peaks at low frequencies, corresponding to say three periods or less over the duration of the time series, must be regarded with some caution. One danger is that we are observing spectral leakage from, or harmonics of, a strong frequency component close to $f_1$. Such effects will be observed if there is a trend in the time series, increasing or decreasing steadily throughout the observation window. It is therefore recommended to *detrend* the time series before spectral analysis, by subtracting the linear regression line.

Another almost philosophical question is whether we can reasonably claim the presence of a recurrent oscillation if we have observed only one or two cycles.

If the amplitude or frequency of a sinusoid varies from one cycle to the next, this will result in the widening or even splitting of the spectral peak, with a corresponding drop in the height of the peak. Such *modulation effects* are quite common in real data.

To conclude, the interpretation of a spectrogram involves some subtle considerations and is not entirely straightforward.

## 12.1.1 Discrete Fourier transform

Spectral analysis of sampled data is a large and well-developed field, and there are many different approaches to power spectrum estimation. The simplest way is probably to find the inner products (proportional to correlation coefficients) between the discrete time series $F_n$ (here assumed to be sampled at equal intervals) and a harmonic series of sines and cosines. This is known as the discrete Fourier transform (DFT), and it will produce the spectral coefficients $a_i$ and $\omega_i$ directly:

$$c_i = \frac{1}{N}\sum_{j=1}^{N} F_j \cos\frac{2\pi i (j-1)}{N}$$

$$d_i = \frac{1}{N}\sum_{j=1}^{N} F_j \sin\frac{2\pi i (j-1)}{N}$$

$$a_i = \sqrt{c_i^2 + d_i^2}$$

$$\omega_i = \tan^{-1}\frac{d_i}{c_i}$$

In many texts, the DFT is presented using complex numbers, which simplifies the notation greatly. For very long time series, these calculations can be slow. Much faster implementations are collectively known as fast Fourier transform (FFT) (see Press et al. 1992 for details).

The DFT assumes that the data points are evenly spaced along the time axis, and time values do not therefore enter the expressions. Unevenly spaced data could of course be interpolated at constant intervals, but this can lead to some distortion in the spectrogram in the form of energy loss at high frequencies. A better approach is to use the Lomb periodogram method, as described by Press et al. (1992).

It can be shown that direct Fourier methods are not optimal with respect to maximizing resolution and/or minimizing variance when estimating a spectrum with a noisy background. More sophisticated approaches include the Blackman-Tukey method, which smooths the spectrum in order to reduce variance, and the so-called parametric methods that work by optimizing model parameters (including the autoregressive [AR] method). Proakis et al. (1992) give a short but thorough introduction.

## 12.1.2 Spectral analysis with the REDFIT procedure

Modern spectral analysis of paleoclimatic and paleontological time series has developed beyond the straightforward application of Fourier analysis and a white-noise null hypothesis, as described earlier. We will describe two of the commonly used procedures: REDFIT and the multitaper method.

REDFIT (Schulz and Mudelsee 2002) was originally implemented as a stand-alone FORTRAN program, but is now also available in other software, including PAST and R. It is based on the Lomb periodogram method (Press et al. 1992) which allows it to process time series with uneven sampling. REDFIT includes several options that are used generally in spectral analysis, in addition to testing against a red-noise null hypothesis.

### 12.1.2.1 Window function

Above, we discussed the phenomenon of spectral leakage, which happens when the underlying periodicities in the signal do not match the analysis frequencies at integer multiples of the fundamental frequency (harmonics). This causes spurious "side lobes" in the spectrum. One way of reducing the amplitude of these annoying side lobes is to multiply the original signal with a *window function*, which typically has a maximum towards the middle of the analysis window and tapers towards zero near the edges. However, this procedure reduces the spectral resolution and gives some information loss at the beginning and end of the time series. Different window functions provide different trade-offs between side lobe attenuation and resolution. The window functions available in REDFIT are, in order of increasing side lobe attenuation but decreasing resolution: Rectangular, Triangle, Welch, Hanning, and Blackman-Harris. The rectangular window is a constant function (i.e., no tapering). Typically, the latter windows are used for longer and more densely sampled time series where we can afford a lower resolution to gain a clearer spectrum.

### 12.1.2.2 Segment averaging

One efficient way to reduce noise in the spectrum is to split the time series into several segments and average their individual spectra. This will tend to amplify the stable periodicities relative to random variation. However, the smaller sequence lengths imply a reduction in spectral resolution, as discussed earlier.

### 12.1.2.3 Red-noise model and significance curve

As the name implies, REDFIT includes fitting to a red-noise model and testing spectral peaks for significance using this model. Above, we discussed a white-noise model. A white-noise spectrum is expected from a sequence of random numbers where each number is independent of the others. However, this is not a good null hypothesis for most time series. Most natural sequences show some degree of smoothness, usually because the system reacts slowly. A simple way to model this is with a so-called autoregressive (AR) model, where the output of the system is a linear combination of previous outputs in addition to a white noise term (the residual). The simplest case is the AR(1) model, which uses only the most recent previous output:

$$y_t = c + \rho y_{t-1} + \varepsilon_t$$

where $\varepsilon_t$ is white noise and $c$ is just a constant, often set to zero. The interesting parameter, which must be fitted to the data, is $\rho$ (rho), typically in the range 0 to 1 (but can also be negative). With $\rho = 0$, there is no serial dependence, and the AR model reduces to white noise. With, e.g., $\rho = 0.9$, the output depends strongly on the previous value, and the signal will be smoother.

A signal obeying the AR(1) equation is often called a red-noise (or random walk) signal. It will have a particular spectral shape, depending on $\rho$, with higher amplitudes at low frequencies because of the smoothness. It turns out that the AR(1) model often gives a good representation of the "noise" part of natural time series, and it therefore usually provides a better null hypothesis than the white-noise model.

---

**Example 12.1**

The Fossil Record database compiled by Benton (1993) contains counts of fossil families per stage throughout the Phanerozoic and also the number of extinctions and originations per stage. We will use spectral analysis to investigate the putative 26 million years periodicity in extinction rates during the last 250 million years (Raup and Sepkoski 1984). We extend the analysis window to 356 million years here. The data points are unevenly spaced in time, which is handled well by the Lomb periodogram method.

The extinction curve from the Carboniferous (Tournaisian) to the Holocene as derived from Benton (1993) is shown in Fig. 12.3, and the power spectrum calculated with the Lomb method is given in Fig. 12.4A. Frequencies up to 0.08 cycles per million years are shown. A strong peak is found at a frequency of 0.0055 cycles/myr. This corresponds to a period of $1/0.0055 = 182$ myr, which is comparable to the length of the time series (356 myr). Periodicities with such low frequencies should be disregarded. In fact, this period simply reflects the single time interval from the P/T to the K/Pg extinction events.

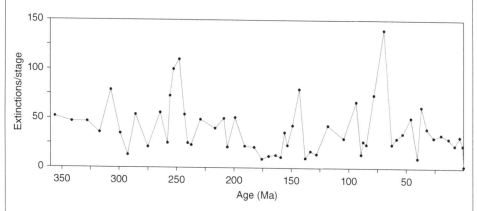

**Figure 12.3** Extinctions per stage from the Carboniferous to the present, at the family level. Data from Benton (1993).

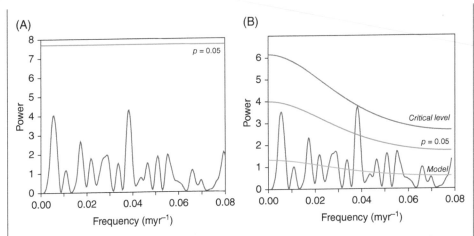

**Figure 12.4** (A) Power spectrum of the extinction curve computed with the Lomb periodogram method and detrending. The frequency axis is in units of cycles per million years. The $p = 0.05$ white noise line is found at a spectral power value of 7.7. (B) The REDFIT analysis, with the fitted red-noise model (orange), $p = 0.05$ (red), and critical significance (green) levels.

The next substantial peak is found at a frequency of 0.038 cycles/myr, corresponding to a period of 1/0.038 = 26 myr. This reproduces the result of Raup and Sepkoski (1984) and indicates the presence of a "cycle of doom." However, the peak does not reach the white noise line for a significance level of $p = 0.05$.

One technical detail to be noted here is that the spectrogram is seemingly drawn with a higher resolution than the 1/356 = 0.00281 cycles/myr that we would expect from the discussion above. This allows the low-frequency peak to be localized to 0.0055 cycles/myr, which is more than the fundamental frequency of the analysis but less than the frequency of the second harmonic. This degree of precision is achieved by a procedure equivalent to interpolation in the spectrogram, and it does not improve the resolution of the analysis.

The REDFIT algorithm produces the same spectrum (Fig. 12.4B) as the Lomb method (in fact the same algorithm is used), but in addition, it fits a red-noise model (orange line) and uses it to estimate significance levels. The "critical level" is corrected for multiple testing. In contrast with the white-noise model, the red-noise model leads to a significant peak for a periodicity of 26 myr.

### 12.1.3 Spectral analysis with the multitaper method

Above, we discussed the trade-offs involved in the selection of window functions for spectral analysis. In the *multitaper* method, several window functions (tapers) are applied to the time series, and the resulting spectra are averaged (Thomson 1982). Some of the tapers have high amplitude at the beginning and end of the window, reducing information loss near the edges. In addition, the tapers are chosen to be orthogonal, giving an improved signal-to-noise ratio in the averaged spectrum.

Multitaper spectral analysis has become a popular method, especially in paleoclimatology. However, the standard algorithms do not support unevenly sampled data, which would have to be interpolated prior to analysis (with some loss of accuracy). Also, the multitaper method is often less successful for very short time series, at least with the default parameters, giving a very smooth spectrum with little resolution.

Several programs, including PAST, include variants of a multitaper code first described by Mann and Lees (1996). This implementation includes red-noise modeling which is more advanced than the one in REDFIT. A possible argument against the REDFIT approach is that the fitting procedure is applied to the original time series, which is usually not a red-noise signal but contains the periodic components that we want to test for significance. Schulz and Mudelsee (2002) suggest that this is usually not a concern, because the spectral peaks are narrow and do not strongly influence the red-noise fitting. Still, the Mann and Lees (1996) algorithm addresses this problem by trying to identify strong peaks in the spectrum, which are then removed. The resulting "reshaped" spectrum is then subjected to red-noise fitting, and confidence lines are constructed.

## 12.1.4 Evolutive spectral analysis

So far, we have subjected the whole time series to one big spectral analysis, giving a single spectrum. This spectrum hopefully gives a faithful representation of the information in the signal. However, the periodicities that we are interested in may vary in amplitude and frequency through the time series. This is the case for climate curves partly controlled by Milankovitch cycles, where the amplitudes of different cycles (eccentricity, obliquity, precession) vary greatly through time. Another example is a sample series collected through a stratigraphic sequence. Since the sedimentation rate is not constant, the frequencies measured in cycles per meter will vary even for cycles of constant temporal duration. For such non-stationary series, the modulation of amplitude and frequency will not be obvious in the spectrum but will give rise to effects such as peak widening and sidebands that are very difficult to interpret.

A way forward is a hybrid between analysis in the time domain and the frequency domain called *evolutive* (or short-time) spectral analysis. The idea is to split the time series into a sequence of pieces, typically of length 64–1024 samples (these are powers of 2, for efficient implementation with the FFT). These pieces are subjected to spectral analysis separately. The pieces are made to overlap strongly, usually being positioned at increments of a single sample.

---

**Example 12.2**

---

We will use the oxygen isotope data from a composite sequence based on the deep-sea cores V19-30 and ODP 677 (Shackleton and Pisias 1985; Shackleton et al. 1990). The evolutive spectrum for the last one million years of this sequence is given in Fig. 12.5. The values in the time series have an equal spacing of 3000 years. A strong band is visible around a periodicity of 0.01 cycles/kyr, i.e., the 100,000 years Milankovitch cycle

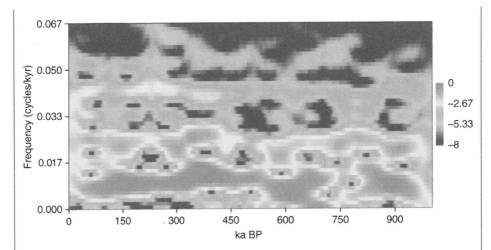

Figure 12.5   Evolutive spectral analysis of the oxygen isotope data, with a Welch window of size 64 samples, i.e., 192 kyr. The color coding of power is logarithmic. Note the high power around 0.01 cycles/kyr (100 kyr eccentricity cycle), around 0.025 cycles/kyr (41 kyr obliquity cycle), and 0.043 cycles/kyr (23 kyr precession).

(eccentricity), and also around 0.025 cycles/kyr, close to the 41,000 years obliquity cycle. Finally, a weaker band around 0.04 cycles/kyr may be interpreted as the 23,000 years precession. All these cycles are varying in amplitude and frequency. Such non-stationarity of cycles is easier to see in the evolutive spectrum than in a spectrogram, which has no time axis.

## 12.2   Wavelet analysis

With evolutive spectral analysis, we select one window size, e.g., 1024 samples. Larger window sizes give better frequency resolution, but poorer resolution along the time axis. Low temporal resolution is particularly annoying at the higher frequencies, where we might otherwise hope to pick up faster transitions. In contrast, wavelet analysis (e.g., Torrence and Campo 1998; Walker and Krantz 1999) is a method for studying a time series at all possible scales simultaneously, providing a kind of adjustable magnifying glass. Thus, fast, small-scale fluctuations can be separated from larger-scale trends. Wavelet analysis can also be regarded as a variant of evolutive spectral analysis, where the higher frequencies can be observed with higher temporal resolution. The result is conventionally shown as a *scalogram*, with time along the horizontal axis and scale up the vertical. The "strength" (energy) of the time series around a particular time and when observed at a particular scale is shown with a color or a shade of gray (Fig. 12.6).

The scalogram can bring out features that are difficult to see in the time series directly, where for example large-scale, long-wavelength features can visually "drown" in the presence of small-scale, short-wavelength oscillations. Intermediate-scale features can be particularly difficult to see in the original time series but will come out clearly at intermediate scales in the scalogram. For example, within sequence stratigraphy, geologists operate with sea-level cycles at different scales (first-order cycles of 200–400 million years duration, down to fourth- and fifth-order cycles of 10–400,000 years durations). With the possible exception of the smallest-scale cycles, which may be connected to orbital forcing, these cycles do not have fixed durations, and spectral analysis is therefore not suitable. With wavelet analysis, the overall sea-level curve can be decomposed into different "orders."

Wavelet analysis derives its name from the fact that it involves correlating the time series with a short, localized, oscillatory mathematical function (a "small wave") which is stretched in time to provide a range of observation scales and translated in time to provide a range of observation points along the time axis. There are many possible choices for the "mother wavelet," and in general we recommend that it is selected according to the nature of the time series. In practice, the so-called *Morlet* wavelet often performs well.

The wavelet transform $w(a, b)$ of a time series $y(t)$ using a mother wavelet $g(t)$ at scale $a$ and time $b$ is defined as follows:

$$w(a,b) = \frac{1}{\sqrt{a}} \int_{-\infty}^{\infty} y(t) g\left(\frac{t-b}{a}\right) dt$$

This can be interpreted as the correlation between $y(t)$ and the scaled and translated mother wavelet.

The Morlet wavelet is a complex sinusoid multiplied with a Gaussian probability density function. It has a parameter $w_0$ that can be arbitrarily chosen – in PAST it is set to 5:

$$g(t) = e^{iw_0 t - t^2/2}$$

The scalogram is the modulus squared of the complex $w$, calculated at a sampled grid over the $(a, b)$ plane.

While the wavelet transform can be computed by direct correlation between $y(t)$ and the translated and dilated wavelets, the calculations are made immensely more efficient by realizing that correlation in the time domain corresponds to a simple multiplication in the frequency domain. The scalogram can therefore be computed using the FFT.

---

**Example 12.3**

A scalogram of the oxygen isotope data set is shown in Fig. 12.6. The time series is observed at large scales at the bottom of the diagram. Note that the scales are distributed logarithmically on the vertical axis. A strong signal is seen for a scale close to the 100,000 years Milankovitch cycle (eccentricity). The cycle breaks up a little towards the right of the diagram, which corresponds to older samples. A second possible Milankovitch cycle is seen close to the 41,000 years cycle (obliquity), and some energy is also present around 23,000 years (precession).

The black line is the so-called "cone of influence." Below this line, the scalogram is influenced by edge effects near the beginning and end of the time series, and the values are less reliable.

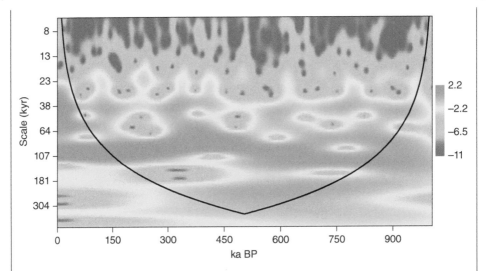

**Figure 12.6** Scalogram of oxygen isotope data from the last one million years before the present, using the Morlet wavelet. Time proceeds from right to left. Periodicities at two or three scales can be identified, corresponding to Milankovitch cycles.

Comparing the wavelet analysis with the evolutive spectrum (Fig. 12.5), the former is perhaps more aesthetically pleasing, but otherwise the information conveyed is comparable.

## 12.3 Autocorrelation

Autocorrelation (Davis 1986) is the comparison, by correlation, of a time series with delayed versions of itself. The result of this procedure is usually shown as an *autocorrelogram*, which is a plot of the correlation coefficient (section 4.5) as a function of delay time (lag time). The method can identify characteristics of the time series such as smoothness and cyclicity.

For a lag time of zero, we are comparing the sequence with an identical copy of itself, and the correlation is obviously equal to 1. If the sequence consists of somewhat smooth data, the correlation will remain relatively high for small lag times, but as the lag time exceeds the interval of interdependency, the autocorrelation will drop. Hence, the autocorrelogram can be used to identify whether there are any dependencies between close points (Fig. 12.7). Totally random, independent data will give small autocorrelation values at all lag times except zero, where the autocorrelation is one by definition. Smooth (autocorrelated) time series will have larger autocorrelation values at small lag times, typically diminishing for larger lag times.

Cyclicity can also be identified in the autocorrelogram as peaks for lag times corresponding to multiples of the period. The reason for this is that when we have delayed the sequence by a time corresponding to the period of the cycle, we are in effect again correlating

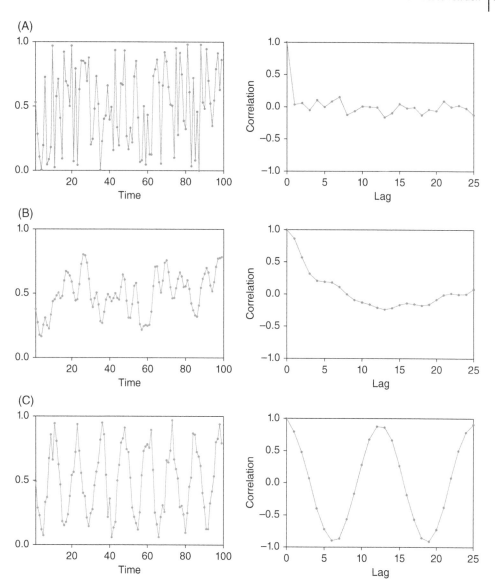

**Figure 12.7** Idealized time series (left) and their autocorrelograms (right). (A) Random, uncorrelated noise has a nonsystematic autocorrelogram with low values, except for a lag time of zero. (B) A random but slightly smoothed time series gives high autocorrelation values for small non-zero lag times. (C) A periodic time series gives rise to a periodic autocorrelogram, with peaks at lag times corresponding to multiples of the period (here around 12).

near-identical sequences, as for a lag time close to zero. Autocorrelation can sometimes show cyclicities clearer than spectral analysis, particularly for non-sinusoidal cycles.

For lag times larger than one-quarter of the sequence length, the variance of the autocorrelation becomes so large that it renders the analysis less useful. It is therefore customary to cut off the autocorrelogram at this point.

Unevenly sampled time series cannot be used directly for autocorrelation analysis. It is recommended to detrend (remove any linear trend in) the time series prior to autocorrelation.

Several different versions of the autocorrelation function have been proposed. The basic expression for the autocorrelation of a time series $x_n$ as a function of discrete lag time $k$ is simply

$$a_k = \frac{1}{N} \sum_{i=1}^{N-k} x_i x_{i+k}$$

In other words, we are taking the inner product of the time series and its delayed copy. The autocorrelation function is symmetrical around $k = 0$. Another similarly defined value is the *autocorrelation coefficient*, which is normalized to the range $-1$ to $+1$:

$$r_k = \frac{\sum_{i=1}^{N-k}(x_i - \bar{x})(x_{i+k} - \bar{x})}{\sum_{i=1}^{N}(x_i - \bar{x})^2}$$

(compare with the correlation coefficient of section 4.5). A slight variant of this expression is used in PAST, attempting to correct for a bias in the normalization that may occur for large lags relative to $N$: the dividend in the expression is replaced with the arithmetic mean of the sum-of-squared-deviations of the first $N - k$ and the last $N - k$ samples.

A simple significance test for one given lag time (Davis 1986) is based on the cumulative normal distribution with

$$z = r_k \sqrt{N - k + 3}$$

This test does not work well for $N < 50$ or $k > N/4$ (Davis 1986).

---

**Example 12.4**

We return to the oxygen isotope data from a composite sequence based on the deep-sea cores V19-30 and ODP 677 (Shackleton and Pisias 1985; Shackleton et al. 1990). The autocorrelogram for the last one million years of this sequence is given in Fig. 12.8. The values in the time series have an equal spacing of 3000 years. The relatively smooth appearance of the graph and the large correlation values for small lags indicate a non-random sequence with strong correlations between close samples. Peaks for lags of 30 and 39 samples indicate periodicities around 90,000 and 117,000 years, respectively. Another peak for a lag of 70 samples (210,000 years) may result from the same periodicity, this time for two periods of the fundamental cycle of about 100,000 years.

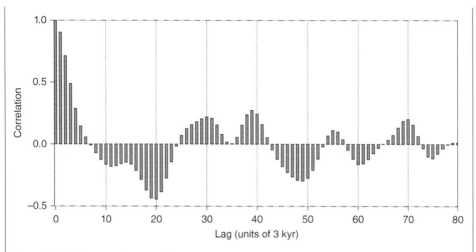

**Figure 12.8** Autocorrelogram of oxygen isotope data for the last one million years before the present.

## 12.4 Cross-correlation

Cross-correlation (Davis 1986) concerns the time alignment of two time series by means of the correlation coefficient ($r$). Conceptually, time series A is kept in a fixed position, while time series B is slid past A. For each position of B, the correlation coefficient is computed (Fig. 12.9). In this way, all possible alignments of the two sequences are attempted, and the correlation coefficient is plotted as a function of the alignment position. The position giving the highest correlation coefficient represents the optimal alignment.

The cross-correlation coefficient for two time series $x$ and $y$ at a displacement $m$ can be calculated as

$$r_m = \frac{\sum_i \left(x_i - \bar{x}\right)\left(y_{i-m} - \bar{y}\right)}{\sqrt{\sum_i \left(x_i - \bar{x}\right)^2 \sum_i \left(y_{i-m} - \bar{y}\right)^2}}$$

Here, the summations and the means are taken only over the region where the sequences are overlapping.

A rough significance test for one given alignment is given by using the Student's $t$ distribution with $n - 2$ degrees of freedom, where $n$ is the length of the overlapping sequence:

$$t = r_m \sqrt{\frac{n-2}{1-r_m^2}}$$

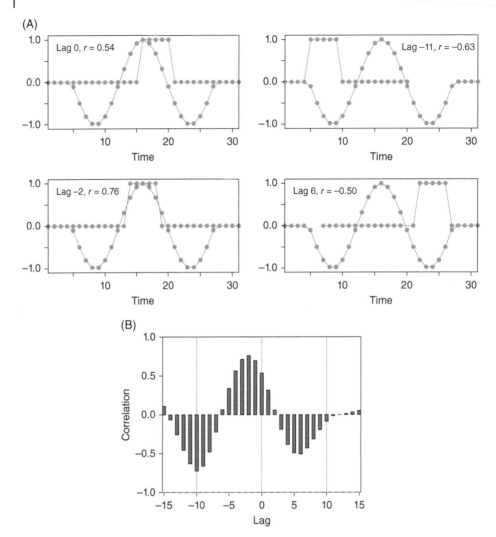

Figure 12.9 Cross-correlation of a "wavy" and a "square" time series. (A) Correlations of the wave (in a fixed position) with the square at different positions or lag times. $r$ = correlation coefficient. (B) The cross-correlogram for a range of lag times. Optimal alignment (maximal $r$) is achieved for a lag time of −2.

---

**Example 12.5**

Hansen and Nielsen (2003) studied the stratigraphic ranges of trilobites in a Floian (Lower Ordovician) section on the Lynna River in western Russia. A log of the proportion of trilobites of the genus *Asaphus* (Fig. 12.10) and the glauconite content through the section indicates a relation between the two curves, but the *Asaphus* curve seems slightly delayed (Fig. 12.11A). Glauconite is often associated with sequence stratigraphic high stands, while Hansen and Nielsen suggested that *Asaphus* is a

Figure 12.10   *Asaphus* with
eyes on stalks from the
Middle Ordovician, Lynna
River, western Russia, ca.
50 mm in length. Image by
Per Aas, NHM, Oslo.

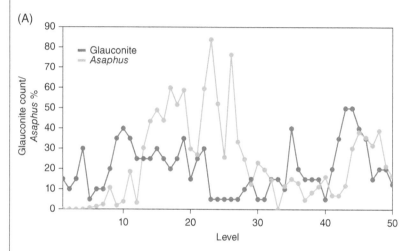

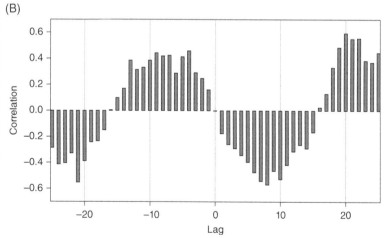

Figure 12.11   (A) Glauconite content and relative abundance of the trilobite genus *Asaphus*
through a Lower Ordovician section at Lynna River, Russia. Data from Hansen and Nielsen
(2003). (B) Cross-correlation of *Asaphus* abundance versus glauconite content. A peak positive
correlation is found for a lag around − 8, in the region around 8 samples delay of *Asaphus*
abundance with respect to glauconite content.

shallow-water indicator. How large is the delay, and what is the optimal alignment? We will try to use cross-correlation for this problem. The assumption of data points being evenly spaced in time may be broken, so we must take the result with a grain of salt. The section represents an interval of almost continuous sedimentation, but the sedimentation rate is likely to vary. However, the samples are taken from consecutive, alternating limestone and shale beds that may represent Milankovitch cycles, so it is possible that the data set does in fact represent an evenly spaced time series.

Figure 12.11B shows the cross-correlation. Maximal match is found around a lag around −8 (although with a broad peak), when the *Asaphus* abundance curve is delayed around 8 samples with respect to the glauconite curve.

## 12.5  Runs test

Consider a binary time series, such as 10011110, coding for the presence or absence of a fossil in a sequence of eight beds, for example. A *run* is defined as a consecutive sequence of the same digit (0 or 1). In this case, we have four such runs: 1, 00, 1111, and 0. If the sequence is totally random, with no interactions between the elements, it can be calculated that the expected number of runs in this example would be 4.75 (see below). In other words, we have fewer runs than expected. On the other hand, the sequence 10101010 has eight runs, much more than expected from a totally random sequence. This idea is the basis of the *runs test* (Wald and Wolfowitz 1942; Bradley 1968; Davis 1986; Swan and Sandilands 1995), also known as the Wald-Wolfowitz test, with the null hypothesis.

$H_0$: the sequence is totally random, with independence between points.

Non-binary data can be converted to binary data so that the runs test can be used. Consider the time series {1, 3, 2, 4, 5, 6, 8, 7}, with a mean 4.5. Write the binary digit 0 for numbers below the mean and 1 for numbers above the mean. This gives the sequence 00001111, with only two runs. In this way, the trend in the original time series causes a low number of runs compared with the expectation from the null hypothesis. Obviously, there could be a pattern in the original sequence that is lost by this transformation, reducing the power of the test.

With the runs test, it is particularly important to understand the null hypothesis. Many time series that we would tend to call random may not be reported as such by the runs test, because of a smoothness that decreases the number of runs. If it is smooth, it does not comply with the null hypothesis of the runs test.

Consider a binary sequence with $m$ zeros and $n$ ones. If the null hypothesis of total randomness holds true, the expected number of runs is

$$R_E = 1 + \frac{2mn}{m+n}$$

The expected variance is

$$\sigma^2 = \frac{2mn\left(2mn - m - n\right)}{\left(m+n\right)^2\left(m+n-1\right)}$$

Given the observed number of runs $R$, we can compute the test statistic

$$z = \frac{R - R_E}{\sigma}$$

that can be looked up in a table for the cumulative standardized normal distribution to give a probability for the null hypothesis. An exact test is also available.

---

**Example 12.6**

We return to the extinction rate data for the post-Permian from section 12.1 (Fig. 12.3). The time series (extinctions per stage) is first converted to binary form by comparing each value with the mean:

111001110100000000011000101000110001011000000

There are 18 runs, against the expected number of 21.62. The probability of the null hypothesis is $p = 0.23$, so we cannot claim a statistically significant departure from randomness.

---

## 12.6 Time Series Trends and Regression

The detection of trends in time series is important in paleontology. Was species richness increasing through the Cenozoic? Was climate cooling through the Ordovician? Do animals tend to get larger through evolution? In chapter 4, we discussed methods for correlation and regression that can be used also for time series, although assumptions of linear trend, normally distributed, and serially independent residuals can be problematic.

The assumption of linear trend can be relaxed by using non-parametric correlation methods, especially Spearman's rank-order correlation (section 4.5). In fact, Spearman's $r$ seems to work well for time series (Yue et al. 2002), but the Mann–Kendall test (see below) is more commonly used.

The assumption of serial dependence is particularly troublesome. Most time series have some kind of serial dependence (autocorrelation), which can be described, e.g., by the AR(1) red-noise model (section 12.1). We will describe some methods that do not assume serial independence, but in this case we must be careful with the interpretation of results. For example, we will see in Fig. 14.18 that a simple "Brownian motion" model with cumulative addition of positive or negative random numbers (i.e., serial dependence) can produce trends over time, although there is no underlying driving mechanism. Special care must then be taken to disentangle the "true" trend from random drift.

### 12.6.1 Mann–Kendall trend test

The Mann–Kendall test (Mann 1945; Kendall 1975; Gilbert 1987) is a classical non-parametric test for trends in time series. It does not assume a linear trend, but in its original form it does assume serial independence. The Mann–Kendall test works by comparing all possible pairs of points, counting the number of pairs where the values are increasing, and subtracting the number of pairs where the values are decreasing. Hence, the test statistic $S$ is calculated as follows.

Data $x_1, \ldots x_n$ are assumed to be ordered in sequence of time or in spatial sequence. Define the indicator function

$$\operatorname{sgn} x = \begin{cases} -1, \text{if } x < 0 \\ 0, \text{if } x = 0 \\ 1, \text{if } x > 0 \end{cases}$$

The $S$ statistic is the sum over all pairs of values:

$$S = \sum_{i=1}^{n-1} \sum_{j=i+1}^{n} \operatorname{sgn}\left(x_j - x_i\right)$$

$S$ will be negative for a negative trend, zero for no trend, and positive for an increasing trend.

For $n \leq 10$, the $p$ value is taken from a table of exact values (Gilbert 1987). For $n > 10$, a normal approximation is used: determine the total number of groups of ties $g$ and the number of tied values $t_j$ within each group, in the sorted sequence. Then estimate the standard deviation of $S$ by

$$SD = \sqrt{\frac{1}{18}\left[n(n-1)(2n+5) - \sum_{j=1}^{g} t_j(t_j-1)(2t_j+5)\right]}$$

The $z$ statistic is then

$$z = \frac{|S|-1}{SD}\operatorname{sgn} S$$

which is used to calculate $p$ from the cumulative normal distribution (z test).

Hamed and Rao (1998) developed extensions of the Mann–Kendall test, allowing for autocorrelated data. As autocorrelation is usually present, the Hamed-Rao test is preferable to the classical test in most cases.

### 12.6.2 Regression in the presence of autocorrelation

Prais and Winsten (1954, see also Wooldridge 2012) developed a method for linear regression in the presence of autocorrelation that can be described with an AR(1) or red-noise model (section 12.1). The fitted Prais-Winsten model is a sum of a linear function and an AR(1) model with the autocorrelation parameter $\rho$ (rho). Although mainly used in econometrics, this procedure has a solid basis and should be applicable to many problems in paleontology.

Several other methods have been developed for regression of autocorrelated time series. Many of them rely on the generalized least-squares (GLM) framework. We will return to GLM in section 14.9.

## 12.7 Smoothing and filtering

Small, rapid fluctuations in a time series, often caused by "noise" (errors and other random influences), can sometimes obscure patterns. Drawing a smooth curve through the points can be a way to emphasize structure. This can be done by hand, but it is preferable to use a more standardized protocol. Keep the following in mind:

1) Smoothing is mainly a cosmetic operation, usually with no estimation of statistical significance.
2) It may not be necessary to smooth. The human brain has an excellent capacity for seeing structure in noisy data without extra guidance (sometimes such pattern recognition can even go a little *too* far!).
3) Always include your original data points on the graph.

### 12.7.1 Moving average

Many different curve smoothing techniques have been described (Davis 1986; Press et al. 1992). The most obvious approach is perhaps smoothing by *moving averages*. For three-point moving average smoothing, each value in the smoothed sequence $y$ is the mean of the original value and its two neighbors:

$$y_i = \frac{x_{i-1} + x_i + x_{i+1}}{3}$$

Higher-order moving average filters using more neighbors are also common and will give stronger smoothing. Although practical and easy to understand, the moving average filter has some far from ideal properties. For the three-point case, it is clear that any cycle with period three such as {2, 3, 4, 2, 3, 4, ...} will completely vanish, while a higher frequency cycle with period two will only lose some of its amplitude – {0, 1, 0, 1, 0, 1, 0, 1, ...} will turn into {1/3, 2/3, 1/3, 2/3, ...}. In other words, the attenuation does not increase with frequency in a simple way.

### 12.7.2 Exponential moving average

It is not difficult to design a simple smoothing filter with better characteristics. One useful form is the *exponential moving average* or AR(1) filter, which is computed recursively as follows:

$$y_i = \alpha y_{i-1} + (1 - \alpha) x_i$$

The degree of smoothing is controlled by the parameter $\alpha$, in the range 0 (no smoothing) to 1. This smoothing method only uses past values at any point in time, puts less weight on values farther back in time, and always attenuates higher frequencies more than lower frequencies.

### 12.7.3 Moving median

An alternative to the moving average filter is to use the median rather than the mean within each sliding window. As with the moving average, a longer window produces stronger smoothing. This is more robust to outliers, because an isolated, extreme value will never be selected as the median. A disadvantage of the moving median is that it tends to produce a staircase appearance of the curve.

### 12.7.4 Non-local means

Non-local means is a powerful smoothing method, originally developed for image denoising but also effective for time series (Tracey and Miller 2012). It may be compared with the moving average method, but the average is not taken over neighboring points but over points in similar regions, which can be far away. This tends to preserve peaks and transitions better than local averaging. The procedure depends on some degree of similarity between short patterns (motifs) across the time series, but this is very common.

The size of the local regions to be compared (patch size) can be selected by the user; it could be set to, e.g., $N = 7$. The search radius can also be specified; it should be large enough to provide analog motifs, but smaller than any large-scale trends. An additional parameter *lambda* controls the amount of smoothing. Tracey and Miller (2012) suggest a value of about 0.6 times the standard deviation of the noise (usually unknown but can be estimated by eye).

### 12.7.5 FIR filtering

The moving average smoother can be modified by assigning different weights to the terms in the sum. The weight factors can be written as a vector – thus, the weight vector for the three-point moving average described earlier is (1/3, 1/3, 1/3). This vector is also known as the *impulse response*, because it is equal to the sequence of numbers produced for an input consisting of a single value of 1 surrounded by zeros (i.e., an impulse). Because the impulse response is of finite length, the resulting smoother (or filter) is called a finite impulse response (FIR) filter. This is in contrast with the recursive exponential moving average described earlier, which has an infinite length, exponentially decaying impulse response (IIR).

Careful design of the weights (i.e., impulse response) of an FIR filter can produce filters with desirable properties. In particular, we can ensure that the phase response is linear, meaning that different frequencies are delayed by equal time through the filter. Many filters do not share this property, leading to distortion of the curve shape. Moreover, we can design the smoother to have a near-monotonic frequency response, in contrast with the simple moving average smoother discussed earlier. Finally, and perhaps surprisingly, we can design filters with an almost arbitrary frequency response. A smoother is basically a low-pass filter, attenuating high frequencies. However, we can also build high-pass (removing slow variation), band-pass, and band-stop filters with the FIR method. The band-pass filter is used to emphasize a narrow frequency band and is commonly used in cyclostratigraphy to identify a single Milankovitch periodicity in a time series.

Designing an FIR filter can be a little tricky. In PAST, we have tried to automate the process using the so-called Parks-McClellan algorithm. An important parameter is the filter order (length of the weight vector) – it should be large enough to give an acceptably sharp and monotonic filter. However, a filter of length $n$ will give less accurate results in the first and last $n/2$ samples of the time series, which puts a practical limit on the filter order for short series.

Another special case of the FIR filter is the Gaussian filter, which is a low-pass filter (smoother) with a Gaussian weight vector.

### 12.7.6  Fitting to models

All the smoothing methods presented so far assume constant time spacing between the values. A very different approach is to fit the data to some smooth mathematical function by minimizing the sum of squared residuals (least-squares fitting). In this sense, even linear regression can be viewed as a smoothing method, but it is usually preferable to use a more flexible function that can follow a complicated time series more closely.

Smoothing by model fitting allows interpolation between the given points and the plotting of confidence intervals by, e.g., bootstrapping.

#### 12.7.6.1  Cubic splines

A popular model choice is the *cubic spline*, one version of which is known as the *B-spline*. The idea here is to divide the time axis into intervals. Within each interval, the spline consists of a third-degree polynomial. The coefficients of these polynomials are chosen to ensure not only that they join together (the spline is continuous), but also that their slopes and second derivatives (curvatures) are continuous. Given these constraints, the coefficients of each piece of the spline can be optimized in order to trace the original data points within its time interval as closely as possible. The result is a visually pleasing smoothed curve. The main question in this procedure is how to choose the intervals – this choice will influence the degree of smoothing and oscillation (Agterberg 1990, section 3.6). The algorithm of de Boor (2001) allows the user to specify a simple smoothing factor.

#### 12.7.6.2  LOWESS/LOESS

The LOWESS smoothing algorithm (LOcally WEighted Scatterplot Smoothing, Cleveland 1979, 1981; Cleveland and Devlin 1988) fits the data to straight lines or second-order polynomials within local windows. Given the number of points $n$ and a smoothing parameter $q$ specified by the user, the program fits the $nq$ points around each given point to a line or quadratic, with a weighting function decreasing with distance. The new smoothed point is the value of the fitted function at the original $x$ position.

LOESS is a newer implementation of the LOWESS algorithm, and the two terms are being used almost interchangeably.

LOESS and splines provide alternative solutions to the same problem and produce similar results. Splines can be smoother and more visually appealing, but they can also suffer from overshooting near discontinuities.

**Example 12.7**

Hammer and Webb (2010) described two cores spanning most of the Holocene in the inner Oslofjord, Norway. For every 10 cm in the cores, 300 foraminiferans were identified to 22 different taxa and counted. From one of these cores ("REF"), we here estimate the total species richness in each sample using the iChao-1 estimator (section 9.1). Fig. 12.12 shows the species richness through the core, with different smoothing methods. A radiometric age of ca. 7300 years (BP, calibrated) was obtained at 411 cm core depth (sample no. 41), close to the peak in species richness. The subsequent steady decrease in richness until today is probably due to the land rising some 200 m by isostatic rebound, causing increasingly restricted conditions in the inner fjord.

The different smoothing methods emphasize different scales and details in the data, and none of them is obviously the "best," although the non-local means smoother (Fig. 12.12D) is perhaps the most visually appealing in this case.

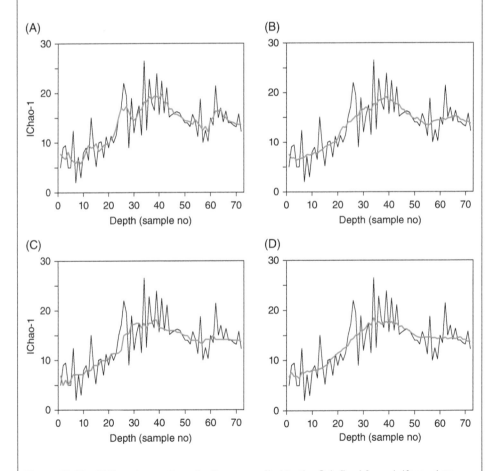

**Figure 12.12**   Different smoothers (red) were applied to the Oslofjord foraminiferan data set (black). (A) Moving average, $n = 5$. (B) Moving average, $n = 11$. (C) Moving median, $n = 11$. (D) Non-local means, $n = 11$, search distance = 9, and $\lambda = 8$ (strong smoothing).

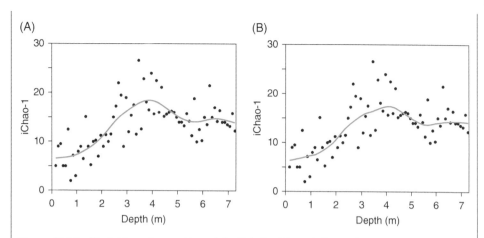

**Figure 12.13** Model-based smoothing of the Oslofjord foraminiferan data set. (A) Cubic spline with an optimized (cross-validated) smoothing factor of 6.3. (B) LOESS, smoothing factor 0.3.

Figure 12.13 shows results from a cubic spline and a LOESS smoother. For the cubic spline, the smoothing factor was selected by cross-validation, i.e., removing one data point at a time and measuring the deviation from the resulting model prediction. The LOESS smoothing factor was selected to give a similar result.

## References

Agterberg, F.P. 1990. *Automated Stratigraphic Correlation.* Developments in Palaeontology and Stratigraphy 13. Elsevier, Amsterdam.

Benton, M.J. (ed.) 1993. *The Fossil Record 2.* Chapman & Hall, New York.

Bradley, J.V. 1968. *Distribution-Free Statistical Tests.* Prentice–Hall, New Jersey.

Cleveland, W.S. 1979. Robust locally weighted fitting and smoothing scatterplots. *Journal of the American Statistical Association* 74, 829–836.

Cleveland, W.S. 1981. A program for smoothing scatterplots by robust locally weighted fitting. *The American Statistician* 35, 54.

Cleveland, W.S., Devlin, S.J. 1988. Locally-weighted regression: an approach to regression analysis by local fitting. *Journal of the American Statistical Association* 83, 596–610.

Davis, J.C. 1986. *Statistics and Data Analysis in Geology.* John Wiley & Sons, New York.

De Boor, C. 2001. *A Practical Guide to Splines.* Springer, New York.

Erlykin, A.D., Harper, D.A.T., Sloan, T., Wolfendale, A.W. 2017. Mass extinctions over the last 500 myr: an astronomical cause? *Palaeontology* 60, 159–167.

Erlykin, A.D., Harper, D.A.T., Sloan, T., Wolfendale, A.W. 2018. Periodicity in extinction rates. *Palaeontology* 61, 149–158.

Gilbert, R.O. 1987. *Statistical Methods for Environmental Pollution Monitoring.* Wiley, New York.

Hamed, K.H., Rao, A.R. 1998. A modified Mann-Kendall trend test for autocorrelated data. *Journal of Hydrology* 204, 182–196.

Hammer, Ø., Webb, K.E. 2010. Piston coring of Inner Oslofjord pockmarks, Norway: constraints on age and mechanism. *Norwegian Journal of Geology* 90, 79–91.

Hansen, T., Nielsen, A.T. 2003. Upper Arenig biostratigraphy and sea-level changes at Lynna River near Volkhov, Russia. *Bulletin of the Geological Society of Denmark* 50, 105–114.

Kendall, M.G. 1975. *Rank Correlation Methods*. 4th edition. Charles Griffin, London.

Mann, H.B. 1945. Non-parametric tests against trend. *Econometrica* 13, 163–171.

Mann, M.E., Lees, J. 1996. Robust estimation of background noise and signal detection in climatic time series. *Climatic Change* 33, 409–445.

Prais, S.J., Winsten, C.B. 1954. Trend estimators and serial correlation. *Cowles Commission Discussion Paper 383*, Chicago.

Press, W.H., Teukolsky, S.A., Vetterling, W.T., Flannery, B.P. 1992. *Numerical Recipes in C*. Cambridge University Press, Cambridge, UK.

Proakis, J.H., Rader, C.M., Ling, F., Nikias, C.L. 1992. *Advanced Digital Signal Processing*. Macmillan, New York.

Rampino, M.R., Caldeira, K., Zhu, Y. 2021. A pulse of the Earth: a 27.5-Myr underlying cycle in coordinated geological events over the last 260 Myr. *Geoscience Frontiers* 12, 101245.

Raup, D., Sepkoski, J.J. 1984. Periodicities of extinctions in the geologic past. *Proceedings of the National Academy of Science* 81, 801–805.

Schulz, M., Mudelsee, M. 2002. REDFIT: estimating red-noise spectra directly from unevenly spaced paleoclimatic time series. *Computers & Geosciences* 28, 421–426.

Shackleton, N.J., Pisias, N.G. 1985. Atmospheric carbon dioxide, orbital forcing, and climate. In: Sundquist, E.T. and Broeker, W.S. (eds.). *The Carbon Cycle and Atmospheric CO2: Natural Variations Archean to Present. Geophysical Monograph* 32, 412–417.

Shackleton, N.J., Berger, A., Peltier, W.R. 1990. An alternative astronomical calibration of the lower Pleistocene timescale based on ODP Site 677. *Transactions of the Royal Society of Edinburgh: Earth Sciences* 81, 251–261.

Smith, D.G., Bailey, R.J. 2018. Discussion on 'A 2.3 million year lacustrine record of orbital forcing from the Devonian of northern Scotland', Journal of the Geological Society, London 173, 474–488. *Journal of the Geological Society, London* 175, 561–562.

Swan, A.R.H., Sandilands, M. 1995. *Introduction to Geological Data Analysis*. Blackwell Science.

Thomson, D. J. 1982. Spectrum estimation and harmonic analysis. *Proceedings of the IEEE* 70, 1055–1096.

Torrence, C., Compo, G.P. 1998. A practical guide to wavelet analysis. *Bulletin of the American Meteorological Society* 79, 61–78.

Tracey, B., Miller, E. 2012. Nonlocal means denoising of ECG signals. *IEEE Transactions on Biomedical Engineering* 59, 2383–2386.

Vachula, R.S., Cheung, A.H. 2023. A meta-analysis of studies attributing significance to solar irradiance. *Earth and Space Science* 10, e2022EA002466.

Wald, A., Wolfowitz, J. 1942. An exact test for randomness in the non-parametric case based on serial correlation. *Annual Mathematical Statistics* 14, 378–388.

Walker, J.S., Krantz, S.G. 1999. *A Primer on Wavelets and Their Scientific Applications*. CRC Press.

Weedon, G.P. 2003. *Time-series Analysis and Cyclostratigraphy*. Cambridge University Press.

Wooldridge, J.M. 2012. *Introductory Econometrics – a Modern Approach*. 5th edition. South-Western Cengage Learning.

Yue, S., Pilon, P., Cavadias, G. 2002. Power of the Mann–Kendall and Spearman's rho tests for detecting monotonic trends in hydrological series. *Journal of Hydrology* 259, 254–271.

# 13

# Quantitative biostratigraphy

The most common approach to biostratigraphy is still that some expert decides what fossils to use for zonation and correlation, until some other expert feels that these fossils are too rare, too difficult to identify, too facies-dependent, or too geographically restricted to be good markers, and selects new index fossils. As with systematics, it can be argued that while such an approach can produce good results in many cases, it is subjective and does not take all the available data into account. We might like some "objective" protocol for zonation and correlation, with the goal of minimizing contradictions and maximizing stratigraphic resolution. This is the aim of *quantitative biostratigraphy*, as reviewed by, e.g., Cubitt and Reyment (1982), Tipper (1988), Armstrong (1999), and Sadler (2004).

In this chapter, we will discuss methods for zonation of a single section or well, estimation of the truncation of stratigraphic ranges due to incomplete sampling, and zonation and correlation across several sections.

## 13.1 Zonation of a single section

When describing the sequence of fossils in a section or core, it is often considered useful to divide the succession into intervals with more or less distinct faunal or floral compositions. Such intervals are usually called biozones, but if they are defined only from a single locality, then this is perhaps a too generous use of the term. In classical biostratigraphy, biozones can be defined by a single taxon (total range zone), but here we will discuss biozonation based on objective methods for delineating units based on the total species composition.

### 13.1.1 Stratigraphically constrained clustering

In chapter 5, we introduced several hierarchical clustering methods, including the UPGMA algorithm. These algorithms work by first grouping together the most similar data points and then proceeding to cluster successively larger groups. A small modification to this procedure is to only allow the clustering of points or groups if they are *stratigraphically adjacent*. This is called stratigraphically constrained clustering (Gordon and Birks 1972; Birks 2012). This method was first made popular by the program CONISS (Grimm 1987). Several variants of stratigraphically constrained clustering are described by Birks (2012).

*Paleontological Data Analysis*, Second Edition. Øyvind Hammer and David A.T. Harper.
© 2024 John Wiley & Sons Ltd. Published 2024 by John Wiley & Sons Ltd.

PAST implements constrained clustering by direct modification of the UPGMA and Single Linkage algorithms. This has the advantage of allowing any similarity or distance measure and also facilitates comparison with unconstrained clustering. However, it can lead to somewhat strange-looking dendrograms when distances are larger between adjacent samples than between samples further apart.

---

**Example 13.1**

Returning to the radiolarian data set of Bjørklund et al. (2019), which we described in section 10.12, a stratigraphically constrained cluster analysis is shown in Figure 13.1. The Younger Dryas samples form a very well-defined cluster, as shown by the large similarities within the cluster and the large distance from other clusters. A single sample at 410-cm core depth is more distant from the Younger Dryas cluster. A cluster from

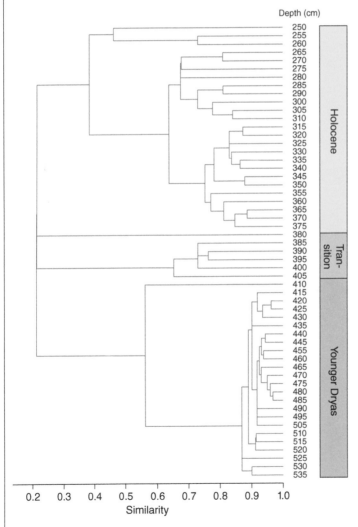

**Figure 13.1** Stratigraphically constrained UPGMA cluster analysis of the radiolarian data. The Bray–Curtis distance was used. Adapted from Bjørklund et al. (2019).

405 to 395 cm represents the transition to the Holocene, also with a fairly distinct fauna. The Holocene part of the core is not clearly resolved into subclusters, with the possible exception of the interval from 375 to 315 cm. It should be clear from this example that the division into biozones based on clustering can be a little arbitrary, although our three zones seem quite robust. A scheme for deciding the number of biozones using a "broken stick" model was suggested by Bennett (1996).

## 13.2   Confidence intervals on stratigraphic ranges

One of the most fundamental tasks of paleontology is to estimate the times or stratigraphic levels of origination and extinction of a fossil taxon. Traditionally, this has been done simply by using the taxon's first appearance datum (FAD) and last appearance datum (LAD) as observed directly in the fossil record. However, doing so will normally underestimate the real range, since we cannot expect to have found fossils at the extreme ends of the temporal or stratigraphic range. What we would like to do is to give *confidence intervals* on stratigraphic ranges. This will enable us to say, for example, that we are 95% certain that a given taxon originated between 53 and 49 million years ago. Obviously, the length of the confidence interval will depend on the completeness of the fossil record. If we have found the taxon in only two horizons, the confidence interval becomes relatively long. If we have found it in hundreds of horizons, we may perhaps assume that the observed FAD and LAD are close to the origination and extinction points, and the confidence interval becomes smaller.

### 13.2.1   Parametric confidence intervals on stratigraphic ranges

A simple method for the calculation of confidence intervals was described by Strauss and Sadler (1989) and by Marshall (1990). They assumed that fossiliferous horizons are randomly distributed within the original existence interval, that fossilization events are independent of each other, and that sampling is continuous or random. These are all very strong assumptions that will obviously often not hold. One example of violation of these assumptions is if we have recorded many randomly distributed fossiliferous horizons in one interval, but this is a truncation of the real range, due to a hiatus or a change in local conditions that may have caused local but not global extinction. The confidence interval will then be underestimated. This and other problems were discussed by Marshall (1990).

Given that the assumptions hold, at least approximately, we can calculate confidence intervals from only the FAD, LAD, and number of known fossiliferous horizons. We also need to specify a confidence level, such as 0.90 or 0.95. The method will give us a confidence interval for the origination time, or for the extinction time, or for the total range. Note that these are independent confidence intervals that cannot be directly combined: the confidence interval for the total range cannot be calculated by summing confidence intervals for either end of the range in isolation.

Let $C$ be the confidence level (such as 0.95), and $H$ be the number of fossiliferous horizons. The confidence interval for either the top or bottom (not both) of the stratigraphic range, expressed as a fraction $\alpha$ of the known stratigraphical range (LAD minus FAD), is given by the equation

$$\alpha = \left(1 - C\right)^{-1/\left(H-1\right)} - 1$$

The confidence interval for the origination ranges from FAD − α(LAD − FAD) to FAD. The confidence interval for the extinction ranges from LAD to LAD + α(LAD − FAD). The confidence interval for the total range is given by the equation

$$C = 1 - 2(1+\alpha)^{-(H-1)} + (1+2\alpha)^{-(H-1)}$$

This must be solved for α by computer, using for example the bisection method. The confidence interval for the total range is then FAD − α(LAD − FAD) to LAD + α(LAD − FAD).

---

**Example 13.2**

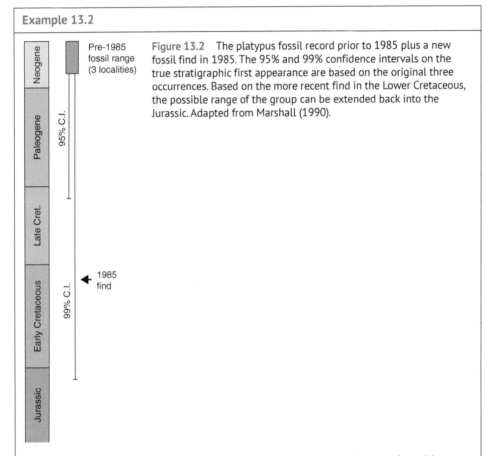

Pre-1985 fossil range (3 localities)

1985 find

Figure 13.2   The platypus fossil record prior to 1985 plus a new fossil find in 1985. The 95% and 99% confidence intervals on the true stratigraphic first appearance are based on the original three occurrences. Based on the more recent find in the Lower Cretaceous, the possible range of the group can be extended back into the Jurassic. Adapted from Marshall (1990).

Marshall (1990) calculated parametric confidence intervals on the stratigraphic range of platypuses. The fossil record of this genus is very limited, and the assumptions of the method do not probably hold, but the example is still illustrative. Prior to 1985, three fossil platypus localities were known, the oldest from the Middle Miocene. This is all the information we need to calculate confidence intervals on the time of origination of platypuses: the 95% confidence interval stretches back to the latest Cretaceous, while the 99% confidence interval goes back to the Jurassic (Fig. 13.2). In 1985, a fourth fossil platypus-like animal from the Albian (late Early Cretaceous) was discovered. This find would perhaps not be expected, lying outside the original 95% confidence interval (but at least inside the 99% interval).

## 13.2.2   Non-parametric confidence intervals on stratigraphic ranges

The method for calculating confidence intervals on stratigraphic ranges given above makes very strong assumptions about the nature of the fossil record. Marshall (1994) described an alternative method that we may call *non-parametric*, making slightly weaker assumptions. While the parametric method assumes a uniform, random distribution of distances between fossiliferous horizons (gaps), the non-parametric method makes no assumptions about the underlying distribution of gap sizes. The main disadvantages of the non-parametric method are that (a) it needs to know the levels of all fossil occurrences; (b) it cannot be used when only a few fossil horizons are available; and (c) the confidence intervals are only estimated and have associated *confidence probabilities* that must be chosen (e.g., $p = 0.95$).

The non-parametric method still assumes that gap sizes are not correlated with stratigraphic position.

Given a confidence probability $p$ (probability of the confidence interval being equal to or smaller than the given estimate), calculate

$$\gamma = \frac{1-p}{2}$$

For $N + 1$ fossil horizons, there are $N$ gaps. Sort the gap sizes from smallest to largest. Choose a confidence level $C$ (such as 0.8). The lower bound on the size of the confidence interval is the $(X + 1)$th smallest gap, where $X$ is the largest integer $X \leq N$ such that

$$\gamma > \sum_{x=0}^{X} \binom{N}{x} C^x (1-C)^{(N-x)}$$

where $\binom{N}{x}$ is the binomial coefficient as defined in section 9.2.

The upper bound of the size of the confidence interval is defined in a similar way as the $(X + 1)$th smallest gap, where $X$ is the *smallest* integer such that

$$1 - \gamma > \sum_{x=0}^{X} \binom{N}{x} C^x (1-C)^{(N-x)}$$

For small $N$ and high confidence levels or confidence probabilities, it may not be possible to assign upper or lower bounds on the confidence intervals according to the above equations.

---

### Example 13.3

Nielsen (1995) studied trilobites in the Middle Ordovician of southern Scandinavia. We will look at the distribution of the asaphid *Megistaspis acuticauda* in the Huk Formation at Slemmestad, Norway (Fig. 13.3). The section was continuously sampled. The correlation between the stratigraphic levels and gap sizes is not statistically

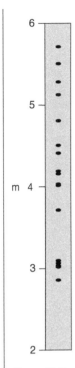

significant ($p > 0.1$ using Spearman's rank-order correlation), but the sequence of gap sizes is not random (runs test, $p < 0.05$, partly due to several close pairs in the section). We therefore use the non-parametric method for estimating confidence intervals on the first occurrence at this locality.

Using a confidence probability of 0.95, the 50% confidence interval is somewhere between 3 and 30 cm wide. Conservatively selecting the upper bound, we can be 50% certain that the real first occurrence of *M. acuticauda* is less than 30 cm below the recorded first appearance. Due to the relatively low number of recorded fossiliferous horizons (17), an upper limit of the confidence interval cannot be computed for 80% or 95% confidence levels (the lower limits are 15 and 30 cm, respectively). In other words, we unfortunately do not have a closed estimate of these confidence intervals. Just one more fossil horizon would in fact be sufficient to produce an estimate for the 80% confidence interval. Informally, we try to add an artificial point, giving an upper bound for the 80% confidence interval of about 60 cm.

**Figure 13.3** Occurrences of the asaphid trilobite *Megistaspis acuticauda* in the Huk Formation, Middle Ordovician, Slemmestad, Norway. Levels in meters. Based on data from Nielsen (1995).

## 13.3 Regional and global biostratigraphic correlation

Finding an overall sequence of fossil occurrences from a number of sections, and correlating horizons based on these sequences, has traditionally been something of a black art. Which taxa should be used as index fossils, and what should we do if correlations using different fossils are in conflict? Biostratigraphy would have been a simple and boring exercise if there were no such contradictions!

It is important to keep in mind the nature of the fossil record with respect to stratigraphy. When time-correlating rocks over large areas, we are to some extent assuming that each taxon originated and went extinct synchronously over the whole region under study. If this is not the case, we are producing diachronous correlations that have more to do with paleoecology and local environmental conditions than with time. Unfortunately, in most cases, the originations and extinctions were probably *not* totally synchronous, for example, if a species lingered on in an isolated refugium long after it went extinct elsewhere. We are therefore normally in more or less serious violation of the assumption of spatial homogeneity. Hence, our first serious problem in biostratigraphy is that the global ranges of taxa through time were not fully reflected by the local faunas or floras.

Our second serious problem is that even the local existence interval in time, from local origination to local extinction, can never be recovered. Since very few of the living specimens were fossilized, fewer of these fossils were preserved through geological time, and fewer still were recovered by the investigator on her field trip, we cannot expect to find the very first and very last specimen of the given taxon that lived at the given locality. Thus, the observed local stratigraphic range is truncated, with respect to the original local range, to an extent that depends on the completeness of the fossil record.

Apart from blatant errors such as reworking or incorrect identification of taxa, it is the truncation of stratigraphic ranges, whether due to diachronous originations/extinctions or incompleteness of the fossil record, which is responsible for biostratigraphic contradictions (Fig. 13.4).

In addition to attempting to summarize the data as a composite sequence, the biostratigraphic method should ideally also report on taxa or occurrences that do not seem to follow the general pattern, allowing the investigator to have a second look at these particular data points in order to correct possible errors or identify problems that might warrant their removal from the analysis. Quantitative biostratigraphy is usually an iterative and time-consuming process, with multiple runs of the algorithm with better data until a stable result is achieved. Such a process will of course involve many more or less subjective decisions, so there is still room for personal opinion (fortunately or unfortunately, depending on your viewpoint).

Once a global sequence of stratigraphic positions has been obtained, either the full sequence or subsets of it can be used to erect a *biozonation*. Horizons at different localities can then be correlated using the sequence. The different biostratigraphic methods more or less clearly separate the processes of zonation and correlation.

At the present time, the four most popular methodologies in quantitative biostratigraphy are graphic correlation, ranking and scaling, unitary associations (UA), and constrained optimization (CONOP). Each of these has strengths and weaknesses, and none of them is an obvious overall winner for all types of data. We therefore choose to present them all in this book.

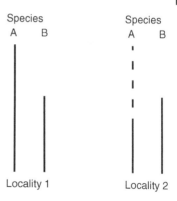

**Figure 13.4** Range charts of two species A and B at two localities. At the first locality, the true global extinctions of both species have been (miraculously!) recovered, with the last appearance of A later than the last appearance of B. At the second locality, the global range of species A has not been preserved, either because it originally disappeared earlier at this locality or because of incomplete preservation or sampling. The order of last appearances is therefore reversed, resulting in a biostratigraphic contradiction between the two localities.

## 13.3.1 Graphic correlation

Graphic correlation is a classical method for stratigraphic correlation of events such as the first and last appearances of species across wells or sections. It is based on a graphical, semiquantitative comparison of two wells at a time (Shaw 1964; Edwards 1984; Mann and

Lane 1995; Armstrong 1999). Graphic correlation is a somewhat simplistic approach, both labor-intensive and theoretically questionable for several reasons, and in these respects inferior to the CONOP, ranking-scaling, and UA methods (see below). Still, its simplicity is itself a definite advantage, and the interactivity of the method encourages a thorough understanding of the data set.

Graphic correlation can only correlate two wells (or sections) at a time. It is therefore necessary to start the procedure by selecting two good wells to be correlated. This will produce a composite sequence, to which a third well can be added, and so forth until all wells are included in the composite standard section. This means that the wells are not treated symmetrically, and the order in which wells are added to the composite sequence can influence the result. In fact, the whole procedure should be repeated several times, each time adding wells in a new order, to check and improve the robustness of the result.

When correlating two wells, the events that occur in both wells are plotted as points in a scatter plot, with the $x$ and $y$ coordinates being the levels/depths in the first and second wells, respectively. If the biological events were synchronous in the two wells, the fossil record perfect, and sedimentation rates in the wells equal, the points would plot on a straight line with a slope equal to one, $y = x + b$, as shown in Fig. 13.5A. If the sedimentation rates were unequal but constant, the points would plot on a straight line with a slope different from one, $y = ax + b$ (Fig. 13.5B). In the real world, the biological events may not have been synchronous in the two wells, and/or the fossil record is not complete enough to show the event at the correct position. The points will then end up in a cloud around the straight line in the scatter plot. The aim of graphical correlation in this simple case is to fit the points to a straight line (the line of correlation [LOC]), either by eye or using a regression method such as reduced major axis (section 4.6). The LOC is hopefully a good approximation to the "true" stratigraphic correlation between the two wells, continuous through the column. Individual events may conflict with this consensus correlation line, due to different types of noise and missing data. The investigator may choose to weigh events differently, forcing the line through some points (e.g., ash layers) and accepting the line to move somewhat away from supposedly low-quality events (Edwards 1984). The procedure for constructing the LOC was reviewed by MacLeod and Sadler (1995).

The situation gets more interesting and complex in the common case of sedimentation rates varying through time and the variations not being synchronous in the two wells. We may first consider a simple example where the sedimentation rate in well $x$ is constant and the rate in well $y$ is piecewise constant. In Fig. 13.5C, note in particular that the sedimentation rate in well $y$ was zero for a period of time, producing a hiatus. The LOC will then be piecewise linear, and the fitting of the line becomes more complex. The situation becomes even more involved if sedimentation rates are varying continuously in both wells (Fig. 13.5D). The fitting of the curve in such cases should probably be done manually, with good graphic tools, and using other available geological information.

Once an LOC has been constructed, it is used to merge the two sections into a composite section, which will then be correlated with a third section, etc. The addition of sections to the composite should be in order of decreasing quality, with the best section first

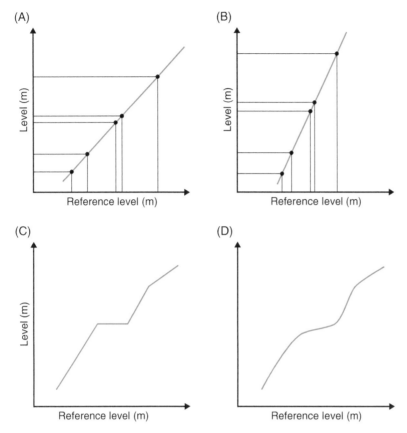

**Figure 13.5** Lines of correlation (LOCs) of four sections versus a reference. (A) Ideal situation where five events plot on a straight line with a slope equal to one. Sedimentation rates are equal in the two sections. (B) The sedimentation rate is faster in the section on the vertical axis. (C) Piecewise constant sedimentation rates. A period of non-deposition in the section on the vertical axis produces a horizontal line. (D) Sedimentation rates are continuously varying.

and the poorest last. The construction of the composite should be based on the principle of maximization of ranges – if the range of a taxon is longer in one section than in another, the latter is assumed to be incomplete. We therefore select the lowest FAD and the highest LAD from the two sections when we transfer the events onto the LOC (Fig. 13.6).

It should be clear from the discussion above that in graphic correlation, the ordering of events and the correlation are not separate processes. In fact, the ordering is erected through the process of correlation, and the two are fully intermingled.

A good example of the application of the graphic correlation method was given by Neal et al. (1994).

Graphic correlation can be carried out by hand but has also been implemented in a number of software packages. These include GraphCor (Hood 1995), StratCor (Gradstein 1996), and Shaw Stack (MacLeod 1994).

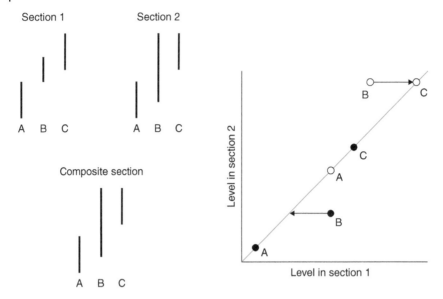

Figure 13.6   The compositing of two sections with three taxa using graphic correlation. In the correlation plot, FADs are marked with filled circles, and LADs with open circles. The FAD and LAD of taxon B fall outside the line of correlation and must be transferred onto it (arrows). This is done by selecting the lowest FAD and the highest LAD in order to maximize the range.

---

**Example 13.4**

We return to the Ordovician trilobite data from Nielsen (1995), introduced in section 13.2. We now focus on the first and last appearances of four species as recorded in three localities: Bornholm (Denmark), southeast Scania (southern Sweden), and Slemmestad (Oslo Region, Norway). The stratigraphic ranges are shown in Fig. 13.7.

The graphic correlation procedure starts by selecting two good-quality sections as our starting point for generating the composite standard reference. We choose the Oslo and the Scania sections. The stratigraphic positions in meters of the first and last appearance events in the Oslo and Scania localities are shown in Fig. 13.8A. We choose to use a straight LOC. When manually placing the LOC, we put less weight on the last appearance of *D. acutigenia*, because this is a relatively rare species. Also, we speculate that the first appearance of *M. simon* in Oslo is placed too high simply because the underlying strata are not yet studied there, so we accept the LOC to diverge also from this point.

The events are then transferred onto the LOC (Fig. 13.8B), always by extending ranges, never shortening them. Other types of events, such as ash layers, can be transferred to the closest point on the LOC. The positions of events on the LOC then constitute the sequence of the Oslo–Scania composite.

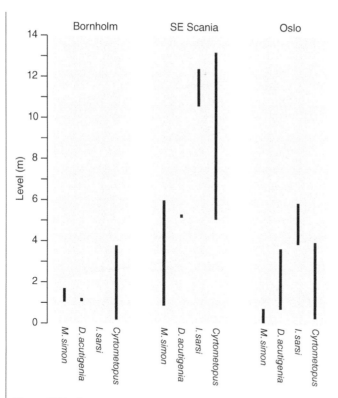

**Figure 13.7** Stratigraphic ranges of four Ordovician trilobite taxa at three Scandinavian localities: Bornholm (Denmark), southeast Scania (Sweden), and Oslo Region (Norway). Data from Nielsen (1995).

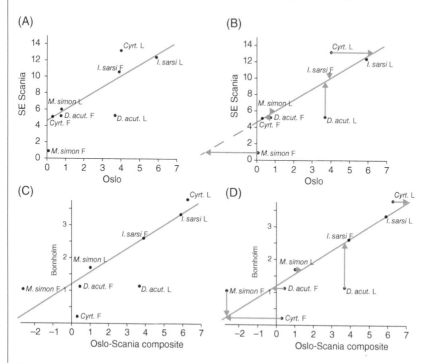

**Figure 13.8** Graphic correlation of the three localities shown in Fig. 13.7. (A) Cross-plotting of the Oslo and SE Scania sections, and construction of the LOC. (B) Transfer of events onto the LOC, constructing an Oslo-Scania composite. (C) Cross-plotting of the Bornholm section and the Oslo-Scania composite, and construction of the LOC. (D) Transfer of events onto the LOC, constructing a Bornholm–Oslo–Scania composite.

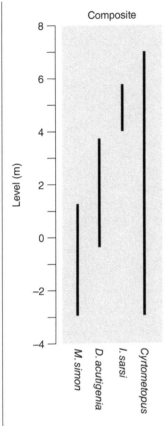

Next, the Bornholm locality is integrated into the composite in the same way (Figs. 13.8C, D). The final composite sequence is shown as a range chart in Fig. 13.9. Ideally, we should have repeated the procedure several times, e.g., by adding the Oslo locality back to the composite, until the sequence stabilizes. By comparing the composite with the original range charts (Fig. 13.7), we see that the Oslo ranges are in good accordance with the composite, but *D. acutigenia* extends farther down in Oslo (as shown by the Bornholm and Scania ranges). Also, *Cyrtometopus* extends much farther up, beyond the range of *I. sarsi*, as shown by the Scania ranges. Armed with this biostratigraphic hypothesis, we are in a position to define biozones and correlate the three sections.

**Figure 13.9** Composite standard reference range chart, based on Fig. 13.8D. Compare with Fig. 13.7.

## 13.3.2 Constrained optimization

The procedure of graphic correlation has several basic problems and limitations. Being basically a manual approach, it is a time-consuming affair full of subjective choices about the construction of the lines of correlation. Also, the localities are integrated into the solution (the composite section) one at a time, and the order in which they are introduced can strongly bias the result.

Constrained optimization (CONOP; Kemple et al. 1989, 1995) can be regarded as a method that overcomes these difficulties. First of all, the sections are all treated equally and simultaneously. In graphic correlation, two sections are correlated at a time, one of them usually being the composite section. This is done by constructing an LOC in the two-dimensional space spanned by the positions of the events in the two sections. In contrast, CONOP constructs a single LOC in $J$-dimensional space, where $J$ is the number of sections (Fig. 13.10). The bias caused by the sequential integration of sections into the solution is therefore removed.

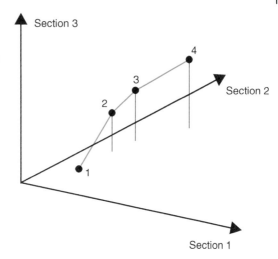

**Figure 13.10** Four events (1–4) in three sections (*J* = 3). The line of correlation (red) is constructed in three-dimensional space. In this case, the sedimentation rate is fastest in section 2.

Obviously, manual construction of the LOC in a high-dimensional space is out of the question for practical reasons and undesirable anyway because of all the subjective and complicated decisions that would have to be made. In CONOP, the construction of the LOC is automatic, based on the principle of *parsimony* ("economy of fit"). An ordering (ranking) and stratigraphic position of events are sought, with the following properties:

1) All known constraints must be honored. Such constraints include the preservation of known co-occurrences (Fig. 13.11), preservation of known superpositions, and obviously that a first occurrence of a taxon must be placed before the last occurrence.
2) A sequence constructed according to property 1 will normally imply an extension of observed ranges in the original sections (Fig. 13.11). A solution is sought that minimizes such extensions, making a minimal number of assumptions about missing preservation.

The name "CONOP " stems from these two properties of the stratigraphic solution.

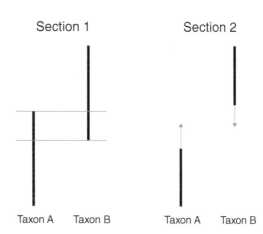

**Figure 13.11** The constraint of observed co-occurrences. In section 1, the two taxa A and B are found together in two samples (red lines). In section 2, the taxa are non-overlapping (black lines). The observed co-occurrence should be preserved in the final ordering of first and last appearance events: FAD(A), FAD(B), LAD(A), and LAD(B). This is achieved by extension of the ranges in section 2 – either taking taxon A up, taxon B down, or both (blue arrows). This is only one of several different situations where range extensions may be enforced by CONOP.

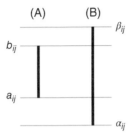

Figure 13.12 Observed and implied ranges for a taxon *i* in section *j*. (A) The observed range, with the first appearance at $a_{ij}$ and last appearance at $b_{ij}$. (B) The implied range in this section. The range is extended to stay consistent with the sequence of events in the global solution.

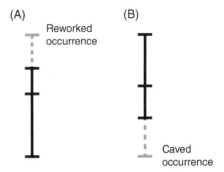

Figure 13.13 Two situations where range contraction should be allowed. (A) Reworking. The observed last occurrence is higher than the real last occurrence. (B) Caving in a well. The observed first occurrence is lower than the real first occurrence.

The search for the most parsimonious solution (property 2) is based on a quantitative measure of the economy of fit, known as a *penalty function*. Following the notation of Kemple et al. (1989), we denote by $a_{ij}$ the observed first occurrence of a fossil *i* in section *j*. Similarly, the observed last occurrence is $b_{ij}$. The estimations of "real" first and last occurrences, as implied by the global solution, are called $\alpha_{ij}$ and $\beta_{ij}$ (Fig. 13.12). A simple penalty function might then be defined as the discrepancy between estimated and observed event horizons, summed over all taxa and sections:

$$p = \sum_i \sum_j \left( \left| a_{ij} - \alpha_{ij} \right| + \left| \beta_{ij} - b_{ij} \right| \right)$$

A solution is sought that minimizes *p*, subject to the constraints. The quantities $a_{ij} - \alpha_{ij}$ and $\beta_{ij} - b_{ij}$ will normally be positive, implying an extension of the range. If the investigator chooses, this may even be enforced through the constraints (property 1). Negative values would mean that the range has been contracted, as might be necessary for data sets with reworking or caving (Fig. 13.13).

This simple penalty function regards all taxa and all sections as of equal stratigraphic value. However, the investigator might choose to put different weights on different events by assigning weight values as follows:

$$p = \sum_i \sum_j \left( w_{ij}^1 \left| a_{ij} - \alpha_{ij} \right| + w_{ij}^2 \left| \beta_{ij} - b_{ij} \right| \right)$$

Let us consider some situations where differential weighting might be useful:

1) Reworking is possible but unlikely. In this case, we would allow $\beta_{ij} - b_{ij}$ to become negative (positive values would not be enforced through the constraints), but then set $w_{ij}^2$ to a large value.
2) Caving (pollution of samples down in a drilled well by material from farther up) is possible but unlikely. In this case, we would allow $a_{ij} - \alpha_{ij}$ to become negative, that is,

positive values would not be enforced through the constraints, but then set $w_{ij}^1$ to a large value.

3) The different fossils occur in different densities in the different sections. If a fossil $i$ has been found densely from FAD to LAD in a section $j$, it is likely that the real range is not much larger than the observed range, and extension of the range should be "punished" by using a relatively large weight. Let $n_{ij}$ be the number of samples (horizons) where a given fossil has been found in a given section. We could then assign weights in inverse proportion to the average gap length $(b_{ij} - a_{ij})/(n_{ij} - 1)$.

4) The investigator has independent information indicating that a fossil has low preservation potential, or a section displays poor or incomplete preservation. She may assign low weight to the events involved with these fossils and sections.

The stratigraphic position of an event, and therefore the range extensions and penalty values, may be given in different types of units:

1) Time. This is rarely possible.
2) Thickness in meters. This was recommended by Kemple et al. (1995) but may give biased results if sedimentation rates differ strongly between sections (weighting with the inverse average gap length can partly solve this problem).
3) Number of horizons within the section. This can be counted as the number of sampled horizons, the number of fossiliferous horizons, or the number of horizons containing an FAD or LAD. Using the number of horizons can to some extent make the method more robust to non-constant sedimentation rates.

Until now, we have measured penalty values in terms of range extensions. This approach is in accordance with common practice in graphic correlation. However, CONOP is a general optimization framework that can use any criterion for minimizing misfits in the solution. For example, the sequence can be found by minimizing the number of reversals of pairs of events, somewhat similarly to the "step model" used in ranking-scaling (see later in the chapter). Another option is to minimize the number of pairwise co-occurrences in the global solution that are not observed in the individual sections. This emphasis on co-occurrences is reminiscent of the philosophy behind the UA method (see below).

The most parsimonious solution cannot be found directly – it must be searched for. Since an exhaustive search is out of the question, it is necessary to search only a subspace, and it is not guaranteed that a global optimum is found. This situation is directly comparable to the procedure of heuristic search in systematics (section 14.4).

The only software that can presently perform this complicated procedure is called CONOP (Sadler 2001), although a minimal version is also provided in PAST. For a given sequence of events that honor all constraints, CONOP attempts to find a minimal set of local range extensions. In an outer loop, the program searches through a large number of possible sequences in order to find the optimal one. The optimization technique used is called *simulated annealing*, which is basically a downhill gradient search in parameter space but where some uphill jumps are permitted at the beginning of the search in order to prevent the program from getting stuck on a suboptimal local minimum. Depending on the data set, the user must specify several parameters to optimize the search. These include the *initial temperature*, deciding the probability of accepting an uphill move at the beginning of the search; a cooling ratio, deciding how fast this probability should drop as the search proceeds; and the total number of steps. The selection of good parameters depends on experience and experimentation.

CONOP will produce a "composite section" in the form of a range chart containing all taxa along an axis of sediment thickness or horizon number. This end result will invariably look impressive and give the impression of a very high stratigraphic resolution. However, as with any data analysis method, it is important to retain a healthy skepticism when confronted with a typical output from CONOP. Even if true first and last appearances were recovered in all sections, different times of local origination and extinction in different localities can distort the result. Incomplete preservation and sampling deteriorate the solution further. That the principle of parsimony applied to many sections and taxa will come to the rescue is only a hope, and it can never correct all errors. Finally, the optimization software is not guaranteed to find the solution with the lowest possible penalty value. In fact, even if an optimal solution is found, it need not be the only one. Other solutions may exist that are as good or nearly as good. We will meet precisely the same problem in chapter 14, when using the principle of parsimony in cladistics.

One way of evaluating the "real" (rather than apparent) resolution of the CONOP solution is to plot a *relaxed-fit curve* for each event. Such a curve shows the total penalty value as a function of the position of the event in the global sequence. The curve should have a minimum at the optimal position of the given event. If this minimum is very sharp and narrow, it means that the event can be accurately positioned and provides high stratigraphic resolution. If the minimum has a broad, flat minimum, it means that there is an interval in the sequence within which the event can be positioned without increasing the penalty value. Such events are not precisely localized, and they do not really provide the stratigraphic resolution they appear to from the composite range chart solution. These "bad" events can be removed from the global solution, producing a more honest *consensus sequence* with lower overall resolution.

To conclude, we regard CONOP as an excellent, flexible method for biostratigraphy, based on simple but sound theoretical concepts and potentially providing high-resolution results. One disadvantage is the long computation times for large data sets, making experimentation with data and parameters difficult. CONOP is closely linked with the concept of biostratigraphic events, in contrast with the method of UA that operate in terms of associations and superpositions between them. The choice between these two philosophies should be made based on the nature of the available data and the purpose of the investigation.

For a recent application of CONOP, see Toro et al. (2023).

---

**Example 13.5**

Palmer (1954) produced a number of range charts for trilobites in the Cambrian Riley Formation of Texas. Shaw (1964) used this data set as a case study for graphic correlation, and Kemple et al. (1995) used it to demonstrate CONOP.

The input data consists of levels in meters of first and last appearances of 62 taxa in seven sections. We specify to the CONOP program that ranges are not allowed to contract (no reworking is assumed) and that the penalty function should use an extension of ranges in meters. A number of other parameters must also be specified, controlling the constraints, the calculation of the penalty function, and details of the optimization algorithm.

After a couple of minutes of computation, CONOP presents an overall range chart of the best sequence. The penalty function of the best solution has the value of 3555. According to Kemple et al. (1995), an even better solution is known, with a penalty value of 3546. The program should always be run again a few times with different starting conditions and optimization parameters to see if a better solution can be found.

Figure 13.14 shows a range chart from one of the sections, with the necessary range extensions as enforced by the global solution. The first and last appearances (observed or extended) are now consistent across all sections, honor all constraints, and can be used directly for correlation.

Morgan Creek

**Figure 13.14** Range chart for the Morgan Creek section of Palmer (1954). The observed ranges of trilobite taxa (thick lines) have been extended (thin lines) according to the global solution, which attempts to minimize the total range extension in all sections. Red lines indicate the inferred ranges of taxa that are not found in this section.

### 13.3.3 Ranking and scaling

The ranking-scaling method, as developed mainly by Agterberg and Gradstein (Gradstein et al. 1985; Agterberg 1990; Agterberg and Gradstein 1999), is based on biostratigraphic events in a number of wells or sections. Such an event can be the first or last appearance of a certain taxon, or an abundance peak (acme). The input data consist of the stratigraphic level of each event in each well or section. It is sufficient that the ordering (ranking) of the events within each well is given – absolute levels in meters are not necessary. Since sedimentation rates have generally been different in different locations, the absolute thickness of sediment between consecutive events can be misleading for stratigraphy.

Given such a data set, a fundamental problem is to try to find the "real," historical sequence of events, that is, a global ordering (ranking) of events that minimizes contradictions. In rare cases, there are no such contradictions: the order of events is the same in all wells. Ranking of the events is then a trivial task. But in the vast majority of cases, it will be observed that some event A is above another event B in one well, but below it in another. Such contradictions can have a number of different causes, but in most cases they are due to inhomogeneous biogeographic distribution of the species or to incomplete preservation or sampling (Fig. 13.4). In many data sets, it will also be found that some events occur in contradictory cycles, such as event A being above B, which is above C, which is above A again.

Ranking-scaling is quite different from the CONOP and UA methods in one fundamental way. CONOP and UA both attempt to recover the *maximal* stratigraphic ranges, from global first appearance to global last extinction. Ranking-scaling, on the other hand, uses *average* stratigraphic positions of events. If a taxon has a particularly long range at one locality but a shorter range at many other localities, the short range will be assumed. Ranking-scaling can therefore be very successful and precise about predicting stratigraphic positions in a majority of localities, at the cost of making occasional errors. Also, the stratigraphic positions of the global first and last occurrence remain unknown. These and other features of the ranking-scaling method have made it useful for applications in the petroleum exploration industry.

Ranking-scaling is available in several software packages, including RASC (Agterberg and Gradstein, 1999) and PAST.

#### 13.3.3.1 Ranking

As the name implies, ranking-scaling is a two-step method. In the first step, the events are ranked (ordered) according to the ranks within wells, with "majority vote" methods used for resolving contradictions and cycles. Several algorithms are available for ranking, but one recommended approach (Agterberg in Gradstein et al. 1985) is to use "presorting" followed by the "modified Hay method." Both algorithms rank the events based on the number of times each event occurs above any other event in any well (ties count as a half unit).

Cycles are resolved by removing the "weakest link." Let us say that A is found below B two times, B is found below A three times, C is found below B five times, and A is found below C four times (Fig. 13.15). In this case, we have a cycle with A below C below B below A. The weakest link in the cycle is obviously the contradictory superposition B below A, and so this link is removed. The resulting ranking is A below C below B.

Before ranking is executed, the user must set a few parameters. The minimum number of wells that an event must occur in for the event to be included in the analysis is called $k_c$. Events that occur in only a couple of wells are of little stratigraphic value and may degrade the result of the analysis. The minimum number of wells that a *pair* of events must occur in for the pair to be included in the analysis is called $m_{c1}$. This parameter is often set to one, but it may be necessary to give it a higher value (but smaller than $k_c$) if there are many cycles. An additional tolerance parameter ($TOL$) can be set to a positive value such as 0.5 or 1.0 in order to remove poorly supported superpositional relationships. This can reduce the number of cycles.

After ranking, all events are sorted from lowest (oldest) to highest (youngest). However, the position of each event in the ranked sequence is often not unique, and a range of possible positions must be given. Such ambiguity can result from ignoring a pair of events due to the threshold parameter $m_{c1}$, the destruction of weakest links in cycles, or simply because of several events occurring at the same position in the inferred sequence.

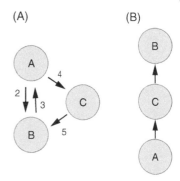

Figure 13.15 (A) The observed numbers of superpositions between three taxa A, B, and C. Arrows point from the taxon below to the taxon above. The contradictory superpositions between A and B are resolved by a majority vote, selecting the direction B below A. This gives a cycle of A below C below B below A. (B) Breaking the weakest link in the cycle gives the ranking A below C below B.

The process of ranking will not necessarily preserve the observed co-occurrences (Fig. 13.16). Since such co-occurrences represent directly observable, trustworthy information, it may be seen as a basic limitation of RASC that they are not used to constrain the solution. However, RASC may give the solution that is most likely to occur in wells or outcrops, which can be a useful feature in industrial applications (RASC computes average ranges, while other methods attempt to find maximal ranges). Moreover, in the petroleum industry, it is common to use drill cuttings, which are prone to caving from higher levels in the well. In these cases, only the last occurrences are considered reliable, and co-occurrence information is therefore not available.

### 13.3.3.2 Scaling

Once events have hopefully been correctly ordered along a relative timeline, we may try to go one step further and use the biostratigraphic data to estimate stratigraphic *distances* between consecutive events. The idea of the scaling step in ranking-scaling is to use the number of superpositional contradictions for this purpose. For example, consider the following two events: A – the last appearance of fusulinacean foraminiferans (globally, end-Permian), and B – the last appearance of conodonts (globally, late Triassic). There may be a few wells with poor data where B appears before A because conodonts for some reason are lacking in the upper part of the well. But in the majority of wells, A is found in the "correct" relative position below B. In this case, we say that there is little stratigraphic *crossover* between the events, and it is inferred that the stratigraphic distance between them is quite large. Now, consider two other events: C – the last appearance of ammonites, and D – the last appearance of rudistid bivalves. On a global scale, both these events are part of the

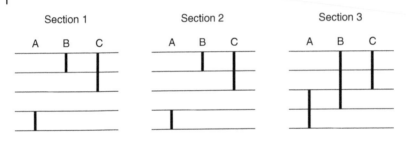

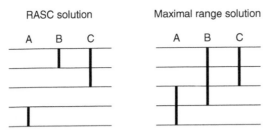

Figure 13.16  The probabilistic nature of the RASC method does not necessarily honor the observed co-occurrences but gives the solution most likely to occur in the next section based on previous data. In this example, there are three taxa (A–C) in three sections (top row). RASC gives the solution at the bottom left, based on a "majority vote" where the identical sections 1 and 2 win over the single section 3. Based on the observed sections, this can be considered the most likely pattern to be observed in a new section 4. However, section 3 proves that A and B lived at the same time. Therefore, the pattern at the lower right might be considered closer to the "true" sequence of the first and last occurrences, implying missing data for A and B in sections 1 and 2. The methods of constrained optimization and unitary associations will produce a "maximal range solution" similar to this.

end-Cretaceous mass extinction, and they were probably quite close in time. You would need very complete data indeed to find C and D in the "correct" order in all wells, and it is much more likely that C will be found above D about as frequently as vice versa. There is a lot of crossover, and it is therefore inferred that the stratigraphic distance between C and D is small.

A problem with this approach is that the distances may be strongly affected by the completeness of the fossil record, which is likely to be different for the different species. For the example of fusulinaceans and conodonts, what happens if the conodont record is very poor? This will produce a lot of crossover, and the distance between the events will be underestimated. In fact, in order to estimate the distance between two events from the degree of crossover, it is assumed that the temporal position of each event has a normal distribution over all wells and that the variance is the same for all events (Fig. 13.17). These assumptions may be difficult to satisfy, although it possibly helps that the *estimate* of inter-event distances (not the real underlying distances) is more Gaussian because of the way it is calculated (Agterberg and Gradstein 1999). Still, this approach seems to work in practice, giving results that are in general accordance with estimates from

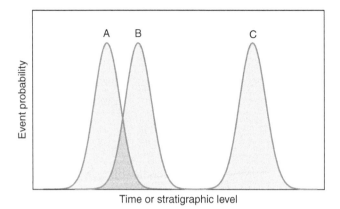

**Figure 13.17** Each of the three biostratigraphic events A, B, and C is found in many wells. Their distributions along the timeline are assumed to be normal, with equal variances. Since the distributions of A and B overlap strongly, there will be many wells where these events are observed in the reverse relationship A above B. Conversely, the stratigraphic distance between two events can be estimated from the degree of inconsistency (crossover) observed in their superpositional relationship: high crossover frequency implies a large overlap between the distributions and hence small stratigraphic distance.

independent data. The available software also contains functions to test the validity of the assumptions.

The distance calculated from one event to the next in the ranked sequence may turn out to be negative. That does not make sense, so then we simply swap the two events in the sequence to get rid of the problem and run the scaling procedure again. This is not an important issue since such events are likely to be very close to each other anyway.

### 13.3.4 Normality testing and variance analysis

Once a ranked or scaled sequence of events has been computed, this "optimal" sequence can be compared with the observed sequence in individual sections or wells. Such a comparison can be very useful for identifying events with particularly stable or unstable stratigraphic positions and to assess the quality of different wells.

Figure 13.19 shows a scatter plot of positions in the optimal sequence versus positions in one particular well. Several statistics are available for quantifying the quality of wells and events versus the optimal sequence:

1) "Step model." Within each well, each event E is compared with all other events in that well. Each reversal with respect to the optimal sequence contributes one penalty point to the event E, while each coeval event contributes 0.5 penalty points. This can be used to evaluate individual events in individual wells. Rank correlation (section 4.5) of the event positions in one particular well versus the global sequence can be used to evaluate the well as a whole.
2) Normality test. For a given event in a given well, the difference between the two neighboring inter-event distances (the second-order difference) is compared with an expected

value to give a probability that the event is out of place with respect to the optimal sequence.

3) Event variance analysis. This procedure is used to calculate a variance in the stratigraphic position for each event, based on the distances between the event and a fitted LOC in each well. Events with small variances are better stratigraphic markers than those with large variances.

### 13.3.5 Correlation (CASC)

As a final step, the different wells can be correlated with each other through the optimal sequence. In the computer program CASC (Gradstein and Agterberg in Gradstein et al. 1985), this is done with the help of spline-fitted (section 12.7) correlation lines. Consider a given event, say event 43, in a given well A. This event has been observed at a certain depth in A (say 3500 m), but this is not necessarily its most likely "real" position in A, both because the well has not been continuously sampled and because of possible contradictions. To find a likely depth, we start by fitting a spline curve through a scatter plot of the well sequence versus the optimal sequence (e.g., Fig. 13.19). This gives us a likely position of event 43 in the well sequence – say at position 45.7. Then a spline is fitted to a scatter plot of the event sequence in the well versus the observed depths. An event sequence position of 45.7 may then correspond to a depth of 3450 m, which is then finally the estimated depth of event 43 in well A. The corresponding depth for event 43 is found in other wells, giving tie points for correlation.

CASC also estimates confidence intervals for the calibrated positions, allowing evaluation of the quality of different events for correlation.

---

**Example 13.6**

We will use an example given by Gradstein and Agterberg (1982), involving one LAD and ten FADs of Eocene calcareous nannofossils from the California Coast Range (Table 13.1).

Table 13.2 shows the ranked sequence of events, from lowest (oldest) at position 1 up to highest (youngest) at position 10. The column marked "Range" shows the uncertainty in the event position. Based on these uncertainties, we can write the sequence of events as (2, 3, 1), (5, 7), 4, (6, 8), and (9, 10), where coeval events are given inside brackets.

Figure 13.18 shows the result of scaling, together with a dendrogram that visualizes the scaling distances between events. This suggests a zonation perhaps as follows: (3, 2, 1), (5, 7, 4), 6, (8, 9, 10).

We can now go back to the original data and see if the global ("optimal") sequence of events is reflected in each well. Figure 13.19 shows the depths of the events in well I plotted against the optimal sequence. There are two reversals: between events 3 and 2 and between events 10 and 9.

Table 13.1 Event names (top) and levels of the events in the nine sections A–I. An empty cell signifies that the event was not observed in the given section.

| Event | Event name |
|---|---|
| 1 | *Discoaster distinctus* FAD |
| 2 | *Coccolithus cribellum* FAD |
| 3 | *Discoaster germanicus* FAD |
| 4 | *Coccolithus solitus* FAD |
| 5 | *Coccolithus gammation* FAD |
| 6 | *Rhaboosphaera scabrosa* FAD |
| 7 | *Discoaster minimus* FAD |
| 8 | *Discoaster cruciformis* FAD |
| 9 | *Discoaster tribrachiatus* LAD |
| 10 | *Discolithus distinctus* FAD |

| Section | 1 | 2 | 3 | 4 | 5 | 6 | 7 | 8 | 9 | 10 |
|---|---|---|---|---|---|---|---|---|---|---|
| A | 1 | 1 | 1 | 1 | 1 | 1 | 2 | 3 | 4 |  |
| B |  | 1 | 1 | 1 | 1 | 1 | 1 |  | 2 | 1 |
| C | 3 | 1 |  |  | 2 |  |  |  | 4 |  |
| D | 2 | 1 |  |  | 4 |  | 3 | 5 | 6 | 7 |
| E | 2 | 1 | 3 | 6 | 1 | 7 | 4 | 5 | 8 |  |
| F | 1 | 3 | 1 | 2 | 2 |  | 4 | 4 | 5 | 6 |
| G | 3 | 3 | 2 | 2 | 3 |  | 1 | 4 | 5 | 4 |
| H | 2 |  |  | 4 | 2 |  | 1 |  | 3 | 2 |
| I | 2 | 1 | 2 | 4 | 3 | 5 |  |  | 6 | 7 |

Table 13.2 Ranked sequence of events, from lowest (position 1) to highest (position 10).

| Pos. | Range | Event |
|---|---|---|
| 10 | −1 to 0 | 10 |
| 9 | 0 to 1 | 9 |
| 8 | −1 to 0 | 8 |
| 7 | 0 to 1 | 6 |
| 6 | 0 | 4 |
| 5 | −1 to 0 | 7 |
| 4 | 0 to 1 | 5 |
| 3 | −1 to 0 | 1 |
| 2 | −1 to 1 | 3 |
| 1 | 0 to 1 | 2 |

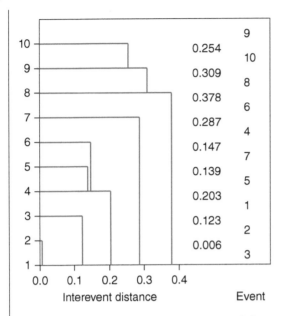

Figure 13.18 Scaling dendrogram. This type of diagram is somewhat complicated to read, but its interpretation is simple. The sequence of events after scaling is shown in the rightmost column, from bottom to top (note slight differences from the ranked sequence). The distance from event to event as calculated by the scaling procedure is given in the next column to the left. For example, the distance from event 1 to 5 is 0.203. These distances are also drawn as vertical lines in the dendrogram, positioned horizontally according to the distance value. This results in a simple dendrogram that can be used for zonation. Note that events 5, 7, and 4 form a cluster, as do events 8, 10, and 9.

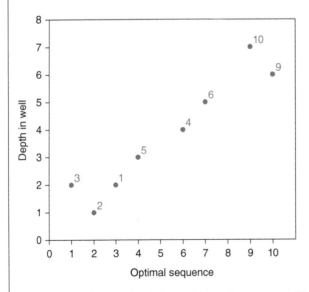

Figure 13.19 Scatter plot of the positions of events in well I versus the optimal sequence. Events 3 and 9 are out of order in this well, and events 7 and 8 are not present.

## 13.3.6   Unitary Associations

One of the key characteristics of the unitary associations (UA) method (Guex 1991; Guex et al. 2016) is that it is based on *associations* rather than events. Instead of using levels of events, it requires fossil presence/absence data in stratigraphic successions of samples from several sections or cores. The method attempts to use information about co-occurring species and tries not to produce a zonation that would involve co-occurrences that have not been observed. The UA method is conservative, in the sense that robustness is weighed heavily at the cost of stratigraphic resolution. There is quite a bit of "philosophy" behind it – one of the premises being that biostratigraphy is regarded as a non-probabilistic problem. Confidence intervals are therefore not calculated, but contradictions and uncertainties are reported in a number of different formats.

There are several alternative ways of describing the UA method (Guex 1991). Here we choose simply to list the steps carried out by the programs Biograph (Savary and Guex 1999), UAgraph (Guex et al. 2016), and PAST (the UA module in PAST provides the most recent implementation).

### 1. *Residual maximal horizons*
The method makes the range-through assumption, meaning that taxa are considered to have been present at all levels between the first and last appearance in any section. Then any sample that contains a taxonomic subset of any other sample is discarded. The remaining samples are called *residual maximal horizons*. The idea behind this throwing away of data is partly that the absent taxa in the discarded samples may simply not have been found even though they originally existed. Absences are therefore not as informative as presences.

### 2. *Superposition and co-occurrence of taxa*
Next, all pairs (A, B) of taxa are inspected for their observed superpositional relationships: A below B, B below A, A together with B, or unknown. If A occurs below B in one locality and B below A in another, they are considered to be "virtually" co-occurring although they have never actually been found together. This is one example of the conservative approach used in the UA method: a superposition is not enforced if the evidence is conflicting. The superpositions and co-occurrences of taxa can be viewed in the *biostratigraphic graph*, where co-occurrences between pairs of taxa are shown as solid lines. Superpositions can be shown as arrows or as dashed lines, with long dashes from the above-occurring taxon and short dashes from the below-occurring taxon (Fig. 13.22).

Some taxa may occur in so-called *forbidden subgraphs*, which indicate inconsistencies in their superpositional relationships. Two of the several types of such subgraphs can be plotted in PAST: $C_n$ *cycles*, which are superpositional cycles (A above B above C above A), and $S_3$ *circuits*, which are inconsistencies of the type "A co-occurring with B, C above A, and C below B." Interpretations of such forbidden subgraphs are suggested by Guex (1991).

### 3. *Maximal cliques*
*Maximal cliques* (also known as initial unitary associations) are groups of co-occurring taxa not being subsets of any larger group of co-occurring taxa. The maximal cliques are candidates for the status of unitary associations but will be further processed below. In PAST,

maximal cliques receive a number and are also named after a maximal horizon in the original data set that is identical to, or contained in (marked with an asterisk), the maximal clique.

### 4. Superposition of maximal cliques

The superpositional relationships between maximal cliques are decided by inspecting the superpositional relationships between their constituent taxa, as computed in step 2. Contradictions (some taxon in clique A occurs below some taxon in clique B, and vice versa) are resolved by a "majority vote," in contrast with the methodological purity we have seen so far. Although it is usually impossible to avoid such contradictions totally, they should be investigated and removed by modification of the data, if appropriate. The contradictions between cliques can be viewed in PAST.

The superpositions and co-occurrences of cliques can be viewed in the *maximal clique graph*. Co-occurrences between pairs of cliques are shown as solid lines. Superpositions are shown as dashed lines, with long dashes from the above-occurring clique and short dashes from the below-occurring clique. Also, cycles between maximal cliques (see below) can be viewed.

### 5. Resolving cycles

It will sometimes be the case that maximal cliques are now ordered in cycles: A is below B, which is below C, which is below A again. This is clearly contradictory. The weakest link (superpositional relationship supported by fewest taxa) in such cycles is destroyed. Again, this is a somewhat "impure" operation.

### 6. Reduction to a unique path

At this stage, we should ideally have a single path (chain) of superpositional relationships between maximal cliques, from bottom to top. This is, however, often not the case, for example, if two cliques share an underlying clique, or if we have isolated paths without any relationships (Fig. 13.20). To produce a single path, it is necessary to merge cliques according to special rules. The maximal cliques graph can often give interesting

(A)    (B)

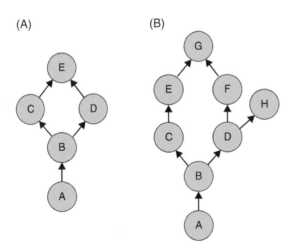

Figure 13.20 Maximal cliques graphs and their reduction to a single superpositional chain or path. Arrows show superpositional relationships. (A) A simple case where the reduction to a single path is clear (cliques C and D will be merged). (B) A graph with parallel paths CE and DF, perhaps due to splitting into different sedimentary basins or facies. H is a "dead end." The reduction to a single path must now proceed in a somewhat *ad hoc* manner, based on the number of shared taxa between cliques, or cliques may simply be removed.

supplementary information. One commonly observed phenomenon is that different geo-graphical regions or facies give rise to parallel paths in the graph.

This merging of cliques is perhaps the most questionable step in the whole procedure. The cliques must usually be merged according to faunal similarity because their strati-graphical relationships are not known. In PAST, there is an option to suppress this merging by simply disregarding cliques that are not part of the longest path. This is a more conserva-tive approach, at the cost of "losing" potentially useful zone fossils that are found in the disregarded cliques.

*7. Post-processing of maximal cliques*
Finally, some minor manipulations are carried out to "polish" the result: generation of the "consecutive ones" property; reinsertion of residual virtual co-occurrences and superposi-tions; and compaction to remove any generated non-maximal cliques. For details on these procedures, see Guex (1991). At last, we now have the unitary associationsUA.

The unitary associations have associated with them an index of similarity from one UA to the next, called *D*, which can be used to look for breaks due to hiatuses or to origination or extinction events:

$$D_i = \left|UA_i - UA_{i-1}\right| / \left|UA_i\right| + \left|UA_{i-1} - UA_i\right| / \left|UA_{i-1}\right|$$

*8. Correlation using the unitary associations*
The original samples are now correlated using the unitary associations. A sample may con-tain taxa that uniquely place it in a unitary association, or it may lack key taxa that could differentiate between two or more UAs, in which case only a range can be given.

*9. Reproducibility matrix*
Some UAs may be identified in only one or a few sections, in which case one may consider merging UAs to improve geographical reproducibility (see below). The reproducibility matrix should be inspected to identify such UAs.

*10. Reproducibility graph and suggested UA merges (biozonation)*
The reproducibility graph (Gk' in Guex 1991) shows the superpositions of UAs that are actually observed in the sections. PAST will internally reduce this graph to a unique maxi-mal path (Guex 1991, section 5.6.3), and in the process of doing so, it may merge some UAs. These mergers are shown as red lines in the reproducibility graph. The sequence of single and merged UAs can be viewed as a suggested biozonation.

---

**Example 13.7**

Drobne (1977) compiled the occurrences of 15 species of the large foraminiferan *Alveolina* (Fig. 13.21) in 11 sections in the lower Eocene of Slovenia. All species are found in at least four sections – otherwise, we might have wanted to remove taxa found in only one section since these are of no stratigraphic value and can produce noise in the result. It may also be useful to identify sections that contain partly endemic

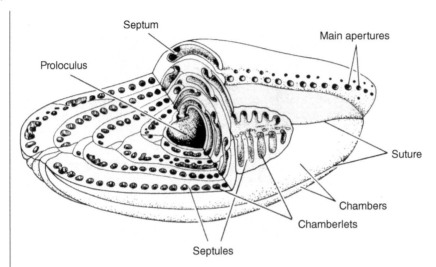

**Figure 13.21** Morphology of a large *Alveolina* foraminiferan. This group has formed the basis for a detailed biostratigraphy of the Paleogene rocks of the Mediterranean region. Adapted from Lehmann and Hillmer (1983).

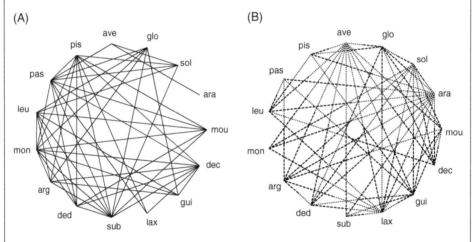

**Figure 13.22** Biostratigraphic graph for the alveolinid data set (15 species). (A) Co-occurrences are shown as solid lines. (B) Superpositions. The dashed line from the species above, dotted line from the species below.

taxa – these may be left out of the analysis, at least initially. For our example, we will use all the sections from the start.

From the original 41 samples (horizons), we are left with only 15 residual horizons.

The biostratigraphic graph (co-occurrences and superpositions between taxa) is shown in Fig. 13.22.

Based on the residual horizons, 12 maximal cliques are found. In Table 13.3, they are given in arbitrary order.

Table 13.3    Maximal cliques from the alveolinid data set, in arbitrary order.

| | sol | leu | Glo | pas | sub | pis | ara | mon | gui | mou | ded | lax | ave | arg | dec |
|---|---|---|---|---|---|---|---|---|---|---|---|---|---|---|---|
| 12 | . | . | . | . | . | ■ | . | . | . | . | . | . | ■ | . | . |
| 11 | . | ■ | ■ | . | . | . | . | ■ | ■ | . | . | . | . | ■ | . |
| 10 | . | . | . | ■ | ■ | ■ | . | ■ | . | . | ■ | . | . | . | ■ |
| 9 | ■ | ■ | . | . | ■ | . | . | . | . | . | . | . | . | . | . |
| 8 | . | . | . | . | ■ | ■ | . | ■ | . | . | . | . | . | ■ | ■ |
| 7 | . | . | . | ■ | . | ■ | . | . | . | . | . | ■ | . | . | ■ |
| 6 | ■ | . | . | ■ | ■ | ■ | . | . | . | . | . | . | . | . | . |
| 5 | ■ | . | . | ■ | . | . | . | . | . | . | . | . | ■ | . | . |
| 4 | . | . | ■ | ■ | ■ | . | . | ■ | . | . | . | . | . | . | . |
| 3 | . | . | . | . | . | . | . | ■ | ■ | . | . | . | . | ■ | ■ |
| 2 | . | ■ | . | ■ | . | . | . | ■ | . | ■ | ■ | . | . | . | . |
| 1 | . | . | . | ■ | ■ | ■ | . | ■ | . | ■ | ■ | . | . | . | . |

After investigating the observed superpositions between the species contained in the maximal cliques, the program arranges the cliques in a superpositional graph. It turns out that in this graph, there are 22 contradictory superpositions that are resolved by majority votes, and four cliques (2, 4, 9, and 10) are involved in cycles that are broken at their weakest links. At this point, the contradictions should be inspected in order to identify any possible problems with the original data (Fig. 13.23 shows the contradictions between two of the cliques). If no such problems are found and corrected, we must continue with the procedure and hope that the "majority votes" have produced reasonable results.

For now, we just continue, perhaps with a slight feeling of unease. First, the program finds the longest possible path through the clique superposition graph. This path turns out to include all the 12 cliques in a linear chain, in the sequence 12-5-6-7-1-10-4-9-2-8-11-3 from bottom to top. This is quite rare, and very fortunate, because there is no need to carry out that questionable merging of cliques.

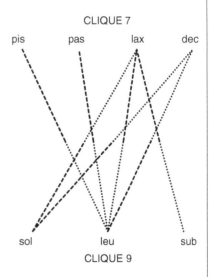

CLIQUE 7

pis   pas   lax   dec

sol   leu   sub

CLIQUE 9

Figure 13.23   Contradictory superpositional relationships between species in clique 7 (top row) and 9 (bottom row). Clique superposition 7 above 9 is indicated by the species superpositions *pis* above *leu*, *pas* above *leu*, *lax* above *leu*, and *lax* above *sub* (four pairs), while clique superposition 7 below 9 is indicated by the species superpositions *lax* below *sol*, *dec* below *sol*, and *dec* below *leu* (three pairs). In this case, the program will decide that 7 is above 9 based on a majority vote.

After some further manipulations (step 7 above), the program still finds it necessary to reduce the stratigraphic resolution, and it merges successive cliques until it ends up with seven UAs, numbered from 1 (bottom) to 7 (top), as shown in Table 13.4.

Given these UAs, the program attempts to correlate all the original samples. For example, the seven samples in the section "Veliko" are assigned to UAs in Table 13.5.

We see that sample 1 is uniquely assigned to UA 1, sample 3 to UA 3, and samples 6 and 7 to UA 7. The other samples can only be assigned to a range of UAs, since they do not contain enough diagnostic taxa.

The program also gives many other reports, the most important ones being the UA *reproducibility table*, showing what UAs are identified in what sections, and the *reproducibility graph*, showing the observed superpositional relationships between UAs. The reproducibility graph can be used to erect a biozonation, as explained in Fig. 13.24.

Table 13.4   Unitary associations from the alveolinid data set, in stratigraphic order.

|   | ara | ave | pas | sol | lax | pis | dec | mou | ded | spy | Mon | glo | leu | arg | gui |
|---|-----|-----|-----|-----|-----|-----|-----|-----|-----|-----|-----|-----|-----|-----|-----|
| 7 | . | . | . | . | . | . | ■ | . | . | . | ■ | ■ | ■ | ■ | ■ |
| 6 | . | . | . | . | . | ■ | ■ | . | . | ■ | ■ | ■ | ■ | ■ | . |
| 5 | . | . | . | ■ | . | ■ | ■ | ■ | ■ | ■ | ■ | ■ | ■ | . | . |
| 4 | . | . | ■ | ■ | . | ■ | ■ | ■ | ■ | ■ | ■ | ■ | . | . | . |
| 3 | . | . | ■ | ■ | ■ | ■ | ■ | . | . | . | . | . | . | . | . |
| 2 | . | ■ | ■ | ■ | . | . | . | . | . | . | . | . | . | . | . |
| 1 | ■ | ■ | . | . | . | . | . | . | . | . | . | . | . | . | . |

Table 13.5   Assignment of samples in the Veliko section to unitary associations.

|           | First UA | Last UA |
|-----------|----------|---------|
| Veliko-7  | 7        | 7       |
| Veliko-6  | 7        | 7       |
| Veliko-5  | 4        | 5       |
| Veliko-4  | 4        | 6       |
| Veliko-3  | 3        | 3       |
| Veliko-2  | 3        | 4       |
| Veliko-1  | 1        | 1       |

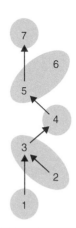

**Figure 13.24**   Reproducibility graph showing the observed superpositional relationships between the seven UAs (arrows pointing from UA below to UA above). This graph suggests a possible biozonation. Since UA 6 is never directly observed above or below any other UA, it should be merged with another UA. It has most taxa in common with UA 5, so we merge UA 5 and 6. UA 2 is also not part of the linear chain. The program suggests merging it with UA 3, with which it shares most taxa. This results in five biozones: UA 1, UAs 2–3, UA 4, UAs 5–6, and UA 7.

### 13.3.7   Biostratigraphy by ordination

So far, we have looked at biostratigraphic methods that put great emphasis on information about superpositional relationships in local sections or wells. Obviously, such relationships can and will greatly constrain the search for an optimal sequence. But in some cases, fossil occurrences are so rare that it is unlikely to find more than one fossiliferous horizon at each locality, and superpositional relationships are then not observable. This is often the case with terrestrial vertebrates, for example. One possible approach is then to assume that taxa come and go in an overlapping sequence, so that samples can be ordered in time by maximizing the number of shared taxa between consecutive samples. The biostratigraphic problem then reduces to an exercise in ordination, as described in chapter 10.

Possible methods for biostratigraphic ordination of isolated samples include correspondence analysis (CA; section 10.8), seriation (section 10.10), principal coordinates analysis (PCoA, section 10.6), and non-metric multidimensional scaling (NMDS; section 10.7). Of these, CA and seriation most directly attempt to localize presence of taxa along the ordination axis and are therefore logical choices for the biostratigrapher.

The method of appearance event ordination (Alroy 1994) bridges to some extent the gap between ecological ordination and biostratigraphy using superpositional information. This method can be compared both with CA and seriation but takes into account a peculiar asymmetry of the biostratigraphic problem: the observation that the first occurrence of taxon A occurs below the last appearance of B is an absolute fact that can never be disputed by further evidence (disregarding reworking, caving, or incorrect identification). The reason for this is that any further finds can only extend down the first appearance of A, or extend up the last appearance of B. Such an observation is called an F/L (First/Last) statement. The method attempts to honor all F/L statements, while parsimoniously minimizing the number of F/L relations that are implied by the solution but not observed. Appearance event ordination can be used for data sets with or without superpositional relationships or a mixture of the two. Despite its interesting theoretical properties, the method has been less frequently used than RASC, CONOP, or UA.

### 13.3.8   What is the best method for biostratigraphic correlation?

It is impossible to give a general recommendation about the choice of biostratigraphic method – it depends on the type of data and the purpose of the investigation. The following comments are provided only as personal opinions of the authors.

If you have only two sections to be correlated, little is gained by using a complicated quantitative method. Graphic correlation, possibly carried out on paper, will probably be an adequate approach.

For well-cutting data with only last occurrence events available, and with a large number of wells and taxa, RASC may be the most practical method. RASC is fast for large data sets, and the problems of constraint violations such as range contractions and breaking of co-occurrences are not a concern if only last occurrences are given or if there is considerable caving or reworking.

For all other event data, in particular if both first and last occurrences are available, and if the data set is not enormous, we recommend CONOP. This method is very reasonably based on the minimization of range extensions, while honoring co-occurrences and other constraints (other optimization criteria and constraint options are also available in existing software).

For data of the taxa-in-samples type, UA is a good choice, although CONOP may also be expected to perform well for such data. The UA method is fast, focuses on observed co-occurrences, and is transparent in the sense that available software gives clear reports on the detailed nature of inconsistencies. The method is therefore a good tool for interactive weeding out of inconsistent data points. The UA method also includes a recommended procedure for the definition of biozones. Finally, the structure of the maximal cliques graph can give valuable insights into paleobiogeography and facies dependence through time.

Cooper et al. (2001) applied graphic correlation, RASC, and CONOP to the same data set and compared the results. They found that all methods gave similar composite sequences of events and concluded that RASC and CONOP should be regarded as complementary methods: RASC giving the expected sequence in new wells, while CONOP giving a sequence corresponding more closely with global maximal ranges and therefore being preferable for comparison with other biozonations and timescales. Graphic correlation was considered unnecessarily labor-intensive compared with the more automated methods. Of course, the congruency between the different methods is expected to deteriorate with more noisy data sets.

## 13.4   Age models

Although absolute dating of rocks and sediments is not part of paleontology, the analysis of fossils through a sedimentary succession is often supported by radiometric dates. Ideally, such dates should be without errors and closely spaced throughout the interval studied. In this case, we could draw a nearly continuous curve through the data points, providing an accurate age for any level in the section or depth in a core.

In practice, however, the radiometric ages will have considerable measurement errors, often assumed to be normally distributed with a certain standard deviation, or, as in the case of calibrated radiocarbon ages, the error distribution can be quite complex. Moreover, the samples may be inaccurately placed in the section or reworked. Finally, because of the analysis cost and the scarcity of dateable material such as ash beds, we usually only have a few points available and need to interpolate between them to obtain an age for any horizon in the section.

It turns out that this is a complex and subtle problem, and many methods have been proposed. We will start with the simplest, naïve (but sometimes useful) approaches and experiment our way toward more advanced methods.

### 13.4.1 Simple interpolation

We return to the Andfjord radiolarian study discussed in sections 10.12 and 11.1 (Bjørklund et al. 2019). In the core JM99-1200, seven radiocarbon dates were available, in addition to the Vedde Ash Bed, which has a fairly precise age known from other studies. The calibrated ages are given in Table 13.6 (for now, we disregard the associated errors).

We saw in section 11.1 that the radiolarian assemblages point to a dramatic temperature increase at 410-cm core depth. We need an age model that allows us to estimate the age at this level. The first thing we can try is to simply draw straight lines between the given points, as in Figure 13.25A. The model implies much higher sedimentation rates in the lower part (older than ca. 11,400 years), which is consistent with the deglaciation history of the area. The age at 410 cm depth can be estimated as 11,990 cal year BP by linear interpolation – slightly older than the currently accepted age of ca. 11,650 years for the base Holocene.

This simple linear method was used for the same data set by Cabedo-Sanz et al. (2013). Although its simplicity and transparency are appealing, it has limitations. First, because of errors and noise, outliers are quite common in such data and superpositional reversals can occur (i.e., older ages are found higher in the sequence). The only way to handle such points with the present method would be to delete them. Second, error bars on the given

Table 13.6   Dated levels in the JM99-1200 core, Andfjord, Norway.

| Core depth (cm) | Calibrated yr BP (median) | Age error (one-sigma) |
|---|---|---|
| 59.0 | 6,323 | 72 |
| 281.8 | 10,873 | 104 |
| 300.5 | 11,429 | 165 |
| 436.5 (Vedde Ash) | 11,980 | 57 |
| 457.5 | 12,590 | 141 |
| 511 | 12,880 | 104 |
| 655.5 | 13,256 | 95 |
| 723.5 | 13,553 | 103 |

From Bjørklund et al. (2019). The one-sigma errors are derived from a simplified error distribution.

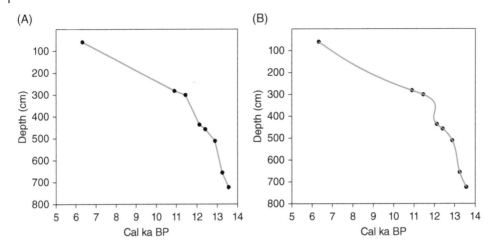

Figure 13.25 Two simple age models for the Andfjord core, based on Table 13.4. (A) Linear interpolation between given points. (B) Cubic spline interpolation, giving a non-monotonic age model.

ages are not taken into account, and third, we get no information about expected errors in the interpolated errors.

Yet another issue with the piecewise linear model is that it implies a constant sedimentation rate between data points, which may be unrealistic. As a partial remedy for this problem, we could try a cubic spline (section 12.7), i.e., interpolating each gap with a third-degree polynomial, giving a smooth curve. As seen in Fig. 13.25B, this fails because the age model becomes non-monotonic.

## 13.4.2 Simple regression and smoothing

Our next step in complexity is to include information about the errors in the given ages and to enforce a monotonic age model. When we accept that the ages have associated errors, we can also accept that the age model does not pass through the given points exactly. The problem then changes from an interpolation to a regression problem.

For very simple problems with few data points, and if we can assume that the sedimentation rate was constant throughout the interval studied, we may consider using simple linear regression (section 4.6). In most cases, the depths are known with precision, and the errors are confined to the ages. The usual OLS regression is then appropriate. We need an OLS program that allows the specification of individual errors, downweighing points with large errors. Figure 13.26A shows such regression for the seven oldest points in the Andfjord data set, where the sedimentation rate seems to vary relatively little.

Extending the cubic spline model, we now turn to a more advanced smoothing spline, allowing the curve to move slightly away from the points (section 12.7), downweighing points with large errors. The smoothing factor is optimized by a cross-validation procedure, constrained to produce a monotonic curve. This procedure is included in PAST. The resulting age model for the complete Andfjord data set is shown in Fig. 13.26B. The interpolated age at 410 cm depth is 11,970 cal year BP.

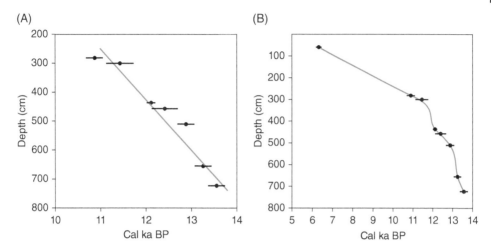

**Figure 13.26** Two more age models for the Andfjord core (cf. Fig. 13.25). (A) Linear regression of the seven oldest dates, giving larger weight to points with smaller errors. (B) Smoothing spline of all points, with weighting. Error bars are 2-sigma.

### 13.4.3   Classical age models with Monte Carlo simulation

We usually want to estimate error bars (confidence intervals) for our age model. For simple linear regression, such confidence intervals can be computed analytically from the standard errors on slope and intercept. For more complex age models, confidence intervals are conveniently estimated by Monte Carlo simulation. Considering piecewise linear interpolation, we can thus produce many random replicates, each with the given dates replaced by random dates selected from normal distributions with the given standard deviations. For each such replicate, the piecewise linear interpolation is computed anew, giving a large number of simulated age models from which confidence intervals can be extracted.

This approach is used, for example, by the program COPRA (Breitenbach et al. 2012), which also includes a spline interpolator. COPRA enforces monotonic solutions by disregarding any replicates with reversals. PAST uses a similar approach, with smoothing splines, automatically increasing the smoothing factor until reversals are resolved. The R script "Clam" (Blaauw 2010) is a popular program in the same class, but it also allows non-Gaussian error distributions as typically produced by radiocarbon calibration.

To illustrate age modeling with Monte Carlo simulation, we will use a set of U-Pb radiometric dates from South China (Ovtcharova et al. 2015). The purpose of the study was to date the Early-Middle Triassic boundary, as identified in the section based on conodont biostratigraphy. The dates with their error bars, and a smoothing spline generated with PAST, are shown in Fig. 13.27A. The data set appears noisy, with large scatter and several serious age reversals. Some of the methods described earlier require that we remove outliers and reversed dates before constructing the age model, alternatively that we increase their error bars until a monotonic model can be fitted. In Figure 13.27A, the algorithm has instead increased the smoothing. This leads to a very high smoothing factor of 4.08. The solution is not unreasonable, however, passing within or close to the error bars of most of the points, except the reversed dates.

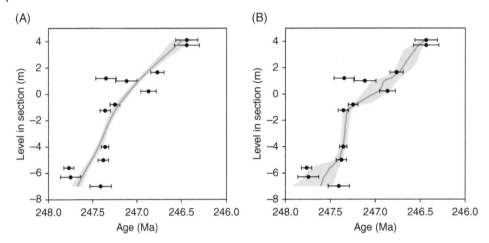

Figure 13.27   Age modes for the Triassic data set of Ovtcharova et al. (2015). (A) Smoothing spline with 95% confidence band (PAST). (B) Bayesian (Bchron) model with 95% credible interval. Adapted from Ovtcharova et al. (2015).

It is important to remember that any age model is constrained by the assumptions we have made about sedimentation rates. For the spline model, the main assumption is that the rate varies smoothly. This can perhaps be viewed as a parsimonious model, not allowing for more complicated model elements such as the existence of unknown sharp changes in sedimentation rate, including hiatuses. In the absence of such information, using a simple model has some merit. The Monte Carlo confidence interval is similarly constrained by these assumptions. This is the reason for the narrow confidence band in Fig. 13.27A, which may seem unreasonable given the spread in the data.

### 13.4.4   Bayesian age modeling

If we are uncomfortable with the smoothness assumptions inherent in piecewise linear and smoothing spline methods, we need to consider age models that allow more abrupt changes in sedimentation rate, also between the given points. This leads to more complex deposition models, with a larger number of free parameters (Bronk Ramsey 2008). Bayesian methods provide a logical, general framework for such age modeling, where we can sample the model parameter space efficiently with, e.g., Markov Chain Monte Carlo methods (section 2.7). Typically, the general parameters of the deposition model are encoded in the prior, while the dates and their error distributions are used to compute the likelihood. This approach was pioneered by Buck et al. (1991) and became popular with the release of powerful Bayesian age modeling software such as OxCal (Bronk Ramsey 1995, 2001, 2008), Bpeat (Blaauw and Christen 2005), Bchron (Parnell et al. 2008), and Bacon (Blaauw and Christen 2011). The latter two programs are available as R packages. These programs have different options for the form of the depositional model and the error distributions. At the moment, OxCal is probably the most popular in the radiocarbon community (e.g., Bjørklund et al. 2019), while Bchron has been used more for U-Pb dating and other deep-time applications (e.g., Ovtcharova et al. 2015). Bacon is rising in popularity, partly because of its flexibility in including prior information on sedimentation rates (Wang et al. 2019).

A Bchron age model for the Ovtcharova et al. (2015) data set is shown in Fig. 13.27B. Two differences from the spline model (Fig. 13.27A) are apparent. First, the deposition model in Bchron allows for a less smooth solution, producing relatively sharp kinks in the curve (i.e., abrupt changes in sedimentation rate). Second, the credible interval band is dramatically wider than the confidence interval of the spline model, especially in the vicinity of age reversals in the data. This is as expected, because the Bchron deposition model has considerably larger freedom than the spline, which is constrained to produce a smooth curve.

## References

Agterberg, F.P. 1990. *Automated Stratigraphic Correlation*. Developments in Palaeontology and Stratigraphy 13. Elsevier, Amsterdam.

Agterberg, F.P., Gradstein, F.M. 1999. The RASC method for ranking and scaling of biostratigraphic events. *Earth Science Reviews* 46, 1–25

Alroy, J. 1994. Appearance event ordination: a new biochronologic method. *Paleobiology* 20, 191–207.

Armstrong, H.A. 1999. Quantitative biostratigraphy. In Harper, D.A.T. (ed.), *Numerical Paleobiology*, 181–226. John Wiley, Chichester, UK.

Bennett, K. 1996. Determination of the number of zones in a biostratigraphic sequence. *New Phytologist* 132, 155–170.

Birks, H.J.B. 2012. Analysis of stratigraphical data. In Birks, H., Lotter, A., Juggins, S., Smol, J. (eds.), *Tracking Environmental Change Using Lake Sediments. Developments in Paleoenvironmental Research*, 5, 355–378. Springer, Dordrecht.

Bjørklund, K.R., Kruglikova, S.B., Hammer, Ø. 2019. The radiolarian fauna during the Younger Dryas–Holocene transition in Andfjorden, northern Norway. *Polar Research* 38. https://doi.org/10.33265/polar.v38.3444

Blaauw, M. 2010. Methods and code for 'classical' age-modelling of radiocarbon sequences. *Quaternary Geochronology* 5, 512–518.

Blaauw, M., Christen, J.A. 2005. Radiocarbon peat chronologies and environmental change. *Journal of the Royal Statistical Society: Series C (Applied Statistics)* 54, 805–816.

Blaauw, M., Christen, J.A. 2011. Flexible paleoclimate age-depth models using an autoregressive gamma process. *Bayesian Analysis* 6, 457–474.

Breitenbach, S.F.M., Rehfeld, K., Goswami, B., Baldini, J.U.L., Ridley, H.E., Kennett, D., Prufer, K., Aquino, V.V., Asmerom, Y., Polyak, V.J., Cheng, H., Kurths, J., Marwan, N. 2012. COnstructing Proxy Records from Age models (COPRA). *Climate of the Past* 8, 1765–1779.

Bronk Ramsey, C. 1995. Radiocarbon calibration and analysis of stratigraphy: the OxCal program. *Radiocarbon* 37, 425–430.

Bronk Ramsey, C. 2001. Development of the radiocarbon calibration program OxCal. *Radiocarbon* 43, 355–363.

Bronk Ramsey, C. 2008. Deposition models for chronological records. *Quaternary Science Reviews* 27, 42–60.

Buck, C., Kenworthy, J., Litton, C., Smith, A. 1991. Combining archaeological and radiocarbon information – a Bayesian approach to calibration. *Antiquity* 65, 808–821.

Cabedo-Sanz P., Belt S.T., Knies J., Husum K. 2013. Identification of contrasting seasonal ice conditions during the Younger Dryas. *Quaternary Science Reviews* 79, 74–86,

Cooper, R.A., Crampton, J.S., Raine, J.I., Gradstein, F.M., Morgans, H.E.G., Sadler, P.M., Strong, C.P., Waghorn, D., Wilson, G.J. 2001. Quantitative biostratigraphy of the Tanaki Basin, New Zealand: A deterministic and probabilistic approach. *AAPG Bulletin* 85, 1469–1498.

Cubitt, J.M., Reyment, R.A. (eds.). 1982. *Quantitative Stratigraphic Correlation.* John Wiley, Chichester, UK.

Drobne, K. 1977. Alvéolines paléogènes de la Slovénie et de l'Istrie. *Schweizerische Paläontologische Abhandlungen* 99, 1–132.

Edwards, L.E. 1984. Insights on why graphical correlation (Shaw's method) works. *Journal of Geology* 92, 583–597.

Gordon, A.D., Birks, H.J.B. 1972. Numerical methods in qaternary palaeoecology I. Zonation of pollen diagrams. *New Phytologist* 71, 961–979.

Gradstein, F.M. 1996. *STRATCOR – Graphic Zonation and Correlation Software – User's Guide.* Version 4.

Gradstein, F.M., Agterberg, F.P. 1982. Models of Cenozoic foraminiferal stratigraphy – Northwestern Atlantic Margin. In Cubitt, J.M., Reyment, R.A. (eds.), *Quantitative Stratigraphic Correlation*, pp. 119–173. John Wiley, Chichester, UK.

Gradstein, F.M., Agterberg, J.C., Brower, J.C., Schwarzacher, W.S. 1985. *Quantitative Stratigraphy.* Reidel, Dordrecht, Holland.

Grimm, E.C. 1987. CONISS: a FORTRAN 77 program for stratigraphically constrained cluster analysis by the method of incremental sum of squares. *Computers & Geosciences* 13, 13–35.

Guex, J. 1991. *Biochronological Correlations.* Springer Verlag.

Guex, J., Galster, F., Hammer, Ø. 2016. *Discrete Biochronological Time Scales.* Springer Verlag.

Hood, K.C. 1995. *Graphcor – Interactive Graphic Correlation Software*, Version 2.2.

Kemple, W.G., Sadler, P.M., Strauss, D.J. 1989. A prototype constrained optimization solution to the time correlation problem. In Agterberg, F.P., Bonham-Carter, G.F. (eds.), *Statistical Applications in the Earth Sciences*, pp. 417–425. Paper 89-9. Geological Survey of Canada.

Kemple, W.G., Sadler, P.M., Strauss, D.J. 1995. Extending graphic correlation to many dimensions: stratigraphic correlation as constrained optimization. In Mann, K.O., Lane, H.R. (eds.), *Graphic Correlation*, 65–82. SEPM (Society for Sedimentary Geology), Special Publication 53.

Lehmann, U., Hillmer, G. 1983. *Fossil Invertebrates.* Cambridge University Press, Cambridge, UK.

MacLeod, N. 1994. Shaw Stack – Graphic Correlation Program for Macintosh Computers.

MacLeod, N., Sadler, P. 1995. Estimating the line of correlation. In Mann, K.O., Lane, H.R. (eds.), *Graphic Correlation*, 51–65. SEPM (Society for Sedimentary Geology), Special Publication 53.

Mann, K.O., Lane, H.R. (eds.). 1995. *Graphic Correlation.* SEPM (Society for Sedimentary Geology), Special Publication 53.

Marshall, C.R. 1990. Confidence intervals on stratigraphic ranges. *Paleobiology* 16, 1–10.

Marshall, C.R. 1994. Confidence intervals on stratigraphic ranges: partial relaxation of the assumption of randomly distributed fossil horizons. *Paleobiology* 20, 459–469.

Neal, J.E., Stein, J.A., Gamber, J.H. 1994. Graphic correlation and sequence stratigraphy of the Paleogene of NW Europe. *Journal of Micropalaeontology* 13, 55–80.

Nielsen, A.T. 1995. Trilobite systematics, biostratigraphy and palaeoecology of the Lower Ordovician Komstad Limestone and Huk Formations, southern Scandinavia. *Fossils & Strata* 38, 1–374.

Ovtcharova, M., Goudemand, N., Hammer, Ø., Guodun, K., Cordey, F., Galfetti, T., Schaltegger, U., Bucher, H. 2015. Developing a strategy for accurate definition of a geological boundary through radio-isotopic and biochronological dating: the Early–Middle Triassic boundary (South China). *Earth-Science Reviews* 146, 65–76.

Palmer, A.R. 1954. The faunas of the Riley Formation in Central Texas. *Journal of Paleontology* 28, 709–786.

Parnell, A.C., Haslett, J., Allen, J.R.M., Buck, C.E., Huntley, B., 2008, A flexible approach to assessing synchroneity of past events using Bayesian reconstructions of sedimentation history. *Quaternary Science Reviews* 27, 1872–1885.

Sadler, P.M. 2001. Constrained Optimization Approaches to the Paleobiologic Correlation and Seriation Problems: A User's Guide and Reference Manual to the CONOP Program Family. University of California, Riverside.

Sadler, P.M. 2004. Quantitative biostratigraphy – achieving finer resolution in global correlation. *Annual Review of Earth and Planetary Sciences* 32, 187–213.

Savary, J., Guex, J. 1999. Discrete biochronological scales and unitary associations: description of the BioGraph computer program. *Memoires de Geologie (Lausanne)* 34, 281p.

Shaw, A.B. 1964. *Time in Stratigraphy*. McGraw-Hill, New York.

Strauss, D., Sadler, P.M. 1989. Classical confidence intervals and Bayesian probability estimates for ends of local taxon ranges. *Mathematical Geology* 21, 411–427.

Tipper, J.C. 1988. Techniques for quantitative stratigraphic correlation: a review and annotated bibliography. *Geological Magazine* 125, 475–494.

Toro, B.A., Sánchez, N.C.H., Goldman, D. 2023. Using Constrained Optimization (CONOP) to examine Ordovician graptolite distribution and richness from the Central Andean Basin and their comparison with additional data from North America and Baltoscandia. *Palaeogeography, Palaeoclimatology, Palaeoecology* 613, 111396.

Wang, Y., Goring, S.J., McGuire, J.L. 2019. Bayesian ages for pollen records since the last glaciation in North America. *Scientific Data* 6, 176.

# 14

# Phylogenetic analysis

It is with some trepidation that we present this short chapter on phylogenetic analysis. Hardly any area of modern biological data analysis has generated so much controversy, produced so many papers, created and destroyed so many careers, or been more energetically marketed. Most professional paleontologists are well acquainted with the principles of contemporary systematics and have strong opinions on both the subject as a whole and on technical details. Moreover, advocates of phylogenetic systematics have suggested that phylogenies alone should form the basis of a whole new way to both classify and name taxa (the phylocode) marking a revolutionary departure from the Linnaean scheme that has served paleontologists for over 250 years.

## 14.1   A dictionary of cladistics

It is far outside the scope of this book to provide a thorough treatment of cladistics, parsimony analysis, and the other paradigms available for tree optimization (see Kitching et al. 1998; Swofford and Olsen 1990; Wiley et al. 1991; or Felsenstein 2003 for in-depth treatments). Rather, in the spirit of the rest of this book, we will give a brief, practical explanation of some important methods and algorithms. So mostly leaving the theoretical background aside, and together with it the whole dictionary of cladistic terms, we will concentrate on the practicalities of parsimony analysis.

Nevertheless, some vital terms cannot be avoided and need to be explained. A cladist differentiates between three types of systematic groups. A *polyphyletic* group is a somewhat random collection of taxa that does not include their most recent common ancestor. Warm-blooded animals (birds and mammals) are a good example: their most recent common ancestor was some Paleozoic reptile, which was not warm-blooded. Polyphyletic groups are defined based on convergent characters and are not acceptable for a systematist. A *paraphyletic* group does contain the most recent common ancestor, but not all its descendants. Here cladistics starts to differ from traditional taxonomy, because a cladist does not accept say "reptiles" (paraphyletic because it does not include all descendants such as birds and mammals) as a valid taxon. The only type of group acceptable to the cladist is the *monophyletic* group (or *clade*), which includes the most recent common ancestor and all descendants. Thus, mammals are a monophyletic group.

*Paleontological Data Analysis*, Second Edition. Øyvind Hammer and David A.T. Harper.
© 2024 John Wiley & Sons Ltd. Published 2024 by John Wiley & Sons Ltd.

An *apomorphy* is a derived (evolutionary new) character state. A *synapomorphy* is a derived character state that is shared between an ancestor and its descendants, and either alone or together with other synapomorphies it can be used to discover a monophyletic group. The presence of a backbone is a synapomorphy for the vertebrates. An *autapomorphy* is a derived character state that is found in only one species. Autapomorphies are necessary for the definition of species but are not otherwise useful in systematics. As an example, the autapomorphy of articulated speech in humans does not help us in the investigation of primate relationships, because this trait is not found in any other species. Finally, a *plesiomorphy* is a "primitive" character state that is found also outside the clade of interest. Plesiomorphies are not very useful in systematics. For example, the presence of a backbone is a primitive character state for vertebrates and does not help us to find out whether cats are more closely related to dogs or to fishes. Note that whether a character state is apomorphic or plesiomorphic depends completely on our focus of interest: the backbone is apomorphic for the vertebrates as a clade within larger clades, but plesiomorphic for the Carnivora.

Character states should be *homologous* (that is, synapomorphic) to be useful. For example, the feathers of hawks, ostriches, and sparrows are homologous; they are thus a homology for birds. The wings of bats, birds, and flying insects perform similar functions but were not derived in the same way; they are thus not homologous but analogous and of little use in phylogenetic analysis. The existence of analogous (convergent) traits and evolutionary reversals is referred to as *homoplasy*.

## 14.2 Parsimony analysis

It is important to discriminate between the terms "cladistics" and "parsimony analysis." They are not at all the same thing: cladistics is a general term for the philosophy of systematics developed by Hennig (1950, 1966) – a philosophy that few people now disagree with. Parsimony analysis is a specific recipe for constructing phylogenetic trees from character states, which paleontologists sometimes need some persuasion to adopt. Nevertheless, parsimony analysis is probably the most implemented approach to phylogenetic analysis at the present time, at least for paleontologists who mainly use morphological characters. Other approaches such as Bayesian methods and maximum likelihood (ML) are popular for molecular data and are becoming fashionable also among paleontologists. We will briefly discuss these alternative methods at the end of the chapter.

The principle of parsimony analysis is very simple: given a number of character states in a number of taxa, we want to find a phylogenetic tree such that the total number of necessary evolutionary steps is minimized. In this way incongruence and mismatches are minimized. This means that we would prefer not to have a tree where a certain character state such as the presence of feathers must be postulated to have evolved independently many times. Five major arguments are often raised against this approach:

1) Why would we expect the most parsimonious solution to have anything to do with true evolutionary history, knowing that parallel evolution and homoplasies (convergent character states) are so common?

The answer to this objection is that parsimony analysis does not actually assume that nature is parsimonious. It is simply a pragmatic, philosophically sound response to our lack of information about true evolutionary histories: given many possible phylogenetic reconstructions that all fit with the observed morphologies, we choose the simplest one. Why should we choose any other? Having a choice, we select the simplest theory – this is the principle of parsimony, or Occam's razor (Sober 1988 gives a critical review of the use of parsimony in systematics). We hope of course that the most parsimonious tree reflects some biological reality, but we do not expect it to be the final truth. It is simply the best we can do.

2) But it is not the best we can do! We have other information in addition to just the character states, such as the order of appearances in the fossil record, which can help us to construct better trees.

It is indeed likely that stratigraphic information could help constrain phylogenetic reconstructions, and techniques have been developed for this purpose within the framework of parsimony analysis, of ML, and of Bayesian analysis. However, this is not entirely unproblematic, because the fossil record is incomplete and because ancestors are impossible to identify.

3) I know this fossil group better than anyone else does, and I know that this tiny bone of the middle ear is crucial for understanding evolutionary relationships. It is meaningless to just throw a hundred arbitrary characters together and ask the computer to produce a tree that does not reflect the true relationships as shown by this bone.

There is no doubt that deep morphological knowledge is crucial for systematics, also when we use parsimony analysis. But we must still ask the question: why are you so certain that this one character is so important? Would you not make fewer assumptions by treating characters more equally and see where that might lead?

4) Parsimony analysis treats all characters equally. This is a problem because many characters can be strongly correlated, such as the presence of legs being correlated with the presence of feet, to give a simplistic example. Such correlated characters, which really correspond to only one basic trait, will then influence the analysis too strongly.

This is a problem that we should try to minimize, but which will never vanish completely. Again, it is the best we can do. It could also be argued that this is a problem not only for parsimony analysis but is shared by all methods in systematics.

5) I get 40,000 equally parsimonious trees, so the method is useless.

This is not a problem of the method, it is, in fact, a virtue! It shows you that your information is insufficient to form a stable, unique hypothesis of phylogeny, and that more and better characters are needed. We don't want a method that seemingly gives a good result even when the data are bad.

Several of the arguments about cladistics really boil down to the following: traditionally, also within cladistics, homologies were primarily hypothesized first, and then the phylogeny was constructed accordingly. The typical modern cladist also tries to identify homologies in order to select good characters and character states but knows that she may often well be wrong. She then performs the parsimony analysis and hopes that her correct decisions about homology will give a reasonable tree, in which her wrongly conjectured homologies will turn out as homoplasies. Some have claimed that this is close to circular reasoning because putative homologies are both used to construct the tree and are identified from it.

Cladograms are, nevertheless, our best effort at understanding phylogeny. Clearly a restudy of the specimens with new techniques, the discovery of new characters, or the addition of new taxa can help confirm or reject what is essentially a hypothesis. When a cladogram is supplemented by a timescale, we can develop a phylogeny; the addition of the timing of the origins of taxa can add shape to the phylogeny and present a more realistic view of the history of a clade.

At present, the most popular software for parsimony analysis in paleontology is probably TNT (Goloboff et al. 2008). TNT has a somewhat high learning threshold, but is flexible, and extremely fast. The latter is important because searching for the most parsimonious trees can be time consuming for large data sets. Other popular parsimony software includes PAUP* and PHYLIP. A basic parsimony analysis package is also included in PAST.

## 14.3 Characters

The data put into a parsimony analysis consist of a *character matrix* with taxa in rows and characters in columns. The cells in the matrix contain codes for the *character states*. It is quite common for characters to be binary, coded with 0 or 1 for the two states of the absence or the presence of a certain trait. Characters can, however, also have more than two states, for example, the number of body segments or the type of trilobite facial suture coded with a whole number. Measured quantities such as length or width cannot be used as characters for parsimony analysis, unless they are made discrete by division into e.g., "small," "medium," and "large." This is only acceptable if the distribution of the measured quantity is discontinuous (see Archie 1985 and several papers in MacLeod and Forey 2002). This type of coding in some respects bridges the gap between cladistics and phenetics, where essentially continuous variables are converted into codable attributes. Unknown or inapplicable character states are coded as missing data, usually with a question mark.

The number of characters and the distribution of character states between taxa will control the resolution of the parsimony analysis. As a very general rule of thumb, it is recommended to have at least twice as many (ideally uncorrelated) characters as taxa.

It is also necessary to specify the method for calculating the evolutionary "cost" of changes in a character state as used in parsimony analysis. So-called *Wagner* characters (Wagner 1961; Farris 1970) are reversible and ordered, meaning that a change from state 0 to 2 costs more than a change from state 0 to 1, but has the same cost as a change from 2 to 0. *Fitch* characters (Fitch 1971) are reversible and unordered, meaning that all changes have an equal cost of one step. This is the criterion with the fewest assumptions and is therefore generally preferable. When using the Fitch criterion, it is not necessary, perhaps not even meaningful, to code the character states according to a theory of polarization; in other words, it is not necessary to code the presumed primitive (plesiomorphic) state with a low number. Character polarity can be deduced *a posteriori* from the topology of the most parsimonious tree, when rooted using an outgroup (next section).

## Example 14.1

Table 14.1 presents a character matrix for the trilobite family Paradoxididae (Fig. 7.6), with 24 characters and nine genera, published by Babcock (1994). These are relatively large trilobites with complex morphologies readily characterized for cladistic analysis. This matrix, nevertheless, contains a few unknown character states, marked with question marks. Some characters have only two states (0 and 1), while others have up to four states. Also note that character state 1 in characters 15, 21, 22, and 23 is *autapomorphic*: it exists in only one taxon and is therefore not phylogenetically informative except for the definition of that one taxon (strictly speaking, the term "autapomorphy" is reserved for character states found in one *species* and should not be used for higher taxa).

Table 14.1  Character matrix for the trilobite family Paradoxididae.

| | 1 | 2 | 3 | 4 | 5 | 6 | 7 | 8 | 9 | 10 | 11 | 12 | 13 | 14 | 15 | 16 | 17 | 18 | 19 | 20 | 21 | 22 | 23 | 24 |
|---|---|---|---|---|---|---|---|---|---|---|---|---|---|---|---|---|---|---|---|---|---|---|---|---|
| *Elrathia* | 0 | 0 | 0 | 0 | 0 | 0 | 0 | 0 | 0 | 0 | 0 | 0 | 0 | 0 | 0 | 0 | 0 | 0 | 0 | 0 | 0 | 0 | 0 | 0 |
| *Centropleura* | 2 | 1 | 0 | 1 | 1 | 2 | 0 | 2 | 1 | 0 | 1 | 1 | 2 | 1 | 0 | 2 | 2 | 1 | 1 | 1 | 0 | 0 | 0 | 1 |
| *Xystridura* | 1 | 0 | 1 | 0 | 0 | 0 | 0 | 0 | 1 | 1 | 1 | 0 | 1 | 0 | 0 | 0 | 1 | 0 | 0 | 0 | 0 | 0 | 0 | 0 |
| *Paradoxides* | 2 | 1 | 2 | 0 | 0 | 1 | 1 | 3 | 0 | 1 | 1 | 0 | 1 | 0 | 1 | 1 | 0 | 1 | 0 | 0 | 0 | 0 | 0 | 1 |
| *Bergeroniellus* | 0 | 0 | 0 | 0 | 0 | 1 | 1 | 1 | ? | 1 | 1 | 0 | 1 | 0 | 0 | 1 | 0 | 0 | 0 | 0 | 0 | 0 | 0 | 0 |
| *Lermontovia* | 0 | 0 | 1 | 0 | 0 | 1 | 1 | 3 | ? | ? | 0 | 0 | 1 | 0 | 0 | 0 | 0 | 0 | 0 | 1 | 1 | 0 | 0 |
| *Anopolenus* | 1 | 1 | 0 | 1 | 1 | 2 | 0 | 1 | ? | ? | 1 | 0 | 3 | 0 | 0 | 2 | 2 | 1 | ? | 1 | 0 | 0 | 0 | 1 |
| *Clarella* | 2 | 1 | 0 | 1 | 1 | 2 | 0 | 1 | ? | ? | 1 | 1 | 3 | 1 | 0 | 2 | 2 | 1 | ? | 1 | 0 | 0 | 0 | 1 |
| *Galahetes* | 2 | 0 | 1 | 0 | 0 | 0 | 0 | 0 | 1 | ? | 1 | 0 | 1 | 1 | 0 | 0 | 1 | 0 | 0 | 0 | 0 | 0 | 1 | 0 |

Adapted from Babcock (1994).
"?" signifies unknown character states.

## 14.4  Algorithms for Parsimony Analysis

Parsimony analysis is used to form a phylogenetic hypothesis (a cladogram) about the relations between a number of taxa, based on a character matrix of discrete character states, with taxa in rows and characters in columns (section 14.3).

The method can also be used for biogeographic analysis, investigating the relations between a number of biogeographic regions or localities. In this case, the method implies the fragmentation of once continuous provinces – the dispersal of taxa can give rise to homoplasies. For parsimony analysis of regions or localities, a taxon occurrence (presence/ absence) matrix is required.

Parsimony analysis involves finding trees with minimal *length*, which is the total number of character changes along the tree as calculated with e.g., a Wagner or Fitch criterion. This does not in itself indicate an evolutionary direction in the tree, and in fact the shortest trees can be arbitrarily rooted, that is, any taxon can be chosen as the most basal one. Since we are presumably interested in phylogeny, we therefore need to root (and thereby polarize) the tree, usually by selecting an *outgroup* among the taxa. The procedure for choosing an outgroup is a difficult and contentious issue.

Unfortunately, there is no method for finding the shortest trees directly, and instead one must search for them by calculating the tree lengths of a large number of possible trees. There are several algorithms available, creating quite a lot of confusion for the beginner. We will go through the classical methods, from the simplest to the most complex.

### 14.4.1 Exhaustive search

An exhaustive search involves constructing all possible trees by sequential addition of taxa in all possible positions and calculating the tree lengths. This approach is, of course, guaranteed to find all the shortest possible trees. However, it is quite impractical, because of the immense number of trees that are searched: for 12 taxa more than 600 million trees must be evaluated! Fortunately, a better strategy is available that speeds up the search considerably (branch-and-bound), so an exhaustive search is never necessary and should be avoided. The only exception is when you have a small number of taxa, say a maximum of 10, and you want to produce a histogram of all tree lengths.

### 14.4.2 Branch and bound

A clever but simple addition to the exhaustive search algorithm can speed it up by several orders of magnitude in some cases. As taxa are added to a tree under construction, the tree length is continuously calculated even before the tree is completed. This takes a little extra computation but allows the construction of the tree to be aborted as soon as the tree length exceeds the shortest complete tree found so far. In this way, "dead ends" in the search are avoided. Branch and bound is guaranteed to find all the shortest trees, and it is recommended for use whenever possible. Unfortunately, even a branch-and-bound search becomes painfully slow for more than say 15 taxa, somewhat dependent on the structure of the character matrix, the speed of the computer and software, and the patience of the investigator.

### 14.4.3 Heuristic algorithms

When even the branch-and-bound search becomes too slow, we have a problem. One way to proceed is to give up the hopeless search for the "Holy Grail" (the most parsimonious trees possible), and only search a subset of all possible trees in some intelligent way. This is called a heuristic search. If we are very lucky, this may find the most parsimonious trees possible, but we have no guarantee. Obviously, this is not entirely satisfactory, because it always leaves us with a concern that there may still be some better tree out there, or at least more trees as good as the ones we found. Still, heuristic search is a necessary and accepted tool in the parsimony analysis of large data sets.

Many algorithms for heuristic searches have been proposed and are in use, but the most popular ones so far are based on so-called "greedy search" combined with tree rearrangement. The idea is to start with any one taxon, and then add each new taxon to the tree in the position where it will minimize the increase in total tree length. One might naively hope that this minimal increase in tree length in each individual step would produce the shortest possible tree overall, but this is generally not the case. To increase the chances of finding good (i.e., short) trees, it is therefore customary to rearrange the tree in different ways to see if this can reduce tree length further. Three such methods of rearrangement are as follows.

### 14.4.3.1 Nearest-neighbor interchange
All possible couples of nearest-neighbor subtrees are swapped (Fig. 14.1).

### 14.4.3.2 Subtree pruning and regrafting
This algorithm is similar to NNI, but with a more elaborate branch-swapping scheme: a subtree is cut off the tree, and regrafting onto all other branches of the tree is attempted in order to find a shorter tree. This is done after each taxon has been added and for all possible subtrees. While slower than NNI, subtree pruning and regrafting (SPR) will often find shorter trees.

### 14.4.3.3 Tree bisection and reconnection (TBR)
This algorithm is similar to SPR, but with an even more complete branch-swapping scheme. The tree is divided into two parts, and these are reconnected through every possible pair of branches in order to find a shorter tree. This is done after each taxon is added, and for all possible divisions of the tree. TBR will often find shorter trees than SPR and NNI, at the cost of longer computation time.

Any heuristic method that adds taxa to the tree sequentially will usually produce different results depending on the order of addition. We can call this phenomenon convergence on a local, suboptimal solution depending on the order. To increase the chance of finding good trees, we usually run the whole procedure repeatedly, say 10 or 100 times, each time with a new, random ordering of the taxa.

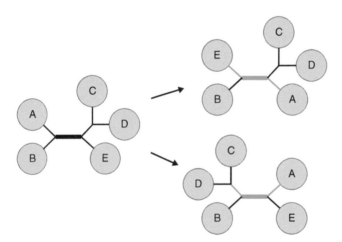

Figure 14.1    The principle of tree arrangement by nearest-neighbor interchange (NNI). In the unrooted tree to the left, focus on the internal branch in bold. Any internal branch will have two nearest-neighbor subtrees on each side – in this case, these are A and B on one side and CD and E on the other. There are then two possible, non-equivalent nearest-neighbor interchanges: we can swap A with E (top right), which is equivalent to swapping B with CD, or we can swap A with CD (bottom right), which is equivalent to swapping B with E. Such rearrangements are attempted for each internal branch in the tree, in order to find shorter trees.

## Examples 14.2

Let us first consider the character matrix for the trilobite family Paradoxididae, given in section 14.3. Because of the small number of genera, we can use the branch-and-bound algorithm, which is guaranteed to find all shortest trees. We also have chosen the Fitch optimization. A single most parsimonious tree is found, with a tree length of 42. This means that a total of 42 character transformations are needed along the tree. In order to root the tree, we chose *Elrathia* as an outgroup (Fig. 14.2; this genus is not a member of the Paradoxididae). The resulting cladogram is shown in Fig. 14.3A. Now that we know what the shortest possible tree is, we can test the performance of a heuristic search on this character matrix. It turns out that even an NNI search with no re-orderings of taxa manages to find the shortest tree of length 42, which is quite impressive.

For an example with more taxa, we will use a character matrix for 20 orders of the Eutheria (mammals), published by Novacek et al. (1988). There are 67 characters, but many of them are uninformative in the sense that their character states either are autapomorphic (exist only in one terminal taxon) or plesiomorphic. In fact, there are only 18 characters with informative distribution of character states on taxa, and we must expect this to generate poor resolution and many "shortest trees." Given the relatively large number of taxa, we are forced to use a heuristic search. We choose the SPR algorithm with 20 reorderings, and Fitch optimization. PAST then finds 790 shortest trees of length 71, but there are probably thousands more of this length and possibly some even shorter ones. One of the shortest trees found is shown in Fig. 14.4.

Figure 14.2    The outgroup, *Elrathia kingii*; the scale bar is 5 mm. Courtesy of Lauren English/ Nigel Hughes.

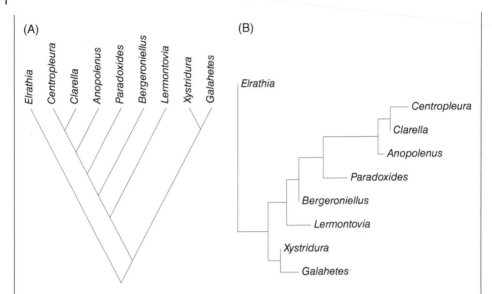

Figure 14.3   The most parsimonious tree for the Paradoxididae character matrix, using the Fitch criterion and *Elrathia* as an outgroup. (A) Cladogram. (B) Phylogram, where branch lengths (number of steps) are indicated by the lengths of horizontal line segments.

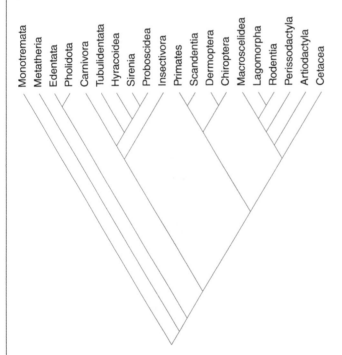

Figure 14.4   One of the 790 most parsimonious trees found with a heuristic search for the Eutheria character matrix, using the Fitch criterion and Monotremata as the outgroup.

## 14.5   Character state reconstruction

Once a tree has been constructed and rooted, we normally want to investigate how a given character changes along the tree. However, in the presence of homoplasy, this is normally ambiguous – there are several ways of placing the character transitions onto the tree, which are all compatible with the tree length, the tree topology, and the character states in the terminal taxa. Normally, one of two approaches is selected. The first approach is to place character transitions as close to the root as possible, and then, if necessary, accept several reversals farther up in the tree. This is known as accelerated transformation or *acctran*. In the other approach, known as delayed transformation (*deltran*), the character transitions are allowed to happen several times farther up in the tree, and then reversals are not necessary. The selection of *acctran* or *deltran* will to some extent be arbitrary, but in some cases, one will seem more reasonable than the other. Anyway, it is important to note that this choice will not influence the search for the most parsimonious tree.

The total number of character state transitions along a given branch in the cladogram is referred to as the *branch length*. A tree where the branches are drawn to scale with the branch lengths is called a *phylogram* (Fig. 14.3B).

---

**Example 14.3**

Returning to our Paradoxididae example, we can plot the transitions of character number 1 onto the shortest tree, using accelerated transformation (Fig. 14.5A). Orange rectangles signify a character change. Note the homoplasy: character states 1 and 2 each evolve twice. Figure 14.5B shows an alternative (delayed) reconstruction.

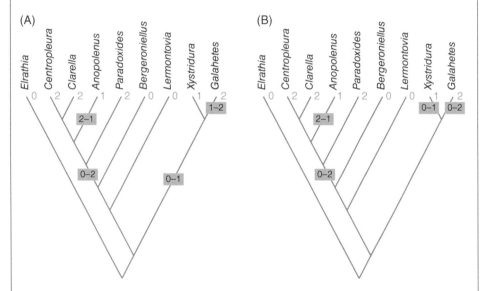

Figure 14.5   (A) Reconstruction of the transformations of character number 1 onto the most parsimonious tree for the Paradoxididae data set, using accelerated transformation (acctran). The character state is given for each taxon. (B) Alternative reconstruction using delayed transformation (deltran).

## 14.6    Evaluation of characters and trees

Parsimony analysis will give one shortest tree or a number of shortest trees of equal length. But the analysis does not stop there, because we need to estimate the "quality" of the result. If we generated several shortest trees, are they similar to each other? Are there many trees that are almost as short? Is the result stable with respect to small changes in the character matrix, such as the removal or duplication of characters? Does the tree involve a lot of homoplasy (convergence or reversal of character states)? A number of indices and procedures have been proposed in order to give some indication about such matters. None of them can be regarded as a perfect measure of the quality of the tree, and even less as some kind of statistical significance value, but they are nevertheless quite useful.

### 14.6.1    Consensus tree

If the parsimony analysis gives many equally parsimonious (shortest) trees, we have several options for the presentation of the result. We could try to present all the trees, but this may be practically impossible, or we can select one or a few trees that seem sensible in the light of other data such as stratigraphy. In addition to these, it can be useful to present a *consensus tree* containing only the clades that are found in all or most of the shortest trees (Fig. 14.6). Clades that are not resolved into smaller subclades in the consensus tree will be shown as a *polytomy*, where all the taxa collapse down to one node. To be included in a *strict* consensus tree, a clade must be found in all the shortest trees. In a *majority rule* consensus tree, it is sufficient that the clade is found in more than 50% of the shortest trees.

   Be a little careful with consensus trees. They are good for seeing at a glance where the individual trees disagree, but they do not themselves represent valid hypotheses of phylogeny. For example, you cannot really plot character changes along their branches.

---

**Example 14.4**

The strict consensus tree for the 790 shortest trees found in our Eutheria example described earlier is shown in Fig. 14.6. Note that many of the clades have collapsed into an unresolved polytomy close to the root. Some clades are however retained, such as the Perissodactyla–Macroscelidea–Rodentia–Lagomorpha clade and its subclades. Given that we have probably not found all possible shortest trees, we might like to double-check the relationships within this clade by including only these taxa in a branch-and-bound search. The topology within this subclade is then confirmed – a more parsimonious subtree does not exist.

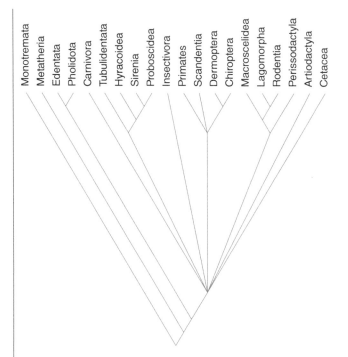

**Figure 14.6** The strict consensus tree for the 790 most parsimonious trees found using a heuristic search of the Eutheria character matrix.

### 14.6.2 Consistency index

The consistency index indicates the degree of homoplasy (convergent evolution) in a given tree, either for one given character or summed for all characters. The consistency index $c_i$ for a given character and a particular tree is the smallest possible number of character changes (steps) for that character on any tree, divided by the actual number of character changes on the given tree (Kluge and Farris 1969). For a binary character, the smallest possible number of character changes is 1, meaning a single change from 0 to 1 or from 1 to 0. The $c_i$ will therefore vary from 1 (for no reversals or homoplasies involving the given character) down towards zero for very "ill-behaved" characters. An ensemble consistency index (CI) can also be computed, which is the smallest possible number of steps summed over all characters, divided by the actual number of steps on the given tree, summed over all characters.

One annoying aspect of the ensemble CI is that it will increase with the addition of any non-homoplastic characters, including uninformative ones (such as autapomorphic characters that have a $c_i$ of 1 whatever the tree topology). Some programs will therefore also calculate a version of CI where uninformative characters have been removed.

---

**Example 14.5**

From Fig. 14.5A, we see that character number 1 has possible states of 0, 1, and 2. The smallest possible number of steps on any tree is therefore 2. The actual number of transitions on this given tree is four (0–1, 1–2, 0–2, and 2–1), so the $c_i$ for character 1 is thus $c_i = 2/4 = 0.5$. The ensemble CI for all characters on this tree is CI = 0.79.

### 14.6.3 Retention Index

The retention index ri for a particular character on a given tree is defined as follows (Farris 1989). Let $M$ be the largest possible number of character changes (steps) for that character on any tree, $m$ is the smallest possible number of steps on any tree, and $s$ is the actual number of steps on the given tree. Then,

$$ri = \frac{M - s}{M - m}$$

Compare this to the consistency index: $ci = m/s$. The retention index can be interpreted as a measure of how much of the similarity in character states across taxa can be interpreted as synapomorphy (Farris 1989). Put another way, if a character transformation occurs far down in the tree and is retained by derived taxa, we have a good, informative synapomorphy and the character will receive a high value for the retention index. If the transformation occurs high up in the tree, the character is more autapomorphic and less informative, and will receive a lower retention index.

The equation for the retention index can perhaps be explained as follows. The observed number of steps $s$ can lie anywhere in the interval between $m$ and $M$. The interval $[m, M]$ is therefore divided by $s$ into two partitions: $[m, s]$ and $[s, M]$. The former, of length $s - m$, represents the excess number of steps, which must be due to homoplasy. The remainder of the interval, of length $M - s$, is considered a measure of retained synapomorphy, and the proportion of the total interval is then the retention index.

An *ensemble* retention index RI can be defined by summing $M$, $m$, and $s$ over all characters. This index will vary from 0 to 1, with values close to 1 being "best" in terms of the cladogram implying little homoplasy and a clear phylogeny. The ensemble retention index seems to be less sensitive than the ensemble CI to the number of taxa and characters.

The rescaled consistency index (Farris 1989) is the product of the consistency index and the retention index, thus incorporating properties from both these indices.

---

**Example 14.6**

From Fig. 14.5A, character number 1 has possible states of 0, 1, and 2. Assuming Fitch optimization, the largest possible number of steps for character 1 on any tree with the given distribution of character states on taxa is five (this is not obvious!), so $M = 5$. The smallest possible number of steps on any tree is 2, so we have $m = 2$. The actual number of transitions on this given tree is 4 (0–1, 1–2, 0–2, and 2–1), so we have $s = 4$. The retention index for character 1 is then

$$ri = (M - s)/(M - m) = (5 - 4)/(5 - 2) = 0.33.$$

The ensemble retention index for all characters is RI $= 0.83$.

---

### 14.6.4 Bootstrapping

Bootstrapping in the context of parsimony analysis (Felsenstein 1985a) indicates the stability of the most parsimonious clades under random weighting of characters. This is achieved as follows. Given that you have $N$ characters, the program will pick $N$ characters at random and then perform a parsimony analysis based on these characters. It may well select one original character two or more times, thus in effect weighting it strongly, while other characters may not be included at all. A 50% majority rule consensus tree is constructed from the most parsimonious trees. This whole procedure is repeated a hundred or a thousand times (you need a fast computer!), each time with new random weighting of the original characters, resulting in a hundred or a thousand consensus trees.

Finally, we return to our most parsimonious tree from the analysis of the original data. Each clade in this tree receives a *bootstrap value*, which is the percentage of bootstrapped consensus trees where the clade is found. If this value is high, the clade seems to be robust to different weightings of the characters, which is reassuring.

Such bootstrapping has been and still is popular in phylogenetic analysis. However, the method has been criticized because a character matrix does not adhere to the statistical assumptions normally made when bootstrapping is carried out in other contexts, such as the characters being independently and randomly sampled (Sanderson 1995). While this may be so, and bootstrapping obviously does not tell you the whole story about the robustness of the most parsimonious solution, nevertheless it still gives some idea about what might happen if the characters had been weighted differently. Given all the uncertainty about possible overweighting of correlated characters, this ought to be of great interest, but the values should not be seen as statistical probabilities allowing formal significance testing of clades.

---

**Example 14.7**

The bootstrap values based on 1000 bootstrap replicates and branch-and-bound search for the Paradoxididae data set are shown in Fig. 14.7. The values are disappointingly low, except perhaps for the *Anopolenus–Clarella–Centropleura* and the *Clarella–Centropleura* clades.

Figure 14.7  Bootstrap values for the most parsimonious tree of the Paradoxididae data set.

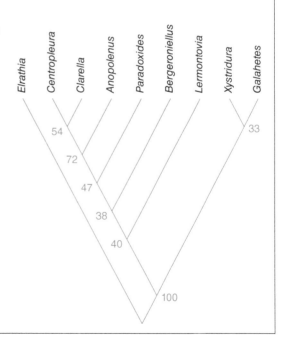

### 14.6.5 Bremer support

Like the bootstrap value, the Bremer support (also known as the *decay index* or *branch support*) is calculated for each clade (Bremer 1994). The idea is to look at not only the shortest tree(s), but also slightly longer trees to see if the clade is still supported in these trees. If the clade collapses only in the sets of trees that are much longer than the shortest tree, it must be regarded as robust. Some programs can calculate the Bremer supports automatically, but in the present version of PAST it must be done manually as follows.

Perform parsimony analysis, ideally using branch-and-bound search. Take note of the clades and the length $N$ of the shortest tree(s) (e.g., 42). If there is more than one shortest tree, look at the strict consensus tree. Clades that are no longer found in the consensus tree have a Bremer support value of 0.

1) In the box for "Longest tree kept," enter the number $N + 1$ (43 in our example) and perform a new search.
2) Additional clades that are no longer found in the strict consensus tree have a Bremer support value of 1.
3) For "Longest tree kept," enter the number $N + 2$ (44) and perform a new search. Clades that now disappear in the consensus tree have a Bremer support value of 2.
4) Continue until all clades have disappeared.

The Bremer support is currently considered a good indicator of clade stability.

---

**Example 14.8**

Again returning to the Paradoxididae example, we find that the clade *Anopolenus–Clarella–Centropleura* has Bremer support 7, while the clade *Clarella–Centropleura* has Bremer support 2. All other clades have Bremer support 1, meaning that they collapse already in the consensus tree of the trees of length 43 (Fig. 14.8). These results reinforce the impression we got from the bootstrap values.

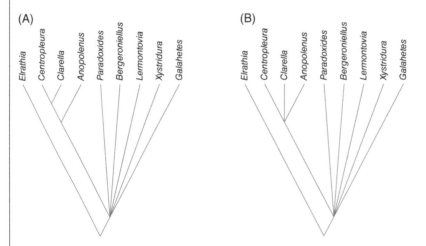

Figure 14.8 (A) Strict consensus of all trees of length 43 or lower. The shortest tree has a length 42. The collapsed clades (all except two) have Bremer support 43 − 42 = 1. (B) Strict consensus of all trees of length 44 or lower. The collapsed clade (*Clarella–Centropleura*) has Bremer support 44 − 42 = 2. The one remaining clade (*Anopolenus–Clarella–Centropleura*) collapses in the consensus of all trees of length 49 or lower (not shown), giving a Bremer support of 49 − 42 = 7. (C) Summary of the Bremer supports for all clades.

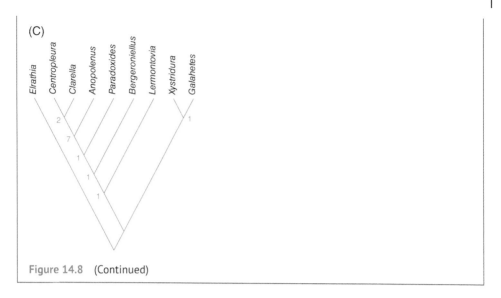

Figure 14.8   (Continued)

### 14.6.6   Stratigraphic congruency indices

Stratigraphic congruency indices assess the degree of congruence between a proposed phylogeny (rooted cladogram or phylogram) and the stratigraphic ranges of the taxa. A rooted cladogram implies a temporal succession of at least some of the phylogenetic branching events. Sister taxa cannot be temporally ordered among themselves, but they must have originated later than any larger clade within which they are contained.

The stratigraphic congruence index (SCI) of Huelsenbeck (1994) is defined as the proportion of stratigraphically consistent nodes on the cladogram and varies from 0 to 1. A node is stratigraphically consistent when the oldest first occurrence above the node is the same age or younger than the first occurrence in its sister group (Fig. 14.9).

The next indices to be described are based on the durations of gaps in the fossil records called *ghost ranges*. A ghost range is a stratigraphic interval where a taxon should have

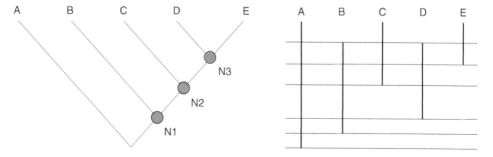

Figure 14.9   The stratigraphic congruence index. Cladogram and range chart of five species A–E. The oldest first occurrence above node N1 (B) is younger than the first occurrence in the sister group (A). Node N1 is therefore stratigraphically consistent. The oldest first occurrence above node N2 (D) is younger than the first occurrence in the sister group (B). N2 is therefore also stratigraphically consistent. However, the oldest first occurrence above node N3 (D) is *older* than the first occurrence in the sister group (C). Hence, N3 is stratigraphically inconsistent. With two out of three nodes consistent, the stratigraphic congruence index is SCI = 2/3.

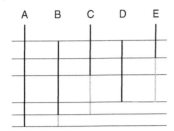

**Figure 14.10** Gray lines show the extensions of the stratigraphic ranges in Fig. 14.9 according to the proposed phylogeny. These extensions are called ghost ranges, and their durations are called *minimum implied gaps* (MIGs).

existed according to the cladogram but is not registered in the fossil record. A ghost range is given as the interval from the first appearance of a taxon down to the first appearance of its sister group (Fig. 14.10).

The relative completeness index (RCI) of Benton and Storrs (1994) is defined as

$$RCI = 100\left(1 - \frac{\Sigma MIG}{\Sigma SRL}\right)$$

where minimum implied gaps (MIGs) are the durations of ghost ranges and SRLs are the durations of observed ranges. The RCI can become negative but will normally vary from 0 to 100.

The gap excess ratio (GER) of Wills (1999) is defined as

$$GER = 1 - \frac{\Sigma MIG - G_{min}}{G_{max} - G_{min}}$$

where $G_{min}$ is the minimum possible sum of ghost ranges on any tree (that is, the sum of distances between successive first appearance datums [FADs]), and $G_{max}$ is the maximum (that is, the sum of distances from first FAD to all other FADs).

These indices can be subjected to permutation tests, where all dates are randomly redistributed on the different taxa say 1000 times. The proportion of permutations where the recalculated index exceeds the original index can then be used as a $p$ value for the null hypothesis of no congruence between the cladogram and the stratigraphic record.

The SCI and RCI were compared by Hitchin and Benton (1997). The two indices describe quite different aspects of the match between stratigraphic and cladistic information: the SCI is a general index of congruency, while RCI incorporates time and gives a more direct measure of the completeness of the fossil record given a correct cladogram. Both indices were criticized by Siddall (1998).

These stratigraphic congruency indices may tentatively be used to choose the "best" solution when parsimony analysis comes up with several shortest trees, although this is debatable.

---

**Example 14.9**

Two trees have been developed cladistically for the two suborders of orthide brachiopod, the Orthidina and the Dalmanellidina (Fig. 14.11). The two trees, however, have markedly contrasting tree metrics (Harper and Gallagher 2001). The three main indices for the two suborders are presented in Table 14.2.

The SCI values for the two orders are similar but there are marked differences between the RCI and GER metrics. Clearly the consistency of the stratigraphical order of appearance of taxa in both groups is similar but the stratigraphical record of the dalmanellidines is much less complete. It is possible that the sudden appearance of some groups without clear ancestors has exaggerated the problem (Harper and Gallagher 2001) or in fact some members of the dalmanellidines have been misclassified as orthidines since the key apomorphy of the group (punctuation) can be difficult to recognize on poorly preserved specimens.

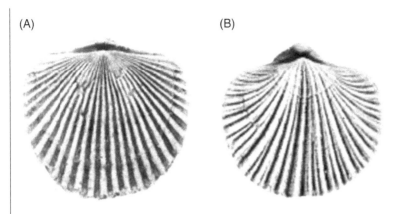

Figure 14.11   The Ordovician orthide brachiopods A, *Orthambonites* (ca. 20 mm long) and B, *Wysogorskiella* (ca. 5 mm long), both from Treatise on Invertebrate Paleontology, part H (revised). Both are superficially similar, but A is an impunctate orthidine and B is an endopunctate dalmanellidine. Lack of the identification of punctae can lead to an erroneous assignment.

Table 14.2   Stratigraphic congruency indices for two brachiopod suborders.

|  | SCI | RCI | GER |
| --- | --- | --- | --- |
| Orthidina | 0.375 | 78.79 | 0.830 |
| Dalmanellidina | 0.350 | 48.47 | 0.395 |

## 14.7   Case study: the systematics of heterosporous ferns

The methodology of parsimony analysis will be demonstrated using a character matrix given by Pryer (1999), concerning the phylogenetic relationships within heterosporous ferns (Fig. 14.12). For illustration, we will concentrate on a subset of the character matrix, and the results below are therefore not directly comparable to those of Pryer (1999).

The reduced character matrix contains seven living heterosporous taxa including members of the marsileaceans:

**Family Marcileaceae**
  *Marsilea quadrifolia*
  *Marsilea ancylopoda*
  *Marsilea polycarpa*
  *Regnellidium diphyllum*
  *Pilularia americana*

**Family Salviniaceae**
  *Azolla caroliniana*
  *Salvinia cucullata*

In addition are included the Late Cretaceous heterosporous fern *Hydropterus pinnata* and an outgroup taxon generated by combining the character states of 15 homosporous ferns.

Figure 14.12  The heterosporous "water fern" *Marsilea quadrifolia*. O. W. Thomé 188/Kurt Stueber/ public domain.

After removing characters that are invariant within the nine taxa, we are left with 45 characters. The matrix contains a number of unknown and inapplicable character states, coded with question marks.

### 14.7.1  Parsimony analysis

With this small number of taxa, we can easily use the branch-and-bound algorithm, which is guaranteed to find all shortest trees. This results in two most parsimonious trees of length 69 steps, as shown in Fig. 14.13. The bootstrap values are also given. The strict consensus tree with Bremer support values is shown in Fig. 14.14. Neither the bootstrap nor the Bremer support values indicate very robust clades.

The two trees agree on the monophyletic Salviniaceae (*A. caroliniana* and *S. cucullata*) as a sister group to monophyletic Marsileaceae, and also give the fossil *H. pinnata* as a basal taxon to the heterosporous clade (but see Pryer 1999). In addition, the genus *Marsilea* is monophyletic. However, the two trees diverge on the question of the relationships between the three genera within the Marsileaceae.

The ensemble consistency indices CI for the two trees are 0.83 and 0.81, while the retention index RI is 0.96 for both trees. Cladogram lengths are shown in Fig. 14.15.

### 14.7.2  Comparison with the fossil record

Figure 14.16 shows the stratigraphic ranges for the six heterosporous genera, as given by Pryer (1999). For the first tree, three out of seven nodes are stratigraphically consistent, giving a stratigraphic congruency index SCI of $3/7 = 0.43$ (permutation $p = 0.42$). For the second tree, four out of the seven nodes are consistent, giving SCI $= 4/7 = 0.57$ ($p = 0.26$).

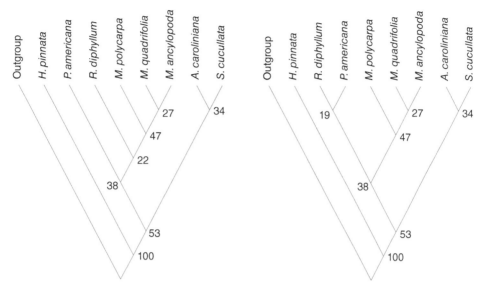

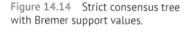

**Figure 14.13** The two most parsimonious trees from a branch-and-bound parsimony analysis of heterosporous ferns, with bootstrap values (1000 replicates).

**Figure 14.14** Strict consensus tree with Bremer support values.

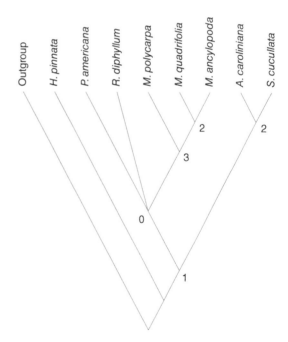

The relative completeness indices RCI are 59.4 ($p = 0.08$) and 67.0 ($p = 0.01$). This means that the second tree is slightly better supported by the fossil record, and we may perhaps use this as a criterion to choose the second tree, with *Regnellidium* and *Pilularia* as a sister group to *Marsilea*.

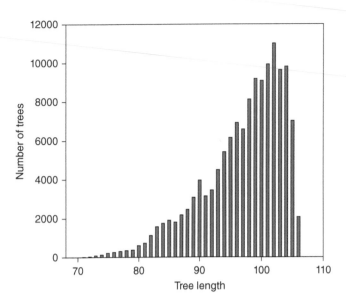

Figure 14.15 Distribution of cladogram lengths, showing that only a few trees are slightly longer than the shortest trees of length 69.

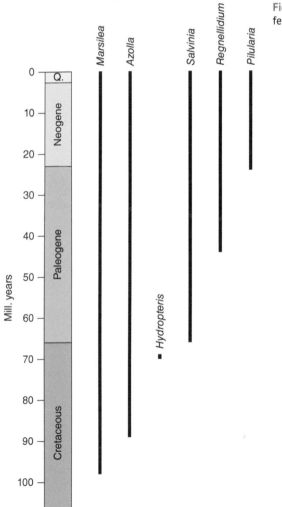

Figure 14.16 Stratigraphic ranges of fern genera. Data from Pryer (1999).

## 14.8   Other methods for phylogenetic analysis

Parsimony has traditionally been the most important criterion for reconstructing the phylogeny of fossil taxa. However, it is not the only way to go, nor necessarily the best. In recent years, the pivotal position of parsimony has been challenged by other methods.

### 14.8.1   Phylogenetic analysis with maximum likelihood

Although maximum likelihood (ML) methods can be very complicated in their details, the fundamental concept is simple. Given a tree A and an evolutionary model (see below), we can calculate the likelihood L(A) of that tree, which is the probability of getting the observed data (e.g., a character matrix and/or stratigraphy) given the tree. A preferred tree is then simply one with the maximal likelihood. Unfortunately, it must be located by a slow, iterative search. In general, an ML tree will be somewhat longer than the parsimony tree for the same data set. In addition to the differences in the cost of character change implied by the evolutionary model, the ML approach can accept more reversals along long branches (those with many steps in other characters) than parsimony analysis.

What is meant by "evolutionary model"? For molecular evolution, these models generally take into account the different probabilities of different types of sequence mutations (Felsenstein 1981, 2003). Such models are not well developed for morphological data, and ML methods have therefore been slow to enter paleontology. However, it is possible to construct simple theoretical models also for the evolution of morphological characters, which can be used for inferring phylogeny by ML. Some such models, such as *TS97* (Tuffley and Steel 1997) will generate ML trees equal to those resulting from a parsimony analysis and are therefore mainly of theoretical interest. A more interesting model is *Mk* with a variant called *Mkv* (Lewis 2001). This model implies that the probability of a character change increases with branch length, which is perhaps not unreasonable, at least if we assume that long branches are associated with long intervals of time and that the mode of evolution is gradualistic rather than punctuated. In contrast, the typical parsimony analysis weighs character transitions equally regardless of branch lengths. There has been a heated discussion in the literature about what method performs best for discrete morphological data – classical parsimony or ML (or Bayesian, see below) analysis with models such as Mk, and the issue is not quite resolved (see, e.g., King 2019 for a useful list of references).

Are there any other kinds of data, in addition to characters, that can help us compute the likelihood of a certain phylogenetic hypothesis? One answer may be stratigraphy. Phylogenetic analysis with ML, based on the combination of a morphological character matrix and stratigraphic information, was pioneered by Huelsenbeck and Rannala (1997) and by Wagner (1998). We will briefly go through the method proposed by Wagner (1998) as an example of this way of thinking.

The basic task to be performed is the evaluation of the likelihood of a given tree. This problem is split into two parts: finding the likelihood of the tree given stratigraphic data and finding the likelihood of the tree given morphological data. The likelihood of the tree given all data is simply the product of the two:

$$L\big(\text{tree given all data}\big) = L\big(\text{tree given stratigraphy}\big) \times L\big(\text{tree given character matrix}\big)$$

We will first consider how to estimate the likelihood of the tree given stratigraphy. Given a tree topology and the stratigraphic levels of observed first occurrences of the taxa, we can compute the sum of MIGs as described in section 14.6. This sum is also known as *stratigraphic debt*. The total stratigraphic debt and the number of taxa appearing after the first stratigraphic unit can be used to estimate the mean sampling intensity $R$, which is the probability of sampling a taxon that existed in the given stratigraphic unit. Finally, thanks to an equation devised by Foote (1997), we can estimate the likelihood of the estimated sampling frequency given the distribution of observed taxon range lengths. This tortuous path has led us to $L$(tree- given stratigraphy).

Second, we need to estimate $L$(tree given character matrix). This is more in line with standard ML methods for non-fossil taxa and can be carried out in different ways. Wagner (1998) suggested using another likelihood: that of having a tree of the length of the given tree, given the length of the most parsimonious tree. Say that the tree under evaluation has 47 steps. How likely is this to occur, given that the length of the most parsimonious tree for the given character matrix is 43? This likelihood is estimated by simulating a large number of phylogenies using an evolutionary model with parameters taken from the data set (frequency of splits, frequency of character changes, extinction intensity, sampling intensity).

Wagner (1998) carried out this procedure for a published data set of fossil hyaenids, and also tested it in simulation studies. As others have also found, the ML method seemed to outperform parsimony analysis. However, the whole idea of using stratigraphy in reconstructing phylogenies has also been strongly criticized (e.g., Smith 2000).

## 14.8.2  Bayesian phylogenetic analysis

Making a phylogenetic tree from morphological or genetic data is basically a form of model fitting, and as such it is well suited to Bayesian analysis. According to Bayes' theorem (section 2.7), we then need a prior, and a way to quantify the likelihood of a tree given the data. Often, no prior information is available, so we use a non-informative, or "flat" prior. In this case, Bayesian phylogenetic analysis is somewhat similar to ML analysis as described earlier, but using MCMC algorithms and with emphasis on the distribution of the posterior rather than the single maximal value (point estimate) of the likelihood. Hence, the support for each clade can be reported as clade credibility (i.e., posterior probability) instead of the bootstrap value commonly used in parsimony analysis.

In paleontology, the most popular program for Bayesian phylogenetic analysis is probably MrBayes (Huelsenbeck and Ronquist 2001; Ronquist et al. 2012). For morphological data, MrBayes uses an evolutionary model based on the Mk model (Lewis 2001) discussed earlier. MrBayes also includes additional functions useful to paleontologists, such as estimation of ancestral character states, dating of nodes based on known ages of fossils, and testing of alternative phylogenetic hypotheses with Bayes factors (section 2.7).

The software BEAST was originally designed for genetic data but later versions (BEAST 2; Bouckaert et al. 2014) support morphological characters and evolutionary models. BEAST is used increasingly by paleontologists, especially for estimating node ages and evolutionary rates together with tree topology, and also for even more complex analyses involving phylogeography.

### 14.8.3 Phylogenetic analysis with distance methods

The first computerized methods for classification of biological taxa were developed in the 1960s (Sokal and Sneath 1963) under the name of "numerical taxonomy." Numerical taxonomy worked by comparing characters and grouping more similar forms using clustering methods similar to those we discussed in section 5.4. This so-called *phenetic* approach was largely discounted and denounced through the rise of parsimony analysis and other phylogenetic methods in the 1980s, which are based on the premise that classification should reflect evolutionary relationships.

A minor comeback for clustering in systematics started with the publication of the neighbor joining clustering algorithm by Saitou and Nei (1987). Like the clustering methods of chapter 5, it is based on a distance matrix comparing all pairs of taxa. The algorithm is somewhat complex, starting with a completely unresolved tree (a "star tree") and iteratively splitting out branches joining the most similar pairs.

The dendrograms made by UPGMA and similar algorithms are *ultrametric*, meaning that all the branch tips end up at the same level. This implies that the total branch length from the root to a terminal taxon is equal for all taxa. This would be fine if all the taxa were living at the same time (such as the Holocene) and they were all evolving at the same rate, e.g., by the assumption of a constant molecular clock. However, these are usually unreasonable assumptions. In contrast, neighbor joining can produce trees with varying total branch lengths, making it more suitable for reconstructing a phylogeny. Moreover, it has been shown that neighbor joining does in fact constitute a greedy search for a tree with minimal length, although in a slightly different sense than in parsimony analysis (Gascuel and Steel 2006). In practice, neighbor joining often gives similar results as other phylogenetic methods.

Still, given the more explicit use of phylogenetic criteria by other methods, and their more exhaustive search algorithms that can produce a collection of shortest trees, neighbor joining will not usually be the primary choice for phylogenetic analysis. But there is one important exception: the analysis of a large number of taxa, say more than 50. In such cases, most phylogenetic methods will be painfully or even prohibitively slow, especially if the tree search is executed many times for bootstrapping or simulation. Neighbor joining is an extremely fast algorithm and will produce a good tree in milliseconds.

---

**Example 14.10**

The tree resulting from a neighbor-joining analysis of the Paradoxididae data set is shown in Fig. 14.17. The topology of the tree is identical to the shortest tree obtained from parsimony analysis (section 14.4), but the branch lengths are distributed slightly differently in the phylogram.

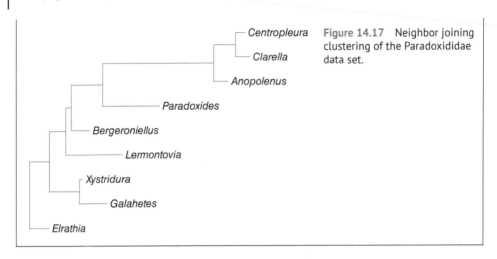

Figure 14.17  Neighbor joining clustering of the Paradoxididae data set.

## 14.9   Phylogenetic Comparative Methods

It is common in paleontology to study interspecific variation, i.e., to compare traits (morphometric data or discrete characters) across taxa. Are wing shapes different in pterosaurs and birds? How does the size of cephalon correlate with the size of pygidium in a set of trilobite species? The purpose of such studies may be to study adaptation to different environments or modes of life, to identify evolutionary trends, or a combination of the two.

We can of course carry out such comparative studies qualitatively. It is quite alright to conclude from casual observation that bats have wings and shrews do not. But in more subtle cases, we need to support our claims statistically, and this raises a problem: our data points are not randomly and independently sampled but are expected to co-vary because of a shared phylogeny. Even under a simple model of random genetic drift through time, we would expect more closely related species to be more similar (which is why traditional taxonomy works quite well also in a phylogenetic context). The methods of *phylogenetic comparative methods* (PCM) have two main purposes: (1) to adjust our statistical methods to take phylogenetic covariation into account, and (2) to try to disentangle the relative contributions of phylogeny and adaptation (including convergence) to interspecific variation.

The philosophy of PCM can be difficult to understand and has been subject to some controversy. Especially, it may seem that "correcting" for phylogeny removes too much of the very information that we want to study. Even if we want to concentrate on the adaptational and environmental aspects of the traits, these are obviously strongly tied to evolution by natural selection, and by removing the phylogenetic source of variation in the data we risk "throwing the baby out with the bath water," leaving us with little or no useful information. While this may be a concern, it may be argued that we have no choice: the assumptions of the standard statistical methods simply do not hold for phylogenetically correlated data, and this must be addressed. However, PCM is more than just an annoying but necessary adjustment of old methods, it is also a newly developing framework for raising and investigating interesting evolutionary theories.

### 14.9.1 Phylogenetic independent contrasts

One of the earliest methods of PCM is phylogenetic independent contrasts or PIC (Felsenstein 1985b; Garland et al. 1992), which converts measurements on $n$ species into $n - 1$ derived quantities (contrasts) in an attempt to eliminate phylogenetic covariation. In its original form, PIC assumes a null model for evolution called the *Brownian motion* (random walk) model, which simply lets the trait drift randomly through time by repeated addition of normally distributed increments (Fig. 14.18). This is analogous to the Brownian motion of a small particle in a fluid, being pushed around by random collisions with molecules in thermal motion.

   Consider the phylogenetic tree in Fig. 14.19, with four taxa A–D. A continuous measurement $x$ has been made on each taxon. First focusing on the pair AB, and under the null hypothesis of Brownian motion, $x_A$ and $x_B$ derive their values from the value $x_{AB}$ at their parent node AB (this part of $x_A$ and $x_B$ is due to their shared evolutionary history) plus a normally distributed increment for each of A and B, with variances proportional to the branch lengths below the sister taxa A and B (assumed to be proportional to time). These two increments are phylogenetically independent. Now the clever trick is to calculate the difference (contrast) $x_A - x_B$, such that the common component $x_{AB}$ cancels out. This is repeated for all terminal nodes, and then by particular rules (Felsenstein 1985b),

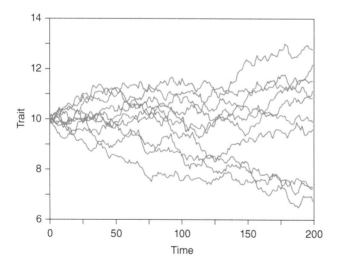

**Figure 14.18**   Ten runs of a Brownian motion simulation of a single continuous trait. The initial value of 10 evolves by the addition of a normally distributed increment with mean zero and variance 0.1 in each time step. Note how the variance across runs increases with time.

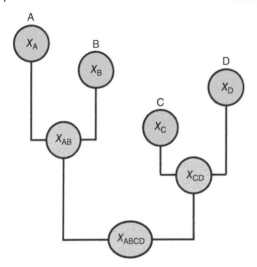

**Figure 14.19**   A tree with four terminal taxa A–D will produce three independent contrasts: $x_A - x_B, x_C - x_D,$ and $x_{AB} - x_{CD}.$ For the latter contrast, the values $x_{AB}$ and $x_{CD}$ in the internal nodes must be estimated from the values in the terminal nodes.

the procedure continues iteratively down the tree. In the end, $n-1$ contrasts have been produced. In addition, the contrasts are standardized by their expected variances under the null model.

Now we are left with a collection of contrasts, and it may not be immediately clear how these can be used. Moreover, the contrasts have been computed with arbitrary signs (positive or negative) because the ordering of pairs (e.g., $x_A - x_B$ or $x_B - x_A$) is arbitrary, making it all seem a bit chaotic. But consider, for example, that we want to carry out a bivariate regression on two measurements $x$ and $y$, fitting to a linear model $y = ax + b$. The contrasts have been computed with the same (arbitrary) pairwise orderings for $x$ and $y$. Now, if $y = ax + b$, then the contrasts $y_A - y_B$ and $x_A - x_B$ will obey the relationship $y_A - y_B = (ax_A + b) - (ax_B + b) = a(x_A - x_B)$. In other words, if we do a linear regression on the contrasts, we will recover the desired slope $a$, as if we regressed on the original variables but without those bothersome phylogenetic correlations between the data points. It is very elegant. One small annoyance is that the intercept $b$ should be forced to zero in the regression of contrasts, as it has been cancelled out by the procedure. However, we are usually primarily interested in the slope.

We have mentioned that this method assumes that branch lengths are proportional to time. However, a phylogenetic parsimony analysis delivers a tree where branch lengths are given as the number of character changes (steps) along the branch, as computed from the character state reconstruction. We can then either (somewhat daringly) assume that the number of changes is proportional to time, or we can try to estimate durations in millions of years based on the tree and the absolute ages of fossil occurrences. This requires estimation of the ages of internal nodes in the tree (e.g., Ruta et al. 2006).

## Example 14.11

Delsett et al. (2023) studied the morphology and evolution of the hyoid ("tongue bone") in ichthyosaurs (Fig. 14.20) and toothed whales, concluding that the mode of feeding evolved along different trajectories in the two groups. As part of this study, the hyoid length (HL) and width (HW) were measured in 29 ichthyosaur species, with a single fossil representing each species.

A standard linear regression (OLS) of the log-transformed measurements gives the model HW = 0.91HL − 0.86. A Model II regression is perhaps more appropriate, and the RMA method gives HW = 0.93HL − 0.91. However, a statistically stringent regression of HL vs. HW should take into account the phylogenetic structure of the data. The phylogeny shown in Fig. 14.21 was used for this purpose (note that some clades are poorly resolved, resulting in polytomies).

The PIC algorithm returns 28 independent contrasts for each of the two variables. An RMA regression with zero intercept then resulted in the slope $a = 0.97 \pm 0.06$ (1 std. error). In this case, the PIC procedure gave a slightly larger slope than simple regression. With $a$ close to 1, no allometry is indicated.

(A)

(B)

(C)

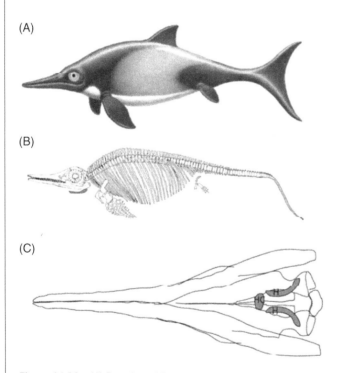

Figure 14.20 (A) Drawing of Early Cretaceous *Keilhauia nui* by Esther van Hulsen. (B) Line drawing of *Ophthalmosaurus icenicus*, with hyoid apparatus colored. Adapted from Moon and Kirton (2016). (C) Ophthalmosaurid skull in ventral view. Adapted from McGowan and Motani (2003).

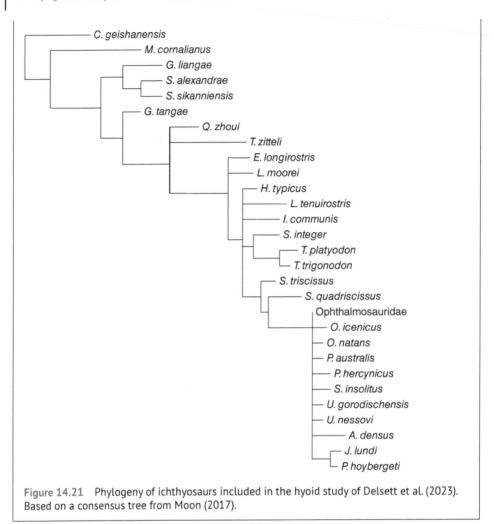

**Figure 14.21** Phylogeny of ichthyosaurs included in the hyoid study of Delsett et al. (2023). Based on a consensus tree from Moon (2017).

### 14.9.2 Phylogenetic generalized least squares

Regression with PIC has now to some extent been superseded by a newer method known as phylogenetic generalized least squares (PGLS; Grafen 1989; Pagel 1999; Symonds and Blomberg 2014). While formal statistical ordinary least-squares regression assumes independence between data points, a method has also been developed for the situation where the covariances are non-zero and can be estimated. This is called *generalized least squares* (GLS). For GLS, we need to produce a variance-covariance matrix $\mathbf{C}$ of size $n \times n$ (where $n$ is the number of data points). In a phylogenetic context, and under the Brownian motion model for evolution, the variance of a trait in a terminal node is proportional to the total branch length from the root of the tree (Fig. 14.18). These variances enter in the main diagonal of $\mathbf{C}$. The expected covariance between a trait in two taxa is proportional to their total shared branch length, entering in the off-diagonal elements of **C**. PGLS can also be used with other evolutionary models, such as the Ornstein-Uhlenbeck model that includes some provision for more directed evolution by selection pressure.

The GLS regression for a single independent variable $x$ and a dependent $y$ is then computed as follows:

$$\beta = \left(\mathbf{X}^T\mathbf{C}^{-1}\mathbf{X}\right)^{-1}\mathbf{X}^T\mathbf{C}^{-1}\mathbf{y}$$

where $\beta$ is a two-vector containing the intercept and slope, **X** is an $n \times 2$ matrix with ones in the first column and the values of the first trait ($x$) in the second column, and $y$ is an $n$-vector with the values for the second trait ($y$). A few things should be noted about this equation. First, unlike the phylogenetic independent contrasts method, the regression model can include an intercept term. Second, GLS regression is a generalization of ordinary least squares, meaning it is a Model I type regression that only minimizes errors in the dependent variable. Third, the equation easily extends to the situation with two or more independent variables (multiple regression) by adding columns to the matrix **X**.

---

**Example 14.12**

Returning to the hyoid example from Delsett et al. (2023), PGLS resulted in the regression model HW = 0.97HL − 0.99. The standard error on the slope is 0.06. PGLS therefore gives basically the same slope as PIC with RMA regression. However, PGLS operates with the original data points, rather than contrasts. This allows us to present a scatter plot of the data together with the PGLS regression line (Fig. 14.22).

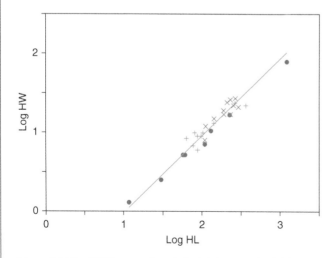

Figure 14.22  PGLS regression of the ichthyosaur hyoid data: HW = 0.97HL − 0.99. Purple dots: Triassic species. Red crosses: Early to Middle Jurassic. Blue crosses: Late Jurassic to Cretaceous.

### 14.9.3 PGLS and phylogenetic signal

Under the pure Brownian motion model, the evolution of a trait is fully controlled by the pattern of phylogeny. At the other extreme, we can imagine that the trait is completely determined by adaptation and natural selection, with no phylogenetic component. In this case, we should probably use a regression method that is more like ordinary least squares, not assuming any covariation between closely related taxa. With PGLS, we can easily accommodate these different situations by multiplying the off-diagonal elements in **C** by a coefficient called *Pagel's lambda* ($\lambda$). With $\lambda = 1$, we have the full PGLS regression as described earlier, while $\lambda = 0$ gives an ordinary linear regression with weighting of the data points depending on the total branch lengths, but without any expected covariation (Pagel 1999; Symonds and Blomberg 2014). Pagel's $\lambda$ is also known as a *phylogenetic signal*, and its value can be estimated by a ML approach. We can estimate $\lambda$ for single traits individually or for a combination of traits (Freckleton et al. 2002), but in the context of regression with PGLS, $\lambda$ should be estimated with respect to the regression residuals (Revell 2010; Symonds and Blomberg 2014).

---

**Example 14.13**

The PGLS regression of the hyoid data shown earlier used $\lambda = 1$. However, the ML value for this data set is $\lambda = 0.40$. This is a relatively weak phylogenetic signal, perhaps indicating that the hyoid shape is a somewhat variable trait, not strongly constrained by phylogeny. Moreover, the regression with this value for $\lambda$ gives HW = 0.89HL – 0.86, i.e., a considerably smaller slope than with $\lambda = 1$.

---

## References

Archie, J.W. 1985. Methods for coding variable morphological features for numerical taxonomic analysis. *Systematic Zoology* 34, 326–345.

Babcock, L.E. 1994. Systematics and phylogenetics of polymeroid trilobites from the Henson Gletscher and Kap Stanton formations (Middle Cambrian), North Greenland. *Bulletin Grønlands geologiske Undersøgelse* 169, 79–127.

Benton, M.J., Storrs, G.W. 1994. Testing the quality of the fossil record: paleontological knowledge is improving. *Geology* 22, 111–114.

Bouckaert, R., Heled, J., Kühnert, D., Vaughan, T., Wu, C.-H., Xie, D., Suchard, M.A., Rambaut, A., Drummond, A.J. 2014. BEAST 2: a software platform for Bayesian evolutionary analysis. *PLoS Computational Biology* 10, e1003537.

Bremer, K. 1994. Branch support and tree stability. *Cladistics* 10, 295–304.

Delsett, L.L., Pyenson, N., Miedema, F., Hammer, Ø. 2023. Is the hyoid a constraint on innovation? A study in convergence driving feeding in fish-shaped marine tetrapods. *Paleobiology* 49, 684–699.

Farris, J.S. 1970. Methods for computing Wagner trees. *Systematic Zoology* 19, 83–92.

Farris, J.S. 1989. The retention index and the rescaled consistency index. *Cladistics* 5, 417–419.

Felsenstein, J. 1981. Evolutionary trees from DNA sequences: a maximum likelihood approach. *Journal of Molecular Evolution* 17, 368–376.

Felsenstein, J. 1985a. Confidence limits on phylogenies: an approach using the bootstrap. *Evolution* 39, 783–791.

Felsenstein, J. 1985b. Phylogenies and the comparative method. *American Naturalist* 125, 1–15.

Felsenstein, J. 2003. *Inferring Phylogenies*. Sinauer, Boston, MA.

Fitch, W.M. 1971. Toward defining the course of evolution: minimum change for a specific tree topology. *Systematic Zoology* 20, 406–416.

Foote, M. 1997. Estimating taxonomic durations and preservation probability. *Paleobiology* 23, 278–300.

Freckleton, R.P., Harvey, P.H., Pagel, M. 2002. Phylogenetic analysis and comparative data: a test and review of evidence. *American Naturalist* 160, 712–726.

Garland Jr., T., Harvey, P.H., Ives, A.R. 1992. Procedures for the analysis of comparative data using phylogenetically independent contrasts. *Systematic Biology* 41, 18–32.

Gascuel, O., Steel, M. 2006. Neighbor-joining revealed. *Molecular Biology and Evolution* 23, 1997–2000.

Goloboff, P.A., Farris, J.S., Nixon, K.C. 2008. TNT, a free program for phylogenetic analysis. *Cladistics* 24, 774–786.

Grafen, A. 1989. The phylogenetic regression. *Philosophical Transactions of the Royal Society B* 326, 119–157.

Harper, D.A.T., Gallagher, E 2001. Diversity, disparity and distributional patterns amongst the orthide brachiopod groups. *Journal of the Czech Geological Society* 46, 87–93.

Hennig, W. 1950. *Grundzüge einer Theorie der phylogenetischen Systematik*. Aufbau, Berlin.

Hennig, W. 1966. *Phylogenetic Systematics*. University of Illinois Press, Chicago, IL.

Hitchin, R., Benton, M.J. 1997. Congruence between parsimony and stratigraphy: comparisons of three indices. *Paleobiology* 23, 20–32.

Huelsenbeck, J.P. 1994. Comparing the stratigraphic record to estimates of phylogeny. *Paleobiology* 20, 470–483.

Huelsenbeck, J.P., Rannala, B. 1997. Maximum likelihood estimation of topology and node times using stratigraphic data. *Paleobiology* 23, 174–180.

Huelsenbeck, J.P., Ronquist, F. 2001. MRBAYES: Bayesian inference of phylogenetic trees. *Bioinformatics* 17, 754–755.

King, B. 2019. Which morphological characters are influential in a Bayesian phylogenetic analysis? Examples from the earliest osteichthyans. *Biology Letters* 15, 20190288.

Kitching, I.J., Forey, P.L., Humphries, C.J., Williams, D.M. 1998. *Cladistics*. Oxford University Press, Oxford, UK.

Kluge, A.G., Farris, J.S. 1969. Quantitative phyletics and the evolution of anurans. *Systematic Zoology* 18, 1–32.

Lewis, P.O. 2001. A likelihood approach to estimating phylogeny from discrete morphological character data. *Systematic Biology* 50, 913–925.

MacLeod, N., Forey, P.L. (eds.). 2002. *Morphology, Shape and Phylogeny*. Taylor & Francis, London.

McGowan, C., Motani, R. 2003. *Ichthyopterygia*. Dr. Friedrich Pfeil, Munich.

Moon, B. C. 2017. A new phylogeny of ichthyosaurs (Reptilia: Diapsida). *Journal of Systematic Palaeontology* 17, 129–155.

Moon, B.C., Kirton, A.M. 2016. Ichthyosaurs of the British Middle and Upper Jurassic. Part 1, *Ophthalmosaurus*. *Monograph of the Palaeontographical Society* 170, 1–84.

Novacek, M.J., Wyss, A.R., McKenna, C. 1988. The major groups of eutherian mammals. In Benton, M.J. (ed.), *The Phylogeny and Classification of the Tetrapods, Volume 2. Mammals*, pp. 31–71. Systematics Association Special Volume 35B. Clarendon Press, Oxford, UK.

Pagel, M. 1999. Inferring the historical patterns of biological evolution. *Nature* 401, 877–884.

Pryer, K.M. 1999. Phylogeny of marsileaceous ferns and relationships of the fossil *Hydropteris pinnata* reconsidered. *International Journal of Plant Sciences* 160, 931–954.

Revell, L.J. 2010. Phylogenetic signal and linear regression on species data. *Methods in Ecology & Evolution* 1, 319–329.

Ronquist, F., Teslenko, M., van der Mark, P., Ayres, D.L., Darling, A., Höhna, S., Larget, B., Liu, L., Suchard, M.A., Huelsenbeck, J.P. 2012. MrBayes 3.2: efficient Bayesian phylogenetic inference and model choice across a large model space. *Systematic Biology* 61, 539–542.

Ruta, M., Wagner, P.J., Coates, M.I. 2006. Evolutionary patterns in early tetrapods. I. Rapid initial diversification followed by decrease in rates of character change. *Proceedings of the Royal Society of London B* 273, 2107–2111.

Saitou, N., Nei, M. 1987. The neighbor-joining method: a new method for reconstructing phylogenetic trees. *Molecular Biology and Evolution* 4, 406–425.

Sanderson, M.J. 1995. Objections to bootstrapping phylogenies: a critique. *Systematic Biology* 44, 299–320.

Siddall, M.E. 1998. Stratigraphic fit to phylogenies: a proposed solution. *Cladistics* 14, 201–208.

Smith, A.B. 2000. Stratigraphy in phylogeny reconstruction. *Journal of Paleontology* 74, 763–766.

Sober, E. 1988. *Reconstructing the Past: Parsimony, Evolution, and Inference*. MIT Press, Cambridge, MA.

Sokal, R.R., Sneath, P.H.A. 1963. *Principles of Numerical Taxonomy*. W.H. Freeman & Co., New York.

Swofford, D.L., Olsen, G.J. 1990. Phylogeny reconstruction. In Hillis, D.M., Moritz, C. (eds.), *Molecular Systematics*, pp. 411–501. Sinauer, Sunderland, MA.

Symonds, M.R.E., Blomberg, S.P. 2014. A primer on phylogenetic least squares. In Garamszegi, L.Z. (ed.), *Modern Phylogenetic Comparative Methods and Their Application in Evolutionary Biology*, 105–130. Springer-Verlag, New York.

Tuffley, C., Steel, M. 1997. Links between maximum likelihood and maximum parsimony under a simple model of site substitution. *Bulletin of Mathematical Biology* 59, 581–607.

Wagner, W.H. 1961. Problems in the classification of ferns. *Recent Advances in Botany* 1, 841–844.

Wagner, P.J. 1998. A likelihood approach for evaluating estimates of phylogenetic relationships among fossil taxa. *Paleobiology* 24, 430–449.

Wiley, E.O., Brooks, D.R., Siegel-Causey, D., Funk, V.A. 1991. *The Compleat Cladist: A Primer of Phylogenetic Procedures*. University of Kansas Museum of Natural History Special Publication 19.

Wills, M.A. 1999. The gap excess ratio, randomization tests, and the goodness of fit of trees to stratigraphy. *Systematic Biology* 48, 559–580.

# Index

*Paleontological Data Analysis*, Second Edition. Øyvind Hammer and David A.T. Harper.
© 2024 John Wiley & Sons Ltd. Published 2024 by John Wiley & Sons Ltd.

Printed in the USA/Agawam, MA
October 15, 2024

874498.004